SEEING THE LIGHT

OPTICS WITHOUT EQUATIONS

WILLIAM L. WOLFE

SPIE PRESS
Bellingham, Washington USA

Library of Congress Control Number: 2022933957

Published by
SPIE
P.O. Box 10
Bellingham, Washington 98227-0010 USA
Phone: +1 360.676.3290
Fax: +1 360.647.1445
Email: books@spie.org
Web: http://spie.org

Front cover design by Erin Wolfe.

Printed in the United States of America.
First printing
For updates to this book, visit http://spie.org and type "PM349" in the search field.

Dedication

This book is dedicated to my family, but especially my loving, patient, and supportive wife of 60 years. We met on the campus of The University of Michigan in 1954. It took me a year to convince her to marry me in 1955. She then followed me to Honeywell in Boston in 1966 and Academia in Arizona in 1969. We traveled the world together – to optical facilities in Sweden, Germany, Poland, Spain, Switzerland, Italy, Israel, Korea, Singapore, Hong Kong, and all over the USA.

Contents

The Author

William L. (Bill) Wolfe is professor emeritus at the James C. Wyant College of Optical Sciences of the University of Arizona. He taught optics there, with an emphasis on infrared techniques and radiometry, for twenty-five years and has been retired for about that long. He spent three years at the Honeywell Radiation Center in Lexington, MA just before that, where he managed the Electro-optics Systems Department and taught part time at Northeastern University. His first real job (not counting delivering papers and pollinating lilies) was at The University of Michigan where he taught in the Electrical Engineering Department and did research at the Environmental Research Institute, then the Willow Run Labs.

He was awarded a BS in Physics by Bucknell University in 1953 an MS in Physics in 1955 and an MSE in electrical Engineering by The University of Michigan in 1965.

He was married for 60 years to his lovely and very supportive wife, Mary Lou. They raised three incredible children who became a doctor, financial planner and aerospace engineer and are now retired. He now has six grandchildren making their way in the world and through college. He has written books on *Infrared System Design*; *Infrared System Design Examples*; *Radiometry*; *Imaging Spectrometers*; *Optics Made Clear*; *Rays, Waves, and Photons, a compendium of historic foundations and emerging technologies of pure and applied optics*, and *40 Fun-filled Float Trips for Fly Fishers*. He co-authored a book much earlier in 1964, *Fundamentals of Infrared Technology* with Marvin Holter, Sol Nudelman and Gwynn Suits. He edited *Optical Engineer's Desk Reference, Handbook of Military Infrared Technology* in 1965 and *The Infrared Handbook* in 1978 with George Zissis. He was a co-editor of the second edition of *Handbook of Optics*.

He is a Fellow of both the Optical Society and SPIE. He served on the Board of Directors of the Optical Society and was President of SPIE in 1989. He received its gold medal in 1999.

He was selected Most Successful in His Chosen Career by Bucknell University in 2004.

He served as an advisor to many government agencies and consultant to many companies. His favorite hobby is fly fishing, and some of that shows up here and there in this book. He solves a crossword puzzle now and then, although he tries many. He is still sharing his enthusiasm for optics with members of the Osher Lifelong Learning Institute, the Tucson Academy, Rotary – and whomever else will listen – by regaling them with some of this information.

He is most proud of the fact that he helped 50 young people attain their graduate degrees in optics at the James C. Wyant College of Optical Sciences.

Thanksgiving

In compiling this treatise, I could not help but remember all the people I worked with and am indebted to for their insight and contributions. This is my way of acknowledging them. I hope I have not left you out, whomever you may be. They are listed in the random order that I remember them.

This is my thanksgiving even though it is not November.

Stan Ballard was instrumental in my understanding of optical materials and scientific writing. I consider him one of my main mentors.

Luc Biberman helped me with all sorts of infrared applications and even ultraviolet ones. He was my other main mentor.

The concept of BRDF and its definition and application are due to my friend Fred Nicodemus, who worked with me on numerous topics in those days of the 1960s.

My ideas of optical design came from my friends Warren Smith, Bob Shannon, and Bob Fischer.

Optical fibers for endoscopy were invented by my colleagues H. H. Hopkins, Narinder Kapany, and Brian O'Brien with whom I had many discussions.

I am indebted to Angus McCloud and Francis Turner, my colleagues at the Wyant College of Optical Sciences, for information about thin films and filters. And to my introduction to them by Phil Baumeister at that other optical place, The Institute of Optics.

My thanks go to Jim Wyant for his interesting, incredible, introspective, imaginative interferometric insight.

Bob Jones was instrumental in my understanding of polarization and infrared detectors. Bob Cashman taught me about lead sulfide detectors, and Henry Levinstein taught me about lead tellurides.

Larry Dunkleman and Manny Goldstein helped me with integrating spheres.

John Strong spent a semester in my lab and helped me understand the VW method, the Palomar telescope, and other optics.

John Jameson was a friend who worked hard on the MIDAS system and shared those insights with me.

Paul Capp and Irv Freundlich of our Radiology Department lead me along the paths of investigating infrared breast cancer detection.

I watched as Aden Meinel and Frank Low developed the first multimirror astronomical telescope.

Harold and Jean Bennett shared their ideas on reflection measurements and polarimetry with me.

Emil Wolf, an almost namesake, was instrumental in my understanding of many theoretical aspects of optics.

Barry Johnson collaborated with me on radiation slide rules and expert witnessing.

Roy Potter explained the techniques of emissivity and absorptivity measurements that he and Don Stierwalt used.

Gerry DeBell was instrumental in running my lab and my contracts. He was both fiscally instrumental and instrumentally instrumental.

Bob Breault and Mitch Ruda were key in the development of what is now the Army Metrological Range in Huntsville. AL.

Eustace Dereniak did the work on infrared and multispectral imaging that led to many other applications.

Mike Nofziger illustrated versatility. He and Eustace, former students, both helped me understand a lot of stuff.

Fred Bartell clarified and extended our knowledge of blackbody simulators.

Larry Brooks did a great job designing our BRDF instrumentation combining optics, electronics, and servomechanisms.

Janet Fender worked hard on the improvements to the Sidewinder guidance optics and did it in a major way.

Mary Turner taught me patience, forbearance, and success over obstacles.

Cesar Sepulveda, Fred Bartell, Ben Platt, Ron Korniski, Jasmin Coté, Sung Muk Lee, Kiebong Nahm, Joanna Schmitt, Amy Lusk, Mike Nagler, Noah Bareket, Boris Venet, DeVon Griffin, Alan Holmes, Larry Brooks, Jim Burge, Jim Palmer, Bob Breault, A. Srinivasan, Maria Skolnik, Marsha Bilmont, and others all confirmed my belief that you can be a successful optical scientist no matter what your age, nationality, religion, ethnicity, or gender. All fifty of them were successful.

I cannot forget my secretaries, Phyllis Miller and Anne Damon, who were the den mothers of the Wolfe pack.

All of my students enabled me to be better at optics and at teaching. I am sure I learned more from them than they did from me.

My neighbor Tom O'Rourke was the non-technical conscience of the book. He helped me de-jargonize and clarify it all.

My son, Doug, helped me with calculations. He was my toughest and therefore best critic. My granddaughter, Erin, designed the front cover; and my granddaughter Melissa posed for many pictures, as did my other grandkids, Elise, Garrett, and Grego, and his wife, Amy. My daughters, Carol and Bardie, provided pictures. They all egged me on during the pandemic.

Preface

I have been involved with optics all my adult life; it has been my entire career. It has been wonderful, enjoyable, illuminating, and fascinating. It has been colorful, delightful and *spectra*cular. I have decided to share this pleasure with those who do not have a mathematical or scientific background. I share their art and literature; I want them to be able to share my fascination with optics. If this book were in another series, it might be called *Optics for Dummies*, but I disagree. It is optics for those who do not have the mathematical tools of calculus and the like to delve through a regular text. Maybe it should be *Optics for Smarties*, those who want to be smarter about our world.

Optics is all around us. I wake in the morning while it is still dark and look at the time projected on the ceiling from the LED clock on my bedstand. As it gets lighter, I check the LED clock on the wall. I flick on a light and go to the bathroom. Next, I go to the full-length mirror to see if I am still all there and to a convex one to check my beard. In the kitchen I see another LED clock and indicators on my microwave oven. There are also lights on my cooktop to show if a burner is hot. Someone turns on the light when I open the refrigerator to get breakfast. I peer through my double pane, insulated windows to see if there is a rainbow, and I notice the beautiful, blue sky. All this happens before I get in my car with its headlights, taillights, turn indicators, night vision sensor, dashboard lights, flip mirrors, and convex *objects closer than they appear* mirrors. There is even more optics the further I go. Traffic lights, retroreflectors, neon signs and white lines painted on the road.

In this book, I have tried to describe optical phenomena without any math. I have also steered away from quantum mechanical concepts and relativity. For those of you who want or need to see the math, it is in the appendices where appropriate.

I have also avoided the somewhat pedantic and crusty writing in the passive voice of the usual scientific texts. This is a topic with vitality, irony and humor that is truly colorful and brilliant. I think writing in the first, second, **and** third person is more appropriate. I think it also is a better way to portray my love and enthusiasm of the topic. I have attempted to make this text gender neutral by using an equal number of feminine and masculine pronouns.

Introduction

And God said Let there be light: and there was light.
And God saw the light. That it was good.

Light is good. It represents Hope, Knowledge, Godliness, Goodness, Charity, Realization, and Evangelism.

Specifically:

Hope: There is light at the end of the tunnel. An old saying.

Knowledge: A man must learn to detect and watch that gleam of light which flashes across his mind from within. Ralph Waldo Emerson

Godliness: I am the light of the world. John 8:1

Goodness: The shining city on a hill. John Winthrop and Ronald Reagan

Charity: As we work to create light for others, we naturally light our own way. Mary Anne Radmacher

Realization: The Inner Light: Self-Realization via Western Exotic Tradition. P. T. Mistberger

Evangelism: Let your light so shine among men that they may see your good works. Matthew 5:16

And do not forget that the good guy is always in the white hat.

In **Buddism**, light is the source of goodness and ultimate reality. In **Hebrew**, the word light (אור) means knowledge. Satan is the prince of darkness. You only do evil if you are ignorant, not bright. According to the **Koran**, God is the light of the heavens and earth. In **Shinto**, much is made of the sun goddess Ameratsu, her isolation in a cave, and her return to light the skies. The origin of the universe, according to the Shinto religion, was when the light separated from the other elements. Light was there first. It has been reported that **Confucius** said that it is better to light a candle than to curse the darkness. **Taoism** (the way), is darkness and light as well as yin and yang. Even **Zoroastrianism** has been characterized as the struggle between good and evil, between the light and the dark. And Darth Vader was on the dark side.

According to my Christian Bible, the world was created in six days, and one of the first occurrences was when God said, *let there be light, and there was light.*[1] According to my scientific education, the universe was created about 14 billion years ago as a big bang. Fortunately, Schroeder has reconciled these two timelines by taking into account the incredible dilation of time due to relativity and gravity.[2]

> Darkness is commonly understood to mean evil and ignorance.
> May you see the light.

I have provided as much information about light as I could for those without the scientific training in physics and mathematics. It is the **concepts** of light in plain language without equations. *Seeing the Light without equations.*

The book is divided into several chapters. The first, **Phenomena**, describes our ideas of the nature of light and its interactions with the world. These include the phenomena of polarization, refraction, reflection, diffraction, interference, and the emission and absorption of light by matter. The second, **Optics in Nature,** describes some of the fascinating things we see around us: haloes, rainbows, blue skies, and sun dogs. The third chapter is **Optical Components**. It includes the individual optical elements such as prisms, polarizers, and detectors. Then comes a major chapter, **Optical Instruments**, which covers over sixty of them. These range from the simple window we peer out of in the morning to the complex devices that save us from ICBM attacks. This chapter is followed by descriptions of several significant and historical **Optical Experiments**, those that helped describe our universe and clarify the nature of light. I have also included a collection of **Appendices** with brief mathematical support of the scientific concepts, as well as one on the foundations of scientific ideas. There is a glossary of special terms. There is no Bibliography, as the internet now suffices for this. It is a rich source of explanations and of references, although not all sites are complete or completely accurate. The references provide the rest.

I suggest that you read the **Phenomena** chapter in sequence, as some of the later concepts such as reflection rely in part on earlier concepts such as polarization. All the other sections may be read in any sequence you like. Some depend upon the phenomena described in Basics but not on each other.

I would say, as Julia Child often did, *Bon appetit*, but I think *bonne lecture* is more appropriate.

Well, maybe you will eat this stuff up!

---

[1]Genesis 1:3.
[2]Schroeder, G. *The Science of God*, Free Press (2009).

Glossary

Aberrations	deviations from perfect imagery
Absorption	the process of light turning into a different form of energy
Absorptivity	the efficiency of radiant absorption, equal to emissivity
Angle	a height divided by a distance, measured in degrees and radians
Ballistic trajectory	unpowered flight paths; parabolic arcs
Blackbody	a theoretical perfect radiator and absorber
c	used almost universally for the speed of light (celeritas)
c_1	first radiation constant $= 2\pi c^2 h$
c_2	second radiation constant $= hc/k$
Camera	a device for obtaining an image of an object
Cat's eye	the eye of a cat and a refractive retroreflector
Centigrade	a temperature scale based on 100
Chief ray	a ray that passes from the edge of an object through the center of a lens, same as principal ray
Coherence	the property of going together as in the same phase and frequency
Coherence length	the distance that two waves are in coherence
Complement	something that is related. A *ones complement* is one minus the quantity
Complementary angle	90 degrees minus the angle
Corner cube	the corner of a cube used as a retroreflector
Cube corner	same as corner cube (corner of a cube)
Cycle	the distance from one point to the next identical one in a wave
Detector	a device that senses radiation
Diffraction	the redistribution of light caused by an aperture or obstruction
Ellipse	one of the conic sections with two internal foci

Ellipsoid	an ellipse of revolution
Emissivity	the efficiency of radiant emission, equal to absorptivity
Emittance	the ratio of the amount of emission to an ideal emitter, a blackbody, and the emitted watts per unit area
Etalon	an element that has two parallel reflecting surfaces like a plane-parallel plate
Eyepiece	the part of an optical instrument near the eye
Etendué	throughput, the $A\Omega$ product of an optical system
Fahrenheit	a temperature scale based on 180 degrees
Field	a region of influence
Flux	the flow of energy, energy per unit time
Function	the relation of one variable to another
Geosynchronous	moving at the same speed the Earth rotates so that the satellite remains over the same spot on the ground.
Geosynchronous altitude	35,786 km or 22,336 miles
h	Planck's constant (helping value) $= 6.26 \times 10^{-34} \text{Ws}^2$
Hertz	frequency, cycles per second, named after Heinrich Hertz
Hyperbola	one of the conic sections with two external foci
Hyperboloid	a hyperbola of revolution
Kelvin	an absolute temperature scale related to centigrade
Illuminance	Lumens per unit area, often in lux
Image space	the region behind a lens in which an image is located
k	Boltzmann's constant $= 1.38 \times 10^{-34} \text{JK}^{-1}$
LED	light emitting diode, a device that emits light generated by electricity
Lens	a refractive device that redirects light by its shape and refractive index
LIDAR	light (or laser) detection and ranging
Light	electromagnetic radiation
Lumen	a light watt
Luminance	a lumen per area and solid angle
Luminous efficacy	sensitivity of the human eye
Luminosity	in astronomy, the total radiant power of an object
Luminous intensity	lumens per solid angle
λ	usually wavelength
Marginal ray	a ray at the outer edge of a beam

Meridional plane	the plane that contains the optical axis and the chief ray
Meridional ray	a ray in the plane of the optical axis and the object point
Micrometer	one millionth of a meter, or a measuring device
Microscope	a device for magnifying the image of objects
Mie scattering	the scattering of light by particles larger than the wavelength
Minimum deviation	minimum angular redirection as in refractive index measurements and rainbows
Mirror	a reflective device that redirects light by its shape and reflection
Nanometer	one billionth of a meter
Normal	a surface normal, a vector perpendicular to a surface.
Incidence	flux density, flux per unit area incident on a surface
Irradiance	power per unit area incident upon a surface, same as incidance
Object space	the space in front of a lens where the object is
Optical axis	a line through the center of a lens or mirror that is perpendicular to both surfaces
Optical path	the product of the refractive index and the true distance
Parabola	one of the conic sections with one focus at infinity
Paraboloid	a parabola of revolution
Photopic efficacy	eye efficiency to normal light levels, in lumens per watt
Polar orbit	an orbit in which a satellite rotates perpendicular to the direction of the rotation of the Earth, over the poles
Polarization	a measure of the direction of vibration of the electric field
Principal ray	a ray that passes from the edge of the object and through the center of the lens. Also called the chief ray
Radiance	power per unit area and unit solid angle
Rayleigh scattering	scattering of light by particles smaller or about the size of a wavelength
Réaumur	an obscene temperature scale! An absolute Fahrenheit scale
Reflection	the redirection of light, usually back towards its origin

Refraction	the redirection of light by materials of different refractive indices
Refractive index	the ratio of the speed of light in vacuum to that in a medium
Rotary polarization	the rotation of the direction of polarization
σ	Stefan Boltzmann constant $= 5.67 \times 10^{-8} \mathrm{Wm}^{-2}\mathrm{K}^{-4}$
Sagittal plane	a plane perpendicular to a meridional plane
Sagittal ray	a ray that is perpendicular to the meridional plane
Scattering	the redirection of light in many directions
Scotopic efficacy	eye sensitivity to low light levels, in lumens per watt
Seidel aberrations	the classical aberrations of spherical, coma astigmatism, etc.
Significant	not by chance, a real result, but not necessarily big
Sine	$\sim\!\sim\!\sim\!\sim\!\sim\!\sim$
Skew ray	a ray that is not in a meridional plane
Solid angle	an area divided by the square of the distance, measured in steradians
Spectrum	frequency distribution of some quantity
Supplementary angle	180 degrees minus an angle
Surface normal	a vector that is perpendicular to a surface. Also, not abnormal
Surface scattering	the redirection of light by the roughness of a surface
T	almost always temperature
Throughput	the product of areas divided by the square of the distance between them
Wavelength	the linear distance between two identical points on a sine wave, usually λ
Zernike polynomial	a means of describing aberrations

Chapter 1
Optical Phenomena

This chapter on the phenomena of optical processes discusses the nature of light and its behavior. We really do not know what light is, but we know what it does. It reflects, refracts, diffracts, interferes, scatters, absorbs, propagates, changes its wavelength with motion, comes and goes, and does it in the shortest time. These topics are covered in this chapter in about that order and in more detail. Their understanding is required for many of the applications discussed in other sections.

Light is a transverse wave motion that vibrates in directions perpendicular to its direction of travel. The direction of that vibration—vertical, horizontal, or otherwise—is its **polarization** direction. **Reflection** is the return of light back in the general direction from which it came. It can be specular or diffuse and high or low. **Refraction** is a change in the direction of propagation of light when it goes from one medium to another, often thought of as bending of light. It is a result of the fact that light goes slower in denser media. **Diffraction** is the spreading of a beam of light when it encounters an aperture or an obstacle. **Interference** is the combining of two or more beams of light that then produce an enhanced or reduced amount of light. **Absorption** of light is the process of transferring the energy of the light to that of the material. It heats it or changes its electronic configuration. **Emission** is the opposite of absorption, that is, the transformation of energy in a material into light. **Propagation** is the way light gets from one place to another. We all know about the speed of light being the fastest thing there is, but it slows down in materials. The change of the wavelength of light with the motion of the source or receiver is called the **Doppler effect** and is more familiar with acoustic waves. In optics and especially in astronomy, the Doppler effect is known as the **red shift** since it is most often observed in stars that are rapidly moving away. And there is a property of light that minimizes its travels. Light is either lazy or efficient, depending upon your point of view. It takes the path of minimum time. Light is fair; it believes in **reciprocity**. What goes out comes back and in much the same way.

These subjects, and **resolution, aberrations, radiometry** and **ray tracing**, are all discussed in this chapter. They are referred to and further explained in later chapters on light in nature and instruments. Mathematical descriptions are provided in Appendix A1.

1.1 Light

Light is perplexing stuff.[1] No one really knows what it is, but we know what it does and how it reacts with its environment. We know its speed, that it refracts, reflects, is polarized, is absorbed, and so on, but we do not know what it is. It is strange in that it acts both as a particle **and** as a wave. The great Albert Einstein showed that it is particulate in nature, but Thomas Young and others showed earlier that it is wavelike. In a letter to his friend Michele Besso in 1951, Einstein wrote that he had *still not come closer* to fully comprehending the nature of light after 50 years of studying it.[2] We now know that it behaves both ways. Fortunately, we do not need to know what it is as long as we know what it does, how it interacts with itself and its surroundings, and how we can use it to our benefit.

We can think of light either as a three-dimensional wave or as clumps of energy, called photons, depending upon the situation.

The following sections describe the early ideas of Newton that were proven wrong and then the two ways we think of light, as waves and as particles.

---

[1]"The Nature of Light, What is a Photon?" *OPN Trends* **3**, 1 (2003).
[2]Einstein, A. "Über einen die Erzeugung und Verwandlung des Lichtes betreffenden heuristischen Gesichtspunkt," *Annalen der Physik* **17**(6), 132 (1905).

Light as Corpuscles[3]

Isaac Newton thought that light consisted of corpuscles, i.e., small particles. These corpuscles (not photons) had kinetic energy when moving, had negligible mass, traveled very fast, and were refracted, reflected, and diffracted (spread out). He ascribed all the properties of regular masses to these corpuscles, although he knew they traveled in straight lines. Ordinary particles travel in parabolic arcs due to the attraction of the Earth's gravity. He believed that these did as well, but the curvature was not observable.

The different colors of light he thought resulted from the fact that the corpuscles were of different sizes. Red ones were larger than blue ones and therefore less refracted. It is also easy to see the analogy of a rubber ball bouncing off a wall to the reflection of light off a mirror. He believed that this was due to the repulsion of the corpuscles by the constituents of the mirror. But Newton assumed that the refraction of light was because they traveled **faster** in a denser medium. He believed the corpuscles were attracted by the particles of the material. But light actually goes **slower** in denser media. He had a somewhat vague idea about polarization in which the corpuscles were not spherical but had sides. His theory could not explain interference or diffraction. He was the first to show us that white light consists of colors, and he led the way for all the improvements of our knowledge.

Three important experiments led to the demise of the Newtonian corpuscular theory: Young's double slit experiment that showed that light interferes,[4] Arago's demonstration that showed light is diffracted to a bright spot behind an occulting disc,[5] and Foucault's and Fizeau's experiments that showed light travels more slowly in denser materials.[6] These experiments are described in Chapter 5.

Light as a Wave

A wave is a periodic disturbance that is in or that moves through a medium such as vacuum, air, glass, or water. A moving ocean wave is a disturbance in the ocean, for instance, an elevated portion of water that moves across the surface, as shown in Figure 1.1.1. There are also circular waves in a pond caused by a rising trout, shown in Figure 1.1.2. Most waves are periodic. They consist of a series of highs and lows. In a trout pond, they are the highs and lows of the water, the crests and troughs. In a standing wave, they move up

---

[3]Newton, I. *Opticks*, London 1704; available from the Gutenberg Project online.

[4]Young, T. "The Bakerian Lecture: On the Theory of Light and Colours," *Philosophical Transactions of the Physical Society of London* **92**, 12 (1802).

[5]Arago, D. "Rapport a fait M. Arago á la Academie le Sciences au nom de commission que avait été de charge été chargée l' examiner le memoir la envoy encores pour le prix de la diffraction," *Annales Chimie et de Physique* **11**, 5 (1814).

[6]Foucault, L. "Sur les vitesses de la lumière dans l'air et dans l'eau," *Ann. de Chim. de Phys.* 3rd series XLI, 129–164 (1851).

Figure 1.1.1 Linear ocean waves.

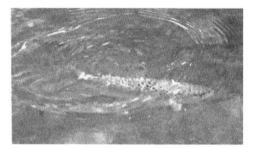

Figure 1.1.2 Circular pond waves.

and down or side to side but not in the direction the wave motion travels. In a traveling wave, they vibrate up and down or side to side, but they move as a group in the direction of travel.

Optical waves are three dimensional. Therefore, a wavefront is the two-dimensional surface of a wave, usually at the maximum. It is a section of a spherical shell that at great distances becomes a plane. It may help to visualize a wavefront as that large wave that surfers ride as it breaks toward the shore. It is almost a flat plane along the length of the shore, and it is surely a long crest, as shown in Figure 1.1.3.

Figure 1.1.3 A wavefront.

A slice through one of those waves on the pond yields a profile that looks like that shown in Figure 1.1.4. This is a single-frequency sine wave. The wavelength is the distance from one crest to another, from one trough to another, or from any two identical parts of the wave. A cycle is one wavelength long. The phase is the position in the cycle. The frequency is how many cycles pass a given point per unit time.

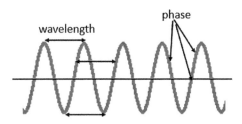

Figure 1.1.4   Simple sine wave.

Since optical waves are three dimensional, the crest of a wave is a spherical surface, until it is very far from its source when it flattens out and becomes a plane. Spheres are shown in blue in Figure 1.1.5 and circles in Figure 1.1.6. The troughs follow the crests, and so do all the other phases in the cycle of the wave. I hope you can visualize the (light blue) plane wavefronts traveling along at the positions of the peaks of the sine waves in Figure 1.1.7.

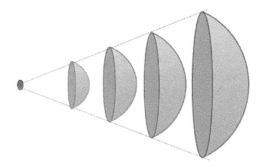

Figure 1.1.5   Expanding spheres.

Figure 1.1.6   Circular representation.

Figure 1.1.7 Plane waves and rays.

Rays are perpendicular to wavefronts. They are a simpler way to show the direction and progress of a wavefront. Wavefronts are real; rays are figments of the optinerd's imagination.

One analogy that might be worth considering to help visualize these three-dimensional waves is a series of spherical balloons of dfferent sizes, one inside the other. They all expand at the same rate as they are inflated. The actual balloons may be considered the crests and the spaces between them the troughs. Another is an onion at a particular instant of time. The skins are the crests, while the spaces between them are the troughs. A time lapse photograph might show the onion expanding as the layers get bigger and bigger as the onion grows.

Light as Photons

In 1900 Max Planck first introduced the idea that the emission of light in the form of infrared radiation was quantized, that it came in little bursts of energy rather than in a continuous stream.[7] That was the birth of quantum mechanics. He considered that the radiators were quantized, but he did not conceive of the radiation as quantized. It was the great Albert Einstein in 1905 who deduced that radiation, light, was quantized.[8] Gilbert Lewis first named them *photons* in 1926.[9]

I prefer to think of **photons** as packets of energy in a sort of Nerf-like ball without well-defined edges that has a wavy structure. One of my colleagues characterized a photon as a corrugated hot dog because it is a particle with waves. Recently two Polish physicists created a hologram of a photon and found that it looks a lot like a Maltese cross.[10] These photons are complicated beasts, but it is enough for us to think of them as very small clumps of very small energy, maybe like tiny golf balls.

---

[7]Planck, M. "Über eine Verbesserung der Wienschen Spektralgleichung," *Verhandlungen der Deutschen Physikalischen Gesellschaft* **2**, 202–204 (1900).
[8]Einstein, A. "Über einem die Erzeugung und Verwandlung des Lichtes betreffenden heuristischen Gesichtspunkt," *Annalen der Physik* **332**(6), 132–148 (1905).
[9]Lewis, G. "The conservation of photons," *Nature* **118**, 874–875 (1926).
[10]Chrapkiewicz, R. et al. "Hologram of a single photon," *Nature Photonics* **10**, 576 (2016).

Photons travel at the speed of light. They have a central wavelength, but each is a wave packet, a small spread of wavelengths around the central one. The energy of a single photon is very small. A yellow photon, right in the middle of the visible range, has an energy of about 10^{-25} kWhr (0.0000000000000000000000001 kilowatt hours). For that same yellow photon, the momentum is about 10^{-21} cm g^{-1} s^{-1}, just enough to move the vanes a Crookes radiometer if it is evacuated enough. The wave packet of a green photon may be represented in two ways. Figure 1.1.8 shows waves of almost the same frequency but different amplitudes. The main wave, the green one, has the greatest amplitude and the nearby ones, just 1% and 2% different in frequency and 10% different in amplitude, combine with it. As the waves propagate, they separate. It is an exaggerated representation. Real quasi-monochromatic waves are tighter than that. That will be discussed further in Section 1.35 Interference.

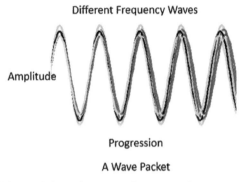

Figure 1.1.8 Conceptual wave packet waves.

A wave packet can also be represented as a spectrum as in Figure 1.1.9.

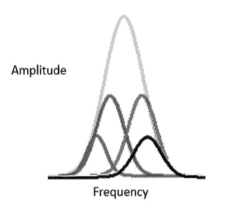

Figure 1.1.9 Wave packet spectrum.

The energy of the photon that is characterized by this wave packet is Planck's constant, *h*, times the central wavelength, which is also the average wavelength of this small spread in wavelengths.

Just remember that there is no such thing as a free lunch or a truly monochromatic wave. The spectrum of a truly single frequency wave is a straight, vertical line. A line has no width. A spectral line with no width can have no energy. There is no such thing as a free lunch or truly monochromatic wave.

It is convenient and correct to think of light in terms of photons when considering detectors and light emitting diodes, but it is equally convenient and correct to think of it as waves when considering interference, refraction and most other phenomena. Fortunately, we do not have to delve further into the rather mysterious world of photon properties.

Light as an Electromagnetic Wave

Light is part of the entire electromagnetic spectrum, the EM, spectrum. One way to think of it is that radio waves are just light with much longer wavelengths or that light is radio waves of much smaller wavelength and higher frequency. The EM spectrum is shown in Figure 1.1.10. It ranges from the very short wavelength gamma and X rays through the ultraviolet and the visible to the infrared. There is a bit of a gap until we reach TV and VHF, UHF and the fm and am bands of radio. The ELF, extra low frequency, is used mostly for submarine communication. That gap, the terahertz region is an area of intense investigation right now.

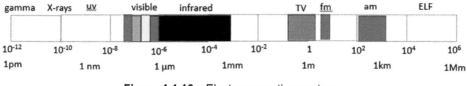

Figure 1.1.10   Electromagnetic spectrum.

This is a short aside on electromagnetism. A simple doorbell, represented in Figure 1.1.11, can be electromagnetic. You push a button that completes a

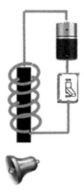

Figure 1.1.11   Electromagnetic doorbell.

circuit that allows current to flow in a wire that is wrapped around an iron rod. The electric current makes the rod magnetic and the rod is attracted to and hits the bell. The battery supplies the electric energy.

Alternately, the doorbell could operate based on the magneto-electric effect. A rod is situated inside a coil of wire. A plunger pushes the rod down through the coil, thereby generating an electric current which rings the bell. The push supplies the energy.

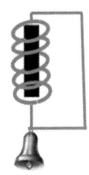

Figure 1.1.12 Magnetoelectric doorbell.

An alternating electric field generates a magnetic field, and an alternating magnetic field generates an electric field.

This explains how light can propagate in a vacuum. It is electromagnetism. The magnetic field changes and generates electricity. The electric field changes and generates a magnetic field. Thus, the changing magnetic field generates a changing electric field that generates a changing magnetic field that generates a changing electric field that... No medium is needed for the propagation of light.

Light as a Duality

One usually thinks of light as **either** a photon or as a wave, and usually one **or** the other concepts is necessary, but there is an exception, the pilot wave theory proffered by Louis de Broglie and David Bohm.[11] In it a wave is associated with the photon so that it can interfere and diffract and otherwise act as a wave even though it is a particle. We will not get into that concept or argument, as it is still a bone (or Bohm) of contention. We will deal with photons when it is convenient to do so and with waves otherwise.

Even if we do not know what light is, we know a great deal about how it interacts with our environment. We know that light is **polarized**, even if it is randomly polarized. Light **refracts.** It changes direction when it is incident at an angle on the surface of a different material. It **reflects**. Some of it returns in

---

[11]Nikoli, H. "Bohmian Particle Trajectories in Relativistic Bosonic Quantum Field Theory," *Foundations of Physics Letters* **17**(4), 363–380 (2004).

the general direction from which it came. It **diffracts.** It spreads out after it encounters an aperture or obstacle. It **interferes** with itself under the proper conditions to either increase or decrease its intensity. Light is **emitted** and **absorbed**. It can change its frequency as a result of the motion of the source or receiver. That is called the Doppler effect or the **red shift**. It is most apparent with acoustic waves and the approach of a train.

Two other principles of light are that it follows the path of least resistance and that it comes and goes. This first principle is known as **Fermat's principle.** It states that light takes the path of minimum time to get from one place to another. It does not take the shortest path; it takes the quickest path. It takes the shortest temporal path not the shortest spatial path.

Another aspect of Fermat's law is minimum deviation. It is the angular version of minimum time and optical path distance. When light is bent by refraction, the minimum path is the angle of minimum deviation. This is important in rainbows and refractometers. It is treated in those sections.

The other principle is that light is reciprocal. That is called the **principle of reciprocity**. One example of reciprocity is a point source of light at the focus of a lens will send out a collimated (parallel) beam of light. If, reciprocally, a collimated beam of light shines on that same lens, it will come to a focus.

The following chapters in this section explain these different optical effects, the way light behaves, more fully. None of them will tell you what light is.

1.2 Polarization

We can start with an analogy. A person is wiggling a rope tied to a tree. She wiggles it up and down as shown in Figure 1.2.1. The vibrations are vertical. We can call this **vertical linear polarization** of the rope. She can also wiggle the rope side to side. We can call that **horizontal linear polarization**. She can also rotate her end of the rope in a circle. That would be **circular polarization**.

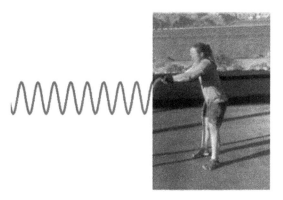

Figure 1.2.1 Vertical polarization.

This is an almost perfect analogy to optical polarization. The electric and magnetic fields of a light wave are always perpendicular to each other and to the direction of travel. Thus, only one needs to be described, and it is conventional to use the electric field direction. If the electric field vibrates up and down as the light travels, it is vertical polarization as in Figure 1.2.2, just like the rope in Figure 1.2.1.

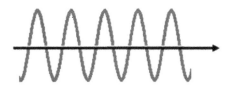

Figure 1.2.2 Vertical polarization.

If it moves side to side, it is horizontal polarization. It can also be at some arbitrary angle and so specified, as in 45 degree, 30 degree and so on.

Figure 1.2.3 Horizontal polarization.

Circular polarization is similar. In that case, the electric field will produce a spiral as it progresses on its way, as shown in Figure 1.2.4.

Figure 1.2.4 Circular polarization.

The usual case, like the light all around us, is that it is unpolarized. That means the direction of the electric field changes randomly from one direction to another. It took some early investigators some time to figure this out.

In general, most light is initially randomly polarized since the charges that cause the emission vibrate randomly in all directions. But much of the light we observe is polarized. The scattering of light in the sky causes it to be polarized. Light reflected from roads and waterways is also polarized.

Biaxial polarization, also known as birefringence, that is, two different refractions, is evident in non-symmetric crystals and in other materials that are not symmetric, notably stressed glass. Rotatory polarization or optical activity occurs in materials that do not have mirror symmetry. It is also discussed elsewhere.

It is difficult to show both states of polarization, parallel to the paper and perpendicular to it. Sometimes it can be shown as in Figure 1.2.5, but sometines not.

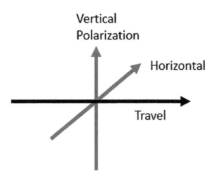

Figure 1.2.5   Polarization indication.

If not, then the convention is that light polarized parallel to the paper is indicated by a vertical arrow, and light polarized perpendicularly is indicated by a small dot to indicate the end-on view of the arrow. Light that is unpolarized is indicated by both in combination. This is shown in Figure 1.2.6.

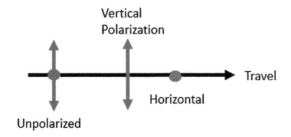

Figure 1.2.6   Polarization indication.

Another convention is that light polarized parallel to the plane of the paper is indicated by a subscript p, whereas the perpendicular component is indicated by a subscript s. We get the s from the German word *senkrecht,* which means "perpendicular."

1.3 Refraction

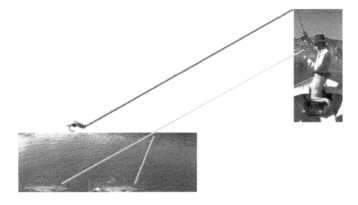

Refraction has several meanings. One is the so-called bending of light as it passes from one medium to another. A second is the apparent altered position of a star by virtue of the bending of light. The third is the evaluation and correction of the seeing ability of a person by an optometrist. We are interested here in the first of these: why light is "bent" as it passes from a medium of one density to another. That is, it changes its direction of propagation. Some of the results, ramifications and interesting features are also included in this chapter.

When light enters a substance with a different density, as from air to water, it is bent, or refracted. The amount of refraction depends upon the difference in densities. It should not be surprising that light travels more slowly in materials with higher density. (It surprised Newton.) An analogy is obvious. We walk more slowly in mud than we do on dry land. We trudge or slog. The slowdown is proportional to the density of the mud. The same is true when we wade in water. The degree of retardation of light is also closely related to the density of the material. This retardation in speed is described by the refractive index, almost always indicated by the letter n. It is the ratio of the speed of light in a vacuum to that in the substance considered.

When light travels from a medium with one density into that of a different density, it is bent; its direction is altered. Figure 1.3.1 shows the geometry of a ray of light incident from a rarer to a denser medium, say, from air to a glass plate. Recall that a ray is a perpendicuar to a wavefront. The incoming ray is bent toward the normal, the perpendicular, to the surface.

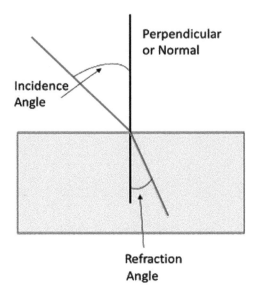

Figure 1.3.1   Rare to dense ray refraction.

One way to understand this is to consider planar wavefronts incident upon an air−water interface at an angle. The leftmost line (representing a plane wavefront) in Figure 1.3.2 is unaffected since it has not entered the denser medium. The next lower line shows a little bending at its bottom edge because that part of the wave is slowed down by the denser material it has entered. The next line shows more of the same effect until the entire wavefront has been refracted.

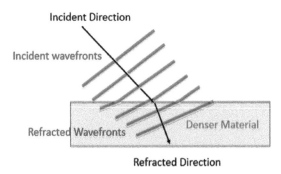

Figure 1.3.2   Refraction by wavefronts.

The degree of refraction is determined by the angle of incidence and the refractive index, as shown in Figure 1.3.3. The red line shows light incident at a rather shallow angle to the surface and refracted at a similar shallow angle, but refracted, nevertheless. The blue line represents light at a 45-degree angle; it is bent at a steeper angle. The green line at the largest angle to the surface but smallest incidence angle with respect to the normal is bent the least. The black line is the normal, and it also represents light incident perpendicular to the surface, which is not bent.

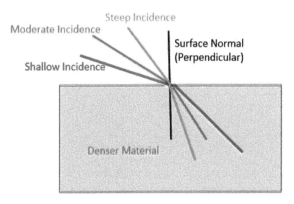

Figure 1.3.3 Refraction at different angles.

The ratio of the refracted angle to the incidence angle is proportional to the refractive index at all but the steepest angles, as shown in Figure 1.3.4. The x axis is the incidence angle; the y axis is the refracted angle. The red line is proportionality; the blue line is reality based on Snell's law of how it really bends (discussed in Appendix A.1.3).

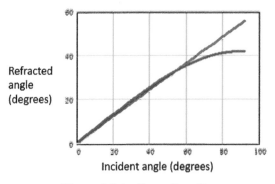

Figure 1.3.4 Proportionality.

When the light travels from a denser medium to a less-dense one, the rays bend away from the surface normal. The more inclined the ray is to the surface, the farther away it bends until it reaches a point where it cannot get out. The green ray shown in Figure 1.3.5 at steepest incidence bends away, the blue ray is more inclined to the surface, and the red ray just goes along the surface. Any ray at a larger angle of incidence, one that is more inclined to the surface, will be reflected into the denser material and obey the law of reflection. The angle at which this total reflection first occurs is called the critical angle; it depends upon the ratio of the refractive indices of the two materials.

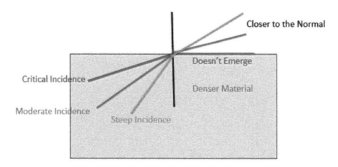

Figure 1.3.5 Dense to less-dense refraction.

The refractive index, and therefore the degree of bending, is a function of the wavelength of the light. It is not a strong variation in the visible, but it is the reason prisms and rainbows display such a range of colors. Figure 1.3.6 shows a typical dependence of the refractive index on the wavelength of the light, called dispersion. It shows the value of the refractive index on the vertical scale as a function of the wavelength in micrometers on the horizontal axis.

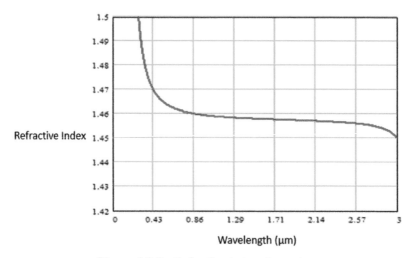

Figure 1.3.6 Refractive index dispersion.

At short wavelengths, less than about 0.45 μm, there is a peak due to an absorption at a shorter wavelength. In these regions of absorption, the refractive index drops precipitously and then increases just as much. They are called regions of anomalous dispersion and are useful in some applications. At a longer wavelength of about 3 μm, there is the beginning of another dip, also due to absorption. In between, there is a gradual decrease in the refractive index toward longer wavelengths. The refractive index is higher at shorter wavelengths (blue) than longer ones (red). In the visible region, a typical glass, BK7, has a small refractive index change of only about 0.01, from 1.47 at 0.43 μm down to 1.46 at 0.86 μm.

An unusual type of refraction is birefringence, meaning two kinds of refraction. In some crystals, light travels in two directions at different speeds because of the crystal asymmetry. As pointed out in Section 1.8, the photons spend a little time at each charged site. In an asymmetric crystal, they have more such sites in one direction than in another. This means that the crystal has two different refractive indices in two different directions.

One example of the importance of understanding refraction is fly fishing, my favorite sport. Assume you are in a boat casting to a wily trout, as shown in Figure 1.3.7. You are six foot above the water (I would have to stand on a seat to do that) and that the fish is about 20 feet away and 3 feet down. Your direct line of sight shows it to be 20 feet away, but it is closer as a result of the bending of the light by refraction. You cast the fly to where you think the fish is, but the fly lands behind it, and the fish does not take it. Oh woe. Better learn optics!

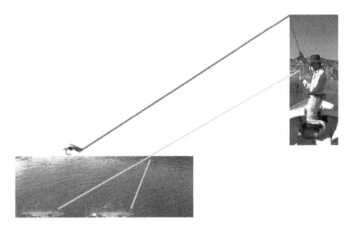

Figure 1.3.7 A practical application of refraction.

There is one other interesting optical effect that should also be of concern to the fly fisher, at least one who uses dry flies. It is the fish's window. As a result of total internal reflection at the proper angles, the fish can only see above the surface in a cone dictated by the refractive index of the water, i.e., about 1.333 and a cone of about 45 degrees. Only the bottom feathers of a dry fly sitting on the surface are visible if the fly is outside that window, not the luscious, tempting body. Obviously, the closer the fish is to the surface, the smaller is the window.

Figure 1.3.8   The fish's window.

But you should still be able to find the cherry in your Manhattan!

Figure 1.3.9   A Manhattan.

1.4 Reflection

When light is incident upon a medium with a different density, such as from air to water, some of it is reflected. It returns in the general direction from whence it came, to that rear hemisphere. Reflections from smooth surfaces are specular, that is, they go back in a single direction. Those from rough surfaces are diffuse; the reflected light scatters back in all directions. A good example of specular reflection is that from an ordinary mirror. All the reflected light from one direction goes in one direction, but some is not reflected. Nothing is 100% in optics or in life. A good example of diffuse reflection is that from an ordinary piece of paper. That too is not 100%, but it is about 90% and in all directions. Some reflections are retro. That is, they return in exactly the same single direction from which they came. Devices that do that are described in Section 3.14. The general case is bidirectional reflection. It is the light reflected in any direction when it comes from any other direction. That is described in Section 4.43.

The amount of light that is reflected specularly depends upon the nature of the surface and the angle of incidence. It is different for insulators such as glass than for metals like silver. Figure 1.4.1 shows how specular reflection changes with incidence angle for an insulator such as glass ($n \approx 1.52$). The horizontal axis is the angle of incidence; the vertical axis is the percent reflection. The blue curve is for light that is polarized parallel to the plane of incidence; the red curve is for perpendicularly polarized light. Because the reflection for perpendicular light is zero at about 60 degrees, only light with parallel polarization is reflected. At 90-degree incidence, grazing angle, all of the light is reflected. For light that is incident perpendicularly on the surface (at zero degrees) the two polarizations are indistinguishable, and the reflection is 4%.

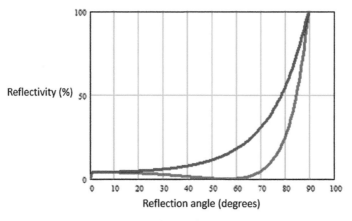

Figure 1.4.1 Surface reflectivity of glass.

Single-surface specular reflectivity is the reflection from a single surface of a material such as glass. Multiple-surface reflectivity occurs when the material is transparent so that light can be transmitted through the first surface to the second surface, reflected back from the second surface and reflected back again from the first surface until it is finally fully transmitted. That is shown in Figure 1.4.2, but at an angle, the only way to show it. Multiple reflectivity, when it occurs, is higher than single-surface reflectivity since all the reflected components add together. This is the case with common window glass.

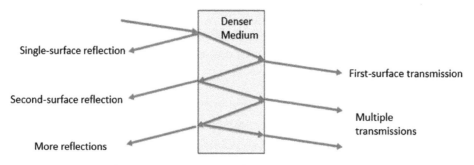

Figure 1.4.2 Multiple-reflection representation.

The magnitude of the single-surface reflection depends upon the ratio of the refractive indices of the two media. The curves in Figure 1.4.1 will be shifted up or down in a material that has a higher or lower refractive index compared to 1.52 for glass. Figure 1.4.3 shows the multiple reflectivity at normal (perpendicular) incidence for refractive index values from 1 to 4, i.e., from no reflectivity to about the maximum for useful insulating materials,

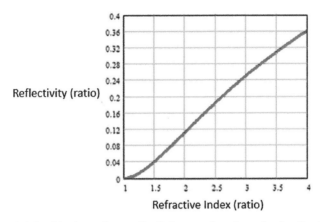

Figure 1.4.3 Single-surface reflectivity as a function of refractive index.

which is germanium in the infrared. Note that the commonly accepted normal incidence reflectivity of glass is 0.04 or 4%

The multiple-surface reflectivity of these materials is shown in Figure 1.4.4, assuming no absorption in the material. The red line is for a single surface as a function of refractive index. The blue line is for light that transmits through the first surface, reflects from the second surface, reflects again off the first and transmits out. The barely visible green line is for a second such pass, and the black line is for them all. It is larger than single-surface reflectivity. For glass with an index of 1.5, it goes from 4% to 7%. For germanium with an index of 4, it goes from 36% to 56%. Each case demonstrates about a 50% increase.

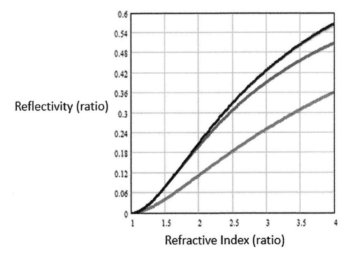

Figure 1.4.4 Multiple-surface reflectivity vs refractive index.

A plot of the parallel and perpendicular reflections for both glass and germanium with refractive indices of about 1.5 and 4, respectively, is shown in Figure 1.4.5.

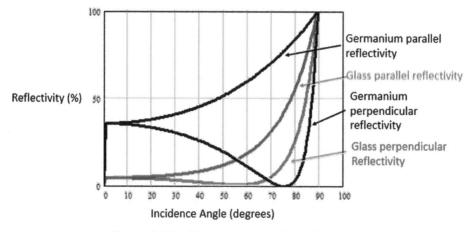

Figure 1.4.5 Glass and germanium reflections.

Diffuse reflection may be considered a form of scattering, but it may also be thought of as specular reflection from a random assortment of very small specular reflectors that are tilted in all sorts of directions, as indicated in Figure 1.4.6. White paper looks white because it has high reflectivity at all visible wavelengths, that is, for all colors of the visible spectrum. It scatters that white light in all directions because it is rough on a minute scale. The variation in surface height of normal writing paper is 5 to 10 μm. Of course, manufacturers can also make red, blue and green diffuse paper with proper dyes. And some papers are smoother or rougher than others.

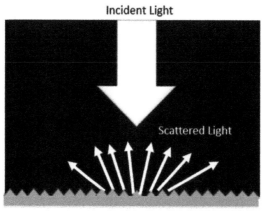

Figure 1.4.6 Schematic of diffuse reflection.

Reflection from metals and other conductors of electrons and electric current is much more complicated. The reflected radiation is often elliptically polarized. The wavefronts are complex. In general, it is messy in detail and extremely complicated to calculate. Fortunately, it is not necessary to do so for almost all optical applications. Metals are mostly used for vanity mirrors, automobile mirrors and astronomical mirrors. They only require high reflectivity.

1.5 Interference

Interference is a wave effect. It is the result of waves combining in a way that they add or subtract with each other to make more or less light in that region. Forget the photons for these concepts.

We can start with an idealized situation for simplicity. Imagine a perfect, single frequency wave, shown in red in Figure 1.5.1. It has a frequency, wavelength and amplitude as described in Section 1.1.

Figure 1.5.1   Single sine wave.

When it combines with another wave, like the one in phase with it shown in blue in Figure 1.5.2, the two amplitudes add, and the result is a wave of twice the amplitude, shown in green. When two waves that are exactly out of phase add, the result is nothing. They cancel each other, as shown in Figure 1.5.3. When the two waves are somewhat out of phase, the result is a diminished, resultant wave, shown in Figure 1.5.4.

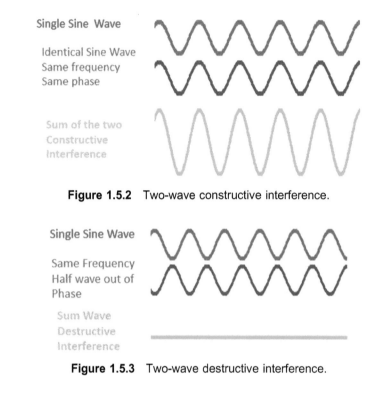

Figure 1.5.2 Two-wave constructive interference.

Figure 1.5.3 Two-wave destructive interference.

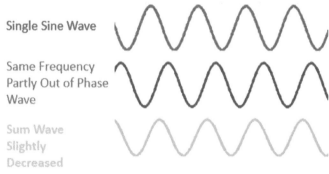

Figure 1.5.4 Two-wave partial interference.

The figure illustrating partial interference is an introduction to the idea of coherence and partial coherence. We have seen that light of the same frequency and phase interferes constructively; and if it is of the same frequency but 90 degrees out of phase, it interferes destructively. This was explained for the hypothetical monochromactic waves. Waves of exactly the same frequency but all starting at slightly different times are shown in Figure 1.5.5. The waves are shown in different colors to differentiate them, but they are of the same frequency. They are all emitted at slightly different times; each has a different phase. The peaks and troughs are not in the same place for the waves. There are only four of them in the figure, but with very many the result is a smear of completely incoherent light. This is **phase incoherence** and the nature of natural light. They are not in step, together, coherent. I think you can imagine that more and more of them added together becomes a blur.

Phase-shifted waves

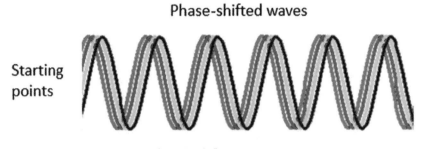

Figure 1.5.5 Phase incoherence.

If many waves with slightly different frequencies are all emitted at the same time, meaning that they are in phase they also get out of synch. Figure 1.5.6 shows that they will be in step, coherent, for a short distance but soon get out of synch. This is **frequency incoherence**. The sum of four waves with slightly different frequencies is shown. They are in phase at the beginning but soon largely cancel out. Enough of them will end up as a jumble some distance from their source.

Different-frequency waves

Figure 1.5.6 Frequency incoherence.

Waves that are phase incoherent are incoherent from the onset and continue with the same offset. Waves that have different frequencies but start in phase stay in phase for a short distance but gradually get out of phase.

Natural light is both phase and frequency incoherent. It is not monochromatic, and it starts at random times. Things all around us emit whenever they feel like it and in any colors they like.

Waves can interfere even if they are not completely phase and frequency coherent, but they must have a reasonable degree of both. The degree of interference depends upon the degree of coherence.

One puzzling aspect of interference is where the light goes when there is destructive interference. Did the energy really get destroyed? The answer is no, but it requires understanding that there really is no such thing as a truly single-frequency, monochromatic wave (again). Thus, as the two waves of slightly different frequencies propagate at a greater distance, they leave the phase relationship that causes destructive interference and then they constructively interfere, as shown in Figure 1.5.7.

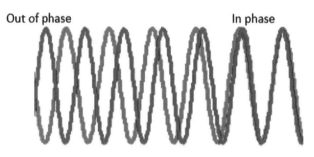

Figure 1.5.7 Constructive interference.

That is where the energy goes. The two waves are out of sync at the left of the figure, where they cancel when added, but they gradually become in phase toward the right, where they add. The combined energy is zero at the left, where they destructively interfere, and gradually increases until they constructively interfere. On average, the energy remains the same; it is not destroyed.

The more monochromatic that waves are, the greater will be their coherence length. Coherence and coherence length are important in optical communication, as is discussed in Section 4.12. **Coherence length** is a measure of the distance over which light waves are reasonably coherent. That is the key, and that is the difficulty. There is a gradual reduction in the degree of coherence over distance, and several measures have been preferred to quantify this, but we need not go into that here.

I have described the concepts of interference in terms of the addition of amplitudes. We actually observe the intensity, which is the square of the combined amplitude. That does not change the concepts of interference that are described here, but it does change some of the calculations.

Interference plays a very large role in the development of physical ideas and in everyday life. It had much to do with the origin of the understanding of the wave nature of light (such as Young's experiment, described in Section 5.4), relativity (such as the Michelson–Morley experiment, discussed in Section 5.7) and the detection of gravity waves. It is used for non-reflective picture glass, thermally insulating windows, in fiber optic gyroscopes and in certain kinds of microscopes.

1.6 Diffraction

When a light wavefront impinges upon an aperture or an obstacle, it diffracts. That is, it spreads out and around. It acts as if each point on the incoming wavefront is a new point source of a new wave that spreads in a spherical manner. This is shown schematically in two dimensions in Figure 1.6.1 for an aperture. An incident plane wave, shown by the vertical red line on the left, is incident upon the hole in the plane shown by the black lines. Each little subsidiary wave that comes from the incident wave in the hole is indicated by the red half circles that represent spherical wavefronts.

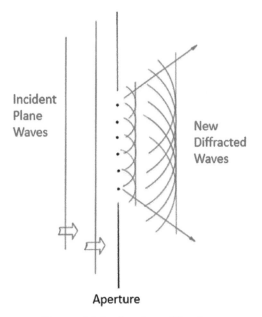

Figure 1.6.1 Aperture diffraction.

The same is true for the obstruction as shown in Figure 1.6.2. Those waves propagate into the space on the right and interfere with each other as they expand from the new sources above and below the obstruction. The crests of these subsidiary waves combine to generate light in the region

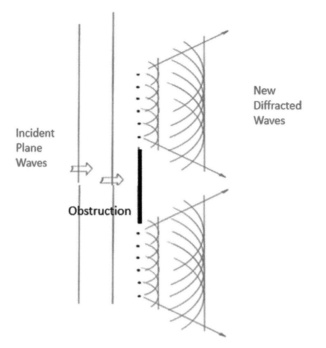

Figure 1.6.2   Obstruction diffraction.

directly behind the obstruction. Therefore, there is a spot that was predicted and measured, and helped prove the wave theory of light. It is explained in Section 5.5.

The angular spread of light in radians caused by diffraction is approximately the wavelength divided by the linear dimension of the aperture or obstruction. The **tightness**, its reciprocal, is the number of wavelengths in that dimension. It makes sense that the more waves you collect, the smaller the spread will be. For many telescopes, the spread is a microradian, a wavelength of about 1 μm divided by a diameter of a meter: a very small spread and a very good resolution. For AM radio with wavelengths of about 300 meters, buildings and hills of about the same size allow a spread of one radian, or 57 degrees. That is why we can receive AM radio in canyons and cities, and conversely why telescopes have such fine resolution. Picture a wave with a wavelength the size of a house flowing and encircling a house before moving on.

One of the applications of diffraction is the **diffraction grating**. It can be used to disburse light into its colors instead of a prism. In several ways, it is better than a prism for this. A diffraction pattern on my ceiling from such a grating is shown in the introductory figure. This application is described further in Section 3.6.

1.7 Scattering

Light is scattered by particles and by rough surfaces, which means that an incident beam of light is broken up into smaller beams that go in many different directions. There are basically two types of particulate scattering: (a) scattering by particles that are about the same size as, or smaller than, the wavelength of the light; and (b) scattering by particles that are much larger than that. They are called, respectively, Rayleigh scattering and Mie scattering after the scientists who developed the way the calculations are made: John William Strutt, known as Lord Rayleigh and Gustav Mie, (pronounced *me*) a German physicist.

The amount of **Rayleigh scattering** is strongly dependent upon the wavelength of the light. It is, in fact, dependent upon the wavelength to the inverse fourth power of the wavelength, λ^{-4}. Blue light has a wavelength about half that of red light. As blue light flies by an atmospheric particle, twice as many waves interact with the particle in a given time than red waves. There are twice as many crests and twice as many troughs in a given time or a given distance with blue light than with red light. Shorter-wavelength light interacts with scattering particles more than longer-wavelength light. This does not explain why it has the fourth-power dependence, but we will leave that to Lord Rayleigh and Appendix A.1.7. It does mean that blue light scatters 2^4 or 16 times more than red light in clear air. This is illustrated in Figure 1.7.1. I could not show the incoming light as white for obvious reasons.

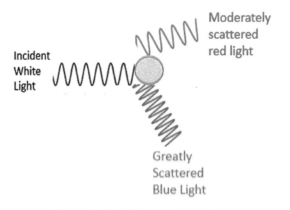

Figure 1.7.1   Rayleigh scattering.

Mie scattering is wavelength independent for particles that are large with respect to a wavelength. The dependence is determined for different shapes by detailed electromagnetic boundary-value calculations much like radar cross-sections. The solutions are usually in terms of advanced functions and/or series approximations, including Bessel functions, Legndre polynomials and other spherical harmonics.

Mie scattering of a metal sphere looks approximately like that shown in Figure 1.7.2. The independent variable is the circumference of the sphere divided by the wavelength of the light. The scattering ratio has an approximate inverse fourth wavelength dependence at low values (when the paticle is smaller than the wavelength of the light, such as Rayleigh scattering) and an exponentially decreasing sinusoidal relationship until it becomes normal reflection with a value of one when the particle circumference is about 30 times the wavelength. The value of the reflectivity is arbitrary in this example.

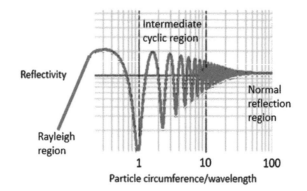

Figure 1.7.2   Mie scattering of a metal sphere representation.

As noted above, **surface scattering** can be considered reflections from a host of randomly arranged small facets of the material. If the average roughness of the surface is a small fraction of a wavelength, the surface is considered smooth. That is a restrictive criterion; only finely polished surfaces such as mirrors and windows meet it. Most materials that we deal with routinely are not smooth according to this criterion. Ordinary paper has an average roughness of about 10 μm, nearly 20 times that of visible light. It is a Mie scatterer. The scattering function for micro-rough scattering surfaces includes an angle factor, an inverse fourth wavelength factor and the spectrum of the surface height distribution. More details are presented in Appendix A.1.7.

1.8 Absorption and Emission

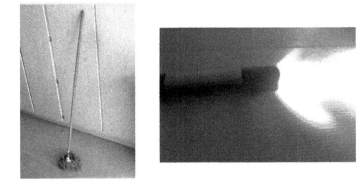

Absorption and emission are reciprocal processes. Gustav Kirchhoff showed many years ago that the efficiency of light emitted from a material is equal to the efficiency of light that is absorbed.[12] That is, the emissivity ε of a material, i.e., the ratio of the amount of light that the material emits compared to that of a perfect radiator, is equal to the absorptivity α, and the same ratio for a perfect absorber. Thus, it is enough to understand emission to get an appreciation of absorption, and vice versa.

One useful, but purely imaginary, way to view the radiation from real objects is that they are all blackbodies, i.e., perfect absorbers and perfect radiators, on the inside. The emissivity and absorptivity limit this perfection so that the objects radiate less than perfectly and in various parts of the spectrum. There is a skin around the perfect radiator that determines this efficiency.

---

[12]Kirchhoff, G. "Über das Verhältniss zwischen dem Emissionsvermögen und dem Absorptions-vermögen der Körper für Wärme und Licht," *Annalen der Physik und Chemie* **109**, 275 (1860).

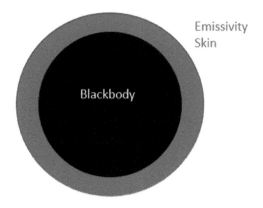

Figure 1.8.1 Imaginary emission model.

Light is absorbed by various materials in different ways, but each one of them involves the transformation of the light (electromagnetic) energy into some other form of energy. Energy is not created or destroyed unless there is nuclear fission or fusion, and those are not considered here.

A good model of **atomic emission** is the Bohr model of the atom, which is adequate for our purposes. It consists of a nucleus of protons and neutrons surrounded by electrons in various orbits. Hydrogen has one electron in a single orbit around its nucleus. Helium has two. For quantum mechanical reasons, the first orbit around a nucleus can have only two electrons. The atom with three electrons is lithium. It has two electrons in the first orbit and one in the second.

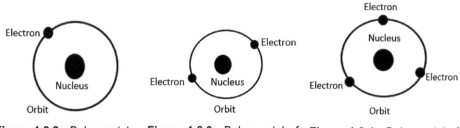

Figure 1.8.2 Bohr model of hydrogen.   **Figure 1.8.3** Bohr model of helium.   **Figure 1.8.4** Bohr model of lithium.

It is followed by beryllium, boron, carbon, nitrogen, oxygen, fluorine and neon, each with one more electron. We do not have to worry about all these elements, but we must be aware that their electrons exist in orbits with increasing radii and increasing numbers of electrons. There are $2n^2$ electrons allowed in the nth orbit, i.e., 2×1 in the first, 2×4 in the second, 2×9 in the third and so on.

The reason the electrons continue to orbit at these distances is wrapped up in quantum mechanics, such as probability amplitudes of wave functions. I will not go there. Suffice it to say that when a photon interacts with an atom, it can transfer its energy to an electron, which then moves to an outer orbit. This is the concept of absorption by atoms, by their electrons. Emission by atoms is the reverse procedure. An electron gives up energy, moves to a lower orbit and emits a photon.

Molecules are collections of atoms. Sodium chloride, common table salt, consists of a cubic arrangement of sodium and chlorine atoms—a tight arrangement. Sodium has a single electron in its outer orbit; chlorine has seven. Both are in the third orbit, which can have eight. They join tightly to make that eight. Copper, a metal, is element number 29. That means it has three electrons in its fifth orbit. Molecular copper is an arrangement of copper atoms with the outermost electrons only loosely bound to the nucleus. These are electrons that are free to roam from molecule to molecule and in between. Copper interacts with light by absorbing photons of any wavelength and accelerating these free electrons. Copper is highly reflective, and its absorption spectrum is quite flat. This is characteristic of most conductors.

The electrons of a conductor are loosely bound to the atom. They can release any amount of energy to a photon generating any color by decelerating. That is perhaps like the Boy Scout at the Jamboree who will go to any tent and help them light a fire. The outer electrons of a semiconductor are only loosely bound to the atom, perhaps like the Boy Scout who expends the energy to go up the hill to another troop to light a fire. A photon of sufficient energy will excite an electron that can participate in current flow if a voltage is applied, but only if it has enough energy, a high enough frequency a short enough wavelength. A material like silicon is a good detector in the visible but is transparent in the infrared, longer wavelength spectrum.

The remaining category is molecules with tightly held electrons, perhaps like the Boy Scout who cuddles with his friends for warmth. Then his shivers generate energy but only a small amount. The energies of the light wave vibrations give rise to molecular vibrations that resonate with them. Carbon dioxide is a good example. It is a linear molecule with carbon in the middle (shown as blue) and one oxygen molecule on each side of it, as pictured in Figure 1.8.5.

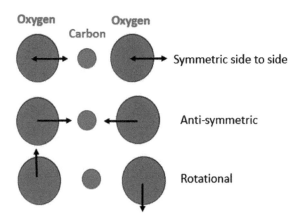

Figure 1.8.5 Representation of carbon dioxide vibrations.

The two oxygens can move in and out together in the same direction as indicated on the first row, or they can move in and out in opposite directions, depicted in the second row. The two oxygens can also go in opposite directions. There are more such motions, (like in and out of the paper), but this is enough to show the principle. These occur at frequencies of 2.2×10^{13} and 8×10^{12} cycles per second (Hertz) (among others), which corresponds to wavelengths of 4.5 and 12 μm, well-known absorption regions of carbon dioxide. The absorptions and emissions occur at optical frequencies that correspond to the natural frequencies of these motions. This is representative of molecular absorption and emission. Other molecules have other motions, molecular and atomic weights, binding energies and frequencies. Complex molecules have complex spectra of vibrations and rotations. It is what makes all of them unique.

When light is incident upon any one of these materials, it interacts with the electrons or molecules. If there is a resonance, there is absorption. If not, the light continues on. This is one way to understand why light goes slower in these media.

I hope you agree that this has been an absorbing discussion!

1.9 Propagation

The propagation of light, that is, radiation transfer, is the process of light traveling from a source through various media to a receiver. It is basic to the calculation of measurements and the performance of cameras, telescopes and other optical receivers—as well as the heat balance of our planet.

It is necessary to start with the definitions of a few terms and their symbols. **Power**, with the symbol P, is the number of watts radiated by some source. It is **energy** U per unit time t. Power per unit area is called power or **radiation (areal) density**, and it is called irradiance or radiant **incidence** E if it is received by a surface (not incidence). It is radiant emittance or **exitance** M if it is emitted. It has units of watts per square meter or per square foot, or equivalent area measurement. Power per unit solid angle is called radiant **intensity** I and has units of watts per steradian (see below for an explanation of solid angle). Power per unit area and per unit **solid angle** is called **radiance** or sterance and have units of watts per unit solid angle and area, and L is used. Radiance is the basic quantity of radiation transfer.[13]

A linear angle, what we normally call an angle, is the length of a line that is perpendicular to the line of sight divided by the length of that line of sight, as shown in Figure 1.9.1. Linear angles are measured in degrees and in radians. A radian is simply the length of that perpendicular distance divided by the length of the line of sight.

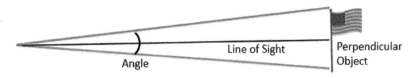

Figure 1.9.1   Linear angle.

Note that it is not the total length of the object but only that which is perpendicular to the line of sight, the projected length. This is illustrated in Figure 1.9.2.

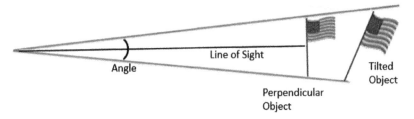

Figure 1.9.2   Projected length.

It is not known who or why someone decided that there should be 360 degrees in a circle. It may have been because there are 365 days in a year or 60 degrees in an equilateral triangle.[14] Nevertheless, it is an arbitrary unit that leads to all sorts of trigonometric complications. A radian angle is simply the height divided by the distance.

Consider this simple example using Figure 1.9.1. Assume the flag is 10 feet high and at a distance of 100 feet. The angle using degrees is the inverse tangent of 10/100. The angle in radians is 10/100. Note that the inverse tangent of 0.1 is 5.68 degrees or 0.0996 radians, which is nearly the 0.1-radians measurement that was obtained directly. The difference is rounding error.

The radian is depicted in Figure 1.9.3. One radian is the angle an arc equal to the radius of the circle makes one radius away. Since the circumference of a

---

[13]Wolfe, W. *Introduction to Radiometry*, SPIE Press (1998) [doi: 10.1117/3.287476].

[14]Blocksma, M. *Reading the Numbers: A Survival Guide to the Measurements, Numbers, and Sizes Encountered in Everyday Life*, Penguin Books (1989).

circle is 2π times its radius, there are 2π radians in a circle. In the familiar but arbitrary system, there are 360 degrees in a circle. Thus, one radian is equal to $360/2\pi = 57.29$ degrees.

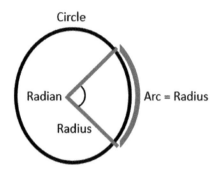

Figure 1.9.3   Radian definition.

A **solid angle** is the area perpendicular to the line of sight divided by the square of the distance, as shown in Figure 1.9.4. The solid angle is important in radiometry because it describes the cones of light that are involved.

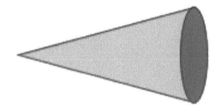

Figure 1.9.4   Solid angle.

The solid angle is also the **projected area** divided by the square of the distance. It is not the light-blue oval area but its projection perpendicular to the line of sight.

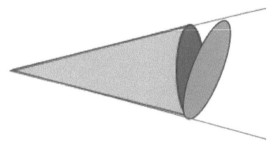

Figure 1.9.5   Projected area.

The solid angle is a unit area on the surface of a sphere divided by the square of the radius. Since the area of a sphere is 4π times the radius squared, there are 4π steradians in a full sphere.

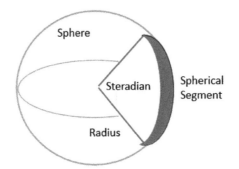

Figure 1.9.6 Steradian definition.

The propagation of light occurs in free space and in optical systems. We deal first with that in free space from point sources and extended sources.

One way to distinguish sources is whether they are extended or "point." There is really no such thing as a true **point** source; it could have no energy. A point is an abstraction with no dimensions. A point source is considered to be one that is smaller than the image of the detector on it. Figure 1.9.7 shows the detector image in green and the source in red. A more accurate but somewhat cumbersome term is **sub-resolution source**, but point source is more commonly used.

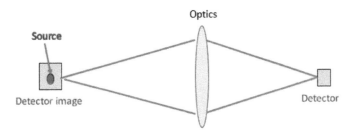

Figure 1.9.7 Point source configuration.

Conversely, an extended source is larger than the image, as shown in Figure 1.9.8. It is a super-resolution source, but it is usually called an extended source. Radiation transfer to a measuring or imaging device is different for the two cases.

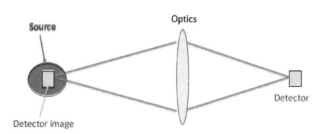

Figure 1.9.8 Extended source configuration.

Probably the simplest radiation propagation rule is the well-known inverse square law. Radiation from a point source expands in spherical wavefronts, as shown in Figure 1.9.9. Each section can be represented as a circle, especially as they move farther from the source. The radii and the distances make similar triangles. Thus, the radii of the circles are proportional to the distance. However, the radiation spreads over an area. It is the power per unit area, and the areas are increasing as the square of the radii which are proportional to the distances. The inverse square law is for small emitters, such as sub-resolution point sources.

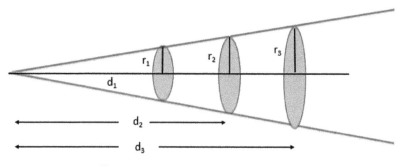

Figure 1.9.9 The inverse square law.

For sources that are not small enough to be considered points but not large enough to be extended, this law must be modified by adding terms that are the square of the ratio of the source linear dimension to the distance. This is shown in Figure 1.9.10. The x axis is the ratio of the source size to the distance, running from 0, meaning negligible (i.e., a true point source), to 1, the same size as the distance. That almost never occurs. Even in this case, the error is only 50% (0.5), and for most sources that fill an appreciable portion of the detector image, it is usually in the range of about 10% (0.1).

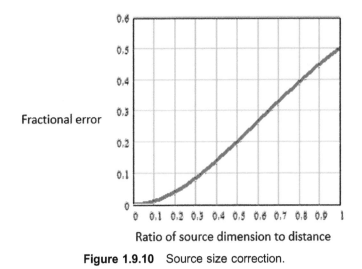

Figure 1.9.10 Source size correction.

The calculation of light from a **point source** starts with the **radiant intensity** of that source, the watts per unit solid angle. If it is a true isotropic source, then the power goes uniformly into a sphere of 4π steradians. The radiant intensity I is $P/4\pi$ watts per steradian. A telescope or other optical device subtends a solid angle of its area divided by the square of its distance from the source. The multiplication of the source intensity and the receiver's solid angle provides the amount of power on the receiver aperture. The rest of the job is to calculate the transfer of power to the detector. In this case, it is simply the power on the aperture times the transmission losses of the system.

The calculation of the received radiation from an **extended source** starts with the **radiance** L of the source, the watts per solid angle and per unit area. In this case, the inverse square rule no longer applies. The source is larger than the detector image. If the distance is doubled, the radiation is reduced by its square, but (the linear dimension of) the image of the detector on the source is doubled such that the area and the power are increased by the square. The power on the receiver is independent of distance for an extended source. It is dependent upon the cone of light, the solid angle that reaches the detector. That is why the F/number in cameras is so important. Almost all camera imagery is of extended objects.

The propagation of optical radiation in optical instruments is subject to two constraints. The power is constant (assuming no transmission losses), and the radiance is constant. These come from the first and second laws of thermodynamics. The first law says that power is not created or destroyed (except for nuclear reactions). Therefore, power through an optical system is conserved. The second law says that it takes energy to improve order. Increasing the radiance without increasing the power means putting more energy in a smaller area or a smaller solid angle. That increases the order.

Every optical system has losses. They are caused by reflectivity, whereby the light is reflected back, or by absorption, whereby the light is converted to another form of energy.

The power through a lossless optical system, by definition of the radiance, is the radiance times the solid angle times the area. This latter product is called the throughput or the $A\Omega$ product (pronounced the **A omega product**). The total power is conserved, and the radiance is conserved. As a result, the throughput—the etendué—is conserved.

The French had a lot to do with this field, not only the term étendue, which means expanse or scope, but also others. The symbol L for radiance comes from *luminosité* and E from *eclairage* for illumination. It could also be that P comes from *puissance* and I from *intensité*.

The F/stop of a camera is a good example of the extended source situation. The F/stop is the ratio of the focal length to the aperture diameter. It is a linear measure of the cone of light on the image plane, which is a solid angle, the omega. The aperture diameter is a measure of the collecting area A. The flux on the film plane of the camera is proportional to the solid angle of light impinging on that plane. The F/number is a linear expression of the solid angle reaching the detector, but the cone is quadratic. This accounts for the strange sets of numbers, such as F/1.4 and F/2.8. If you go from F/2.8 to F/1.4, you do not double the amount of light on the detector; you quadruple it. F/2.8 is twice F/1.4, but that is the linear measure. Square it to get the amount of increased light. To double the amount of light, go from F/2.8 to F/2.0. That is as close as you can get with reasonable numbers (1.96 versus 2.00). It would have to be F/2 to F/2.82845 to be an exact factor of two.

1.10 The Red Shift

This is more generally known as the Doppler effect.[15] It is a shift in the wavelength (and frequency) of a wave as a result of the motion of the source

[15]Doppler, C. "Über das farbige Licht der Doppelsterne und einiger anderer Gestirne des Himmels," in A. Eden, *The Search for Christian Doppler*, Springer-Verlag, Wien (1992).

or receiver. It happens with both sound and light waves. We have all heard the whistle of a train. As it approaches it sounds rather high pitched, and as it recedes the sound changes to a deeper tone. This can be visualized as a squeezing or elongating of the waves by the motion of the source. The speed of light (and sound) is constant. So, as the source approaches us, the crests are squished together. If the source recedes, the crests moves apart. This is illustrated in Figure 1.10.1: as the receiver moves to the right (away from the source on top), the waves need to "reach out" to keep up, but as the receiver moves to the left (towards the source), they bunch up.

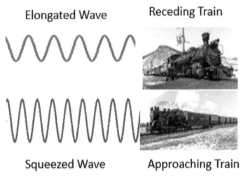

Elongated Wave   Receding Train

Squeezed Wave   Approaching Train

Figure 1.10.1   The red shift.

This is well known and important in astronomy. The amount of red shift is a measure of the speed at which some of the stars and galaxies are expanding in our universe. The light is shifted to the red as the source recedes. It would be shifted to the blue if they approached.

The shift is small, observed mostly in astronomical situations with the recession of stars and galaxies. The ratio of the red-shifted frequency (or wavelength) is the same as the ratio of the speed of the source to the speed of light. If a distant star is moving away at 90% of the speed of light, the yellow light at 0.5 μm will be shifted to 0.95 μm into the near-infrared. The shift of that same yellow line in a car going 100 mph is an undetectable change to 0.500000075 μm. The Doppler shift in sound for that same speeding car is about 13%. We hear it with our ears, but the red shift is not observable with our eyes. Middle C at 256 Hz (cycles per second) becomes 295 Hz if you are moving at 100 mph. It is a clear and discernible difference of one full note, but the yellow line is still the yellow line.

I find it interesting and somewhat curious that Christian Doppler described the effect in terms of double stars: Doppler on Doppelsterne.

1.11 Fermat's Principle

Fermat's principle is a fascinating and useful tool for the development of practical optical rules. I have not found it useful in the solution of any optical problems, but it is essential to include it in a book on optical concepts. It was formulated by Pierre de Fermat in 1662 to explain refraction. Minimum deviation is the angular form and is the basis of the location of the rainbow, described in Section A.2.9.

Fermat's principle states that light is in a hurry, efficient or lazy; it takes the path of least time. It is the optical form of the principle of least action in mechanics and also applies to quantum mechanics. It is an assumed principle that has withstood the test of time and experience. One might infer that God made light efficient.

The time it takes to go a distance d in a material is the distance divided by the speed or d/v. The speed of light in a material is v or c/n. The time it takes to go a distance d is therefore $d/(c/n)$ or nd/c. Thus, since c is a constant, the minimum time is also the minimum optical path nd.

This is a principle that is used intuitively by lifeguards running and swimming to save someone. They run most of the way on the beach and swim part of the way in the water. This minimizes the time because they are going faster most of the time. They run an even shorter distance if they throw a flotation device instead of swimming.

One simple and illustrative example of Fermat's law is the derivation of the law of reflection, which states that the angle of reflection is equal to the angle of incidence. Figure 1.11.1 shows two points and a mirror. A photon is assumed to go from point A to point B after reflection off the mirror. There are many paths it could take. Straight down to the mirror and over to point B (shown in green), or down at any angle to the mirror and up to point B (blue and red), including all the way over to where it goes straight up from the mirror to B (tan). One can do the geometry of all these triangles or simply measure the various distances.

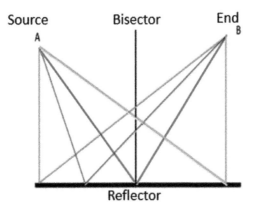

Figure 1.11.1 Possible photon paths.

Figure 1.11.2 compares the measured distances. The shortest distance is from A to the bottom of the bisector of the distance and back up to B (red). This is shown by the measured lengths of the different rays shown in Figure 1.11.2 (you can measure them to be sure I did it right). The green line, not surprisingly, is the longest and is equivalent to the tan one. The blue line is shorter, and the red symmetrical one is the shortest. It is probably clear that as the rays go from the blue angle towards the red one, the rays will get shorter and shorter.

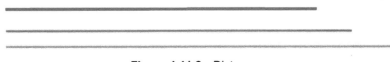

Figure 1.11.2 Distances.

The symmetric arrangement is the shortest. As a result, the angle the ray makes with the bisector upon reflection is the same as the incidence angle. That is the law of specular reflection.

A mechanical equivalent is the bounce of a tennis ball off the court (if it has no spin). Another is the carom of a pool ball off the cushion (again with no spin).

Fermat's law also dictates the path a ray will take to get from A to B if there is some sort of intervening medium in which it travels with a different speed. Figure 1.11.3 shows the paths that light will and will not take to get from A to B when there is a glass plate in between. The shortest path is in red, but the quickest path is in blue since light can travel faster in air than in glass,

50% faster. The longer path is compensated for by the faster speed. I bet Snell and others knew this and pondered how to calculate it. Snell's law of sines was the answer. It is fundamental to refraction. To paraphrase Robert Frost, the direct route is the path not taken.[16]

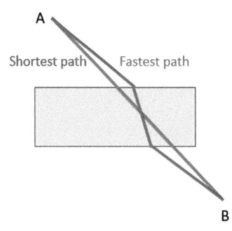

Figure 1.11.3 Shortest and fastest paths.

1.12 Reciprocity

Reciprocity in optics sort of means what goes around comes around. An eye for an eye. If a source is replaced by its receiver and the receiver is replaced by a source in a system, it will just reverse itself, and otherwise be the same.

---

[16]Frost, R. "The Road Not Taken," *Atlantic Monthly* (1915).

A simple example may help. An ideal point source placed at the focus of a perfect lens will produce a perfectly parallel, collimated beam as shown in Figure 1.12.1. If a perfectly parallel, collimated beam is incident upon a perfect lens, it will focus to an ideal point, as shown in Figure 1.12.2.

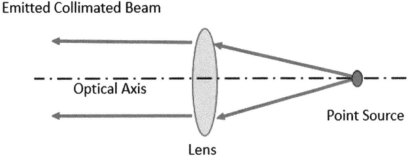

Figure 1.12.1 Point source to collimated beam.

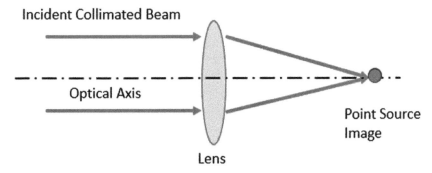

Figure 1.12.2 Collimated beam to point source.

This concept is useful in understanding satellite imaging systems. A telescope projects an image of an individual detector onto the ground. That same spot on the ground radiates back to the detector. The detector radiates its (very few) photons to the ground. The ground patch radiates its (relatively numerous) photons to the detector. They are reciprocal.

Another familiar example is, *If I can see you, you can see me.* Some form of this is often on the back of semis, referring to their rear-view mirrors. Sometimes it is in the negative.

In the theory of diffraction, reciprocity means that apertures and obstacles of the same shape and size produce complementary fields. For instance, where there is a maximum of a sine wave from an obstacle, there will be a minimum for an aperture of the same size and shape.

1.13 Resolution

Resolution is the discernment of things, determining that there are two closely spaced lines and not just one blurry one or that there are two closely spaced audible beeps and not one long one. In optics, there are three types of resolution: spatial, spectral and temporal. Resolution is also the agreement of an argument or a strong intention.

I hope you have the resolution to wade through this.

Spatial resolution is usually determined by the discernment of two or more closely spaced lines or two or more points with a given contrast. The contrast is usually stated either as contrast in a specific sense or as modulation. The input lines or points have full contrast, black on white; the output may be some shades of gray. Contrast has at least two definitions: the difference between the maximum and the minimum divided by either the average or the sum of them. Modulation is the difference divided by the sum.

Telescopes

There are several measures of angular resolution for a telescope. The measure is basically how well it can separate the images of two very small (point) sources. One of the earliest and simplest is that of Lord Rayleigh and is called the **Rayleigh resolution limit** (logically). The central parts of the diffraction patterns of two point sources formed by a telescope are shown in Figure 1.13.1. The Rayleigh criterion states that two images of two closely spaced point sources can be discerned as two separate images if the maximum of the diffraction pattern of one (red) coincides with the first minimum of the diffraction pattern of the other (blue). There is a small dip (about 26%) in the combination of the two patterns as shown by the black line.

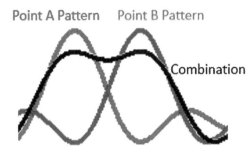

Figure 1.13.1   Rayleigh resolution.

The **Sparrow criterion** states that the two point sources are resolved if there is no dip; i.e., there is a flat area.

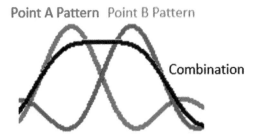

Figure 1.13.2   Sparrow resolution.

Both are arbitrary. In fact, if there is enough signal-to-noise ratio, two point sources can be resolved even if there is an arbitrarily small separation because the normalized sum of the two curves will be different from the single curve. Figure 1.13.3. shows a diffraction curve from a single point in blue and the sum of it with that of a very nearby point (normalized). This may not be resolution, just knowledge that there are two sources.

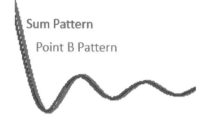

Figure 1.13.3 Arbitrary resolution.

The Rayleigh limit continues to be one of the most reasonable and common ways to specify resolution of astronomical telescopes. The Rayleigh limit for a telescope with a circular aperture is $1.22\lambda/D$, where λ is the wavelength of the light, D is the diameter of the collecting aperture and 1.22 arises because the aperture is circular.

Microscopes

The resolution specification of a microscope is similar. It is also defined in terms of the detection of two, point sources. The Rayleigh criterion and the other comments about telescopes still apply but the consideration is slightly different because many microscopes have immersion lenses. The resolution limit of a microscope is usually specified as λ divided by $2n\sin(\theta)$, where n is the refractive index of the medium, and θ is the angle of the cone of light of the microscope. The medium may be air at 1, or it may be the oil of an immersion lens, at about 1.5.

Imaging Devices

The **point spread function (PSF)** is the simplest way to specify optical resolution in an imager. It is the distribution of light in the image plane of an optical system that is generated by a point source. It is often Gaussian in shape, but not always. Very often, it is specified by its diameter at half its height. This is often called the half width, but it is not half the width; it is the full width at half height. The **circle of confusion** is closely related and is not as good a representation. It is often specified as the diameter of the PSF, but the diameter at the height of the circle is not always well defined. It is too vague for further consideration. Try to not be confused!

The **Strehl ratio**, named after Karl Strehl,[17] is the ratio of the measured intensity peak of the PSF to the peak intensity of the equivalent diffraction-limited PSF.

---

[17]Strehl, K. "Aplanatische und fehlerhafte Abbildung im Fernrohr," *Zeitschrift für Instrumentenkunde* **15**, 362–370 (1895).

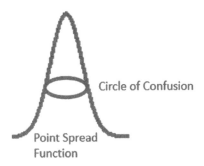

Figure 1.13.4   Point spread function.

The **optical transfer function (OTF)** includes both the **modulation transfer function (MTF)** and the **phase transfer function (PTF)** and is probably the best way to specify the resolution of an imaging system. The MTF is the modulation as defined above as a function of spatial frequency. The PTF is a measure of linear shift of the image and usually of no interest.

The spatial frequency of an object or image can be expressed in terms of the number of lines in a given distance. For instance, 10 lines per millimeter or 20 lines per inch. This is the number of lines in a given length, a linear frequency. As the spatial frequency grows, there are more and more lines per millimeter; more and more lines are squished together; and the optical system will eventually blur them out. The contrast between black and white lines will decrease until there is a gray blur.

The optical transfer function is a plot of how well the imager transfers the 100% modulation of the distinct black and white input lines. The MTF may look something like Figure 1.13.5, with the modulation transfer ratio plotted against the linear frequency in, for instance, lines per millimeter.

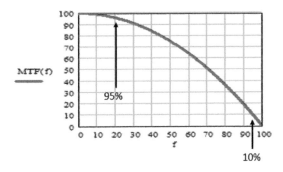

Figure 1.13.5   Representative MTF.

When there are just a few lines per millimeter, 20 near the left of the figure, the transfer is about 95%. As the number per millimeter (the linear spatial frequency) increases, the contrast is reduced; the image becomes blurrier down to about only 10% difference between the gray spaces and the darker gray lines at 95 lines per millimeter; the lines and spaces are smeared. At 100 lines per millimeter, it becomes a gray blur.

In 1958, John Johnson[18] of the US Army Night Vision Labs proposed criteria for detection, recognition and identification. He did so in terms of TV lines, but I will present it in the more modern terms of pixels, since we now almost universally use LED systems. It takes a row of two pixels on an object to detect it as an object as distinct from its background with 50% probability. It takes eight to recognize its type, a car versus a person, for example. It takes twelve to identify the object, say a truck rather than a car, but not a male versus a female. What if he or she wore a coat and a hat? This is also discussed in the book by Lloyd.[19]

In infrared systems, this measurement is replaced by the minimum resolvable temperature difference (MRTD). It is a plot of the minimum temperature difference that can be observed as a function of spatial frequency. In this case, instead of optical contrast as a function of spatial frequency, it is thermal contrast, the temperature difference as a function of spatial frequency.

As an interesting aside, this way of specifying optical image resolution was not readily accepted. Note that it is analogous to our hi-fi systems and auditory perception. We humans sense audio frequencies from about 100 Hz to 20 kHz (depending upon age and other factors). One famous optinerd, who shall remain anonymous to avoid embarrassment, stated in an international symposium, *"The eye is not the ear, and this will never work!"*[20] But it has and is now the preferred method.

Spectral resolution is a measure of how well narrow lines of differing wavelength can be discerned. The resolution is often designated by the wavelength or frequency difference, $d\lambda$ or $d\nu$, where the frequency may be in Hertz (cps) or in wave numbers specified as reciprocal centimeters. Note that a frequency in Hertz is the speed of light divided by the wavelength, but a frequency in wave numbers is just the reciprocal of the wavelength expressed in centimeters. As an example, a wavelength of 10 μm corresponds to a frequency of 3×10^{13} Hertz and 1000 cm^{-1} wave numbers. Although wave numbers can be a convenient way to express a frequency it will not be used in this book.

The quality of a spectral instrument is also specified as the resolving power, the power to resolve spectral lines. It is a measure in which bigger is better (the American way) but also a useful way of discussing these instruments. It is defined as the wavelength divided by the spectral width, $\lambda/d\lambda$. It is easily shown that this is the same as $\nu/d\nu$. Some authors call it resolvance.

---

[18]Johnson, J. "Analysis of image forming systems," in Image Intensifier Symposium, AD 220160 (Warfare Electrical Engineering Department, U.S. Army Research and Development Laboratories, Ft. Belvoir, Va., 1958), pp. 244–273.

[19]Lloyd, J. *Thermal Imaging Systems*, Plenum Press (1945).

[20]I heard him say it. He was a well-respected guy. It was in Paris.

Spectrometers

The resolution of a spectrometer is a measure of how well two closely spaced spectral lines can be observed as two rather than a single smear. Whereas the resolution of telescopes, microscopes and imaging devices is determined by diffraction, the resolution of a spectrometer is determined by interference. The resolution of a spectral line is usually designated as $d\lambda$ or $\delta\lambda$. The resolution is usually specified as resolving power that is defined as the wavelength divided by the spectral width $\lambda/d\lambda$, where λ is the central wavelength of the line. Resolving power can also be expressed the same way in terms of frequencies, f/df or v/dv.

The free spectral range is closely related to resolution and resolving power. It is the range over which there are no spectral lines, the distance between lines. When spectral lines overlap, the results can be confusing or even indecipherable.

Unfortunately, the nomenclature is not used in the same way by everyone. Sometimes the spectral width of a line is designated $\delta\lambda$; sometimes it is the separation between lines, and the free spectral range is usually $\Delta\lambda$.

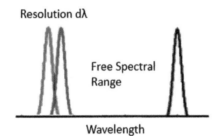

Figure 1.13.6   Resolution and free spectral range.

Interferometers

Interferometers generate interference patterns that are highs and lows of light amplitudes and intensities. They can range from large differences in the highs and lows to no difference at all, i.e., a big smudge. The measure of this is fringe visibility. It is the same as modulation: the difference of the maximum and minimum intensities divided by their sum. It is a measure of the coherence of the interfering waves and was used by Michelson in his stellar interferometer. If the waves are almost monochromatic, coherent waves, then they interfere with very distinct patterns. Otherwise, the pattern is blurry.

Temporal resolution is how well we can determine things that happen in time sequence. In a Google search, it is how long it takes to get an answer. In a home robbery, it is how long it takes for a cop to get there. In detectors, it is their response time. That is the time it takes for an input pulse to raise the output to 63% of its maximum. In LED displays, it is usually specified as the time it takes to go from 10% output to 90% output. Long response times lead to blurred images.

The temporal response of the human eye is very, very complicated. It depends upon age, luminous intensity, waveform, foveal versus peripheral vision, contrast and even whether you are in a good or bad mood. Really! When you are in a good mood, you are more perceptive. When you are in a bad mood, you are more introspective. A nominal value for response time is 0.1 second. But the specification is usually in terms of the flicker fusion rate, the rate at which a varying intensity input becomes constant. Typical flicker fusion rates are 30 to 100 cycles per second.

1.14 Aberrations

According to the online dictionary definition, an aberration is
> *a departure from what is normal, usual, or expected, typically one that is unwelcome.*

That is certainly the case here. An optical or lens aberration is a departure from what we want, unwelcome, unexpected. The ideal is perfect imagery.

There are two major types of aberrations, chromatic and monochromatic.

Chromatic aberrations occur because the value of the refractive index of the lens varies with wavelength. It decreases nonlinearly from the blue to the red. BK-7 glass, for instance, ranges from 1.532 at 400 nm to 1.513 at 700 nm, a change of 1.25%. That is about the change that will occur in the focus, but it is important because many small differences are important in optics.

Chromatic aberration presents itself as on-axis (**axial chromatic aberration**, Figure 1.14.1) wherein the separation is along the axis, or off-axis (**lateral chromatic aberration**, Figure 1.14.2). In the latter case, the spread is sidewise, as well.

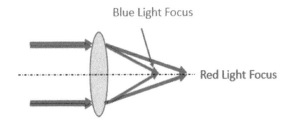

Figure 1.14.1 Axial chromatism.

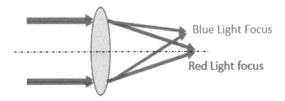

Figure 1.14.2 Lateral chromatism.

The only way to correct for chromatic aberrations is with two or more lenses that have different refractive index profiles, in which the respective refractive indices vary differently with wavelength. This is the case with almost any two different materials. The design issue is choosing the right combination.

Monochromatic aberrations all have a geometrical origin. They combine with the chromatic ones but are considered separately.

Spherical aberration is the first of these. A spherical shape is just not the right one to bring a point source to a sharp focus. The outer portions of the spherical lens bend (refract) the light more than the inner portions. This is shown schematically in Figure 1.14.3.

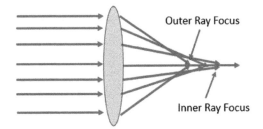

Figure 1.14.3 Spherical aberration.

Coma is the second monochromatic aberration. Even if the lens is corrected for spherical aberration, it still does not work perfectly for rays that are at an angle to the optical axis. One way to think about this is that the shape of the lens is not the same when viewed at an angle as it is when viewed head on. The pattern of the images of a point source is comet-like; ergo, comatic aberration or coma. Figure 1.14.4 is a representation of this.

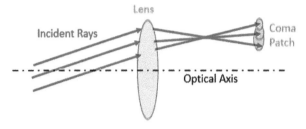

Figure 1.14.4   Coma.

The rays at the outer portion, the thinner potion of the lens, are bent the least and focus the worst. The farther in, the better the imagery and the farther from the axis.

The third (and this is the usual order of presentation, although they occur simultaneously) is **astigmatism**. This is an aberration that occurs when the object is even further off axis. The result is an image that is largely two perpendicular lines. Imagine a lens arranged perpendicular to the plane of the paper. Tilt it to the left about 45 degrees. The collection of parallel rays experiences a tilted lens with effectively a different shape. But the bunch of horizontal rays, arranged perpendicular to the paper, do not sense such a change. It is as if the two sets of rays encounter two different lenses with different magnifications and focal lengths.

Curvature of focus occurs because the constant distance from the lens to an image surface is a radius. The points on a plane image surface are not equally distant from the lens. It is called **curvature of field** when applied to an image.

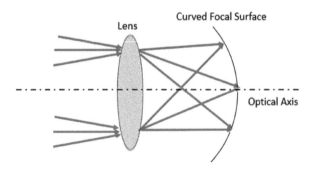

Figure 1.14.5   Curvature of focus.

Distortion can be either an innie or an outie. The image of a rectangle with straight sides can have either concave or convex sides, resembling either some pin cushions or a barrel. It is caused by a variation of magnification across the lens.

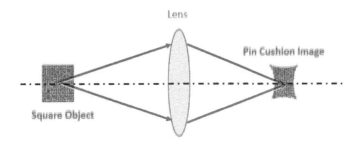

Figure 1.14.6   Pin cushion distortion.

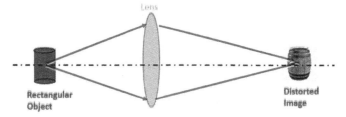

Figure 1.14.7   Barrel distortion.

These are the so-called **third order aberrations**. The calculations of all these images requires the use of the sine function. As with most functions, the sine can be represented as an infinite series of terms. The sineθ function series is $\theta - \theta^3/3! + \theta^5/5! \ldots$ where 3! means $3 \times 2 \times 1$ and so on. The first order wherein only the angle is considered is called the **paraxial approximation** because it is valid for angles close to the axis, only small angles almost **para**llel to the optical axis. The third-order aberrations are the first corrections to the assumption that the angle is small. In most design programs, even the fifth order is considered.

These monochromatic aberrations can be corrected by adjusting the different curvatures and spacings of the lenses and/or mirrors. Curvature of field, for instance, can be corrected by using two lenses with opposite curvatures. Aspherical surfaces are a modern tool for monochromatic aberration correction. They can be considered because we can now make them easily with computer-controlled grinders and polishers.

1.15 Ray Tracing

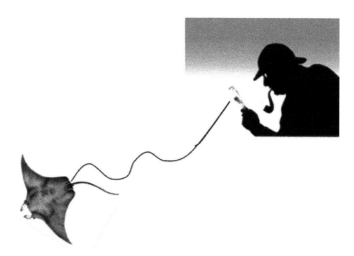

Ray tracing is the process by which optical systems, especially lenses are designed. The process is an exhaustive repetition of simple geometric calculations and Snell's law. This section describes how it is done using a simple spherical lens as an example.

The radius and material of the spherical lens are known, and an incident ray parallel to the optical axis at a height h is assumed. The incident angle can be calculated based on the Rh triangle. It is the same as the angle R makes with the optical axis. It is the inverse sine of h/R.

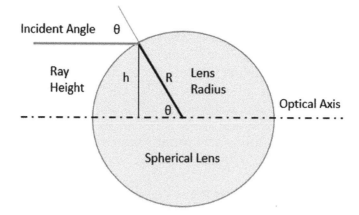

Figure 1.15.1   Step one.

So, as shown in Figure 1.15.2, the refracted angle is θ' by Snell's law. The ray exits the sphere at a radius. So, as shown, there is an equilateral triangle of two radii and two equal angles, which determines the exiting angle. The exiting height h' can be found as R times the sine of $\theta + 2\theta'$ ($\theta + 180 - (180 - 2\theta')$).

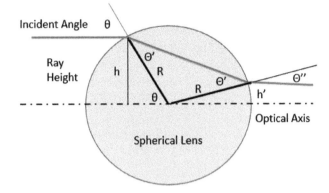

Figure 1.15.2 Next step.

Then do it all over again for the next lens and for more rays at more angles and for more lenses.

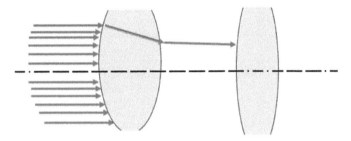

Figure 1.15.3 Step three.

Back in my days as a student we learned how to set up a table to calculate one ray by hand.[21] It took about fifteen minutes. Table 1.15.1 is a representative example.

Table 1.15.1

h	1.5	1.0	0.5	0
log h				
colog r_1				
logsin φ_1				
log n				
colog n'				
logsin $\varphi_{1'}$				
$\varphi_{1'}$				
φ_1				
θ'				
colog sin θ'				
log sin φ_1'				
log r_1				
log $r_1 - r_1'$				
$r_1 - r_1'$				
r_1'				

---

[21]Jenkins, F. and White, H. *Fundamentals of Optics,* McGraw Hill (1937).

I would not venture to take you through that maze; that is why it is empty. It is the way it was done in the early 1950s, adding logarithms and such, before the era of calculators much less computers and Mathcad. Suffice it to say that it is the same calculation that I showed above. It is done for four different incoming ray heights, h. shown in row one.

Later, my friend, Bob Shannon, and his colleagues at Rochester learned how to use their left hands to enter data into Marchant computers so they could pull the lever more efficiently with their rights. Later they entered the programs into computers like ENIAC on stacks of IBM cards. Today we can do all of that and more on our laptops.

The various programs allow us to enter almost any number of lenses of almost any circularly symmetric shape with any separations and any of very many available materials. The programs then calculate a blur circle, a Strehl ratio, an MTF for a host of different field angles. (see Section 1.13. Resolution). Then, *mirabile dictu*, they can optimize the system according to our wishes and hopes.

There are many optical design programs. The leading ones are Zemax, Oslo, Synopsis and Code V. Some are free. I have not tried them all, but I can recommend Zemax. The original version was written by Ken Moore when he was one of our students at the Wyant College of Optical Sciences, and it has been updated ever since. It was originally written to be compatible with modern computer languages and is user friendly. The name is derived from Ze added in front of his dog's name, Max. Max was already copyrighted.

Oslo, Optical Software for Layout and Optimization, was written by Doug Sinclair, one of my colleagues when he was at the Institute of Optics. It was not so user friendly then, back in the early 1970s but it has been updated.

Code V was written by Tom Harris and used internally at Optical Research Associates from 1963 and released for use in 1975. It too has been updated and made user friendly.

I believe they have all been watching each other and saying, *I can do that*. Then they do, and the programs become much more similar in capabilities. A nice review of these three programs is on the net.[22]

Other programs include Tracepro, FRED, ASAP, Optica and probably some others of which I am unaware. They are all listed on the net along with free trial versions.

The design process starts with the user inserting a lens prescription. He specifies the thickness, the material and the curvatures and separations of each element. She then enters a set of field angles. Then a set of parameters to optimize, like four of the six curvatures. The program will change these surfaces a little at a time and assess the result. If it is better, it will go further; if not, go back. This entire process takes maybe a minute. The new

---

[22]"Design software: which package do you need?" optics.org.

result is presented. The user can then specify a new set of parameters to optimize. A reasonable lens can be designed and optimized in just a few minutes. An unreasonable lens with 20 elements will take an unreasonable length of time!

It is possible to start with two or three plane parallel plates and have the program find a solution, but it is not advisable. One should start with a reasonable design. The programs all have libraries of thousands of such lenses. You should not try to optimize too many things all at once. You should have some understanding of aberrations. These are things I have learned by experience and have heard from my friends, Warren Smith,[23] Bob Shannon,[24] and Bob Fischer.[25]

The latest wrinkles in these programs are the techniques for finding global optima and handling non-rotational optics. Imagine a terrain with many hills and valleys. Some hills are taller than others and some valleys deeper than others. The global optimum is the deepest valley. But if you are on the other side of the hill in a shallower valley, you may not get over the hill. Finding the global optimum is like that. There are several techniques, but one is just perturbing the system significantly enough that it gets to the other side of the hill. In this case being over the hill is a good thing! I think I am there.

I find it fascinating that these design programs are used to design the exquisite, complicated lenses of photolithographs that are used to make the computers that the lens design programs use to design the lenses.

---

[23]Smith, W. *Modern Optical Engineering*, McGraw Hill (1990).

[24]Shannon, R. *The Art and Science of Optical Design*, Cambridge University Press (1997).

[25]Fischer, R. and Tadic-Galeb, B. *Optical System Design*, McGraw Hill (2000).

1.16 Radiometry and Photometry

These two optical disciplines deal with the calculation and measurement of radiation and its transfer and detection. Radiometry is more general; photometry is older. They do **not** deal with the measurement of radios and photos!

The radiation terms of Energy, Power, Emittance, Irradiance, Intensity and Radiance are described in Section 1.9 Propagation. Bob Jones unified these in a sense, defining fluometry as the treatment of flux. These are all fluxes, the time rate of energy, of photons and of visible energy. He named them *quantity, fluence, exitance, incidance intensity* and *sterance*. I like to think of them in terms of students. The *quantity* is the number on campus. *Fluence* is the matriculation or graduation rate. *Incidance* is the number per unit area coming at me down the mall or quad. *Exitance* is them running away. *Intensity* might be the spread of them as they race from the door at the end of my class. I leave the *sterance* or radiance of students to you, but it might be their spread from the assembly hall at graduation as they enter the real world. I submit that most students are radiant then.

The calculation of the propagation of radiation is in Section 1.9. Measurements of reflection, emissivity and so on are in those sections.

In radiometric terms these are often specified as power, emittance, irradiance, radiance, and intensity. In photometry they are lumens, lux, illuminance, luminance, and luminous intensity. They are geometrically the same. The difference is the eyeball.

The relationship between radiometry and photometry is the luminous efficacy, the sensitivity curve of our eyes. The two curves are approximately like those shown in Figure 1.16.1. The photopic curve in red is how we sense bright light; the other, the scotopic curve, is when our rods are active in dim light or darkness.

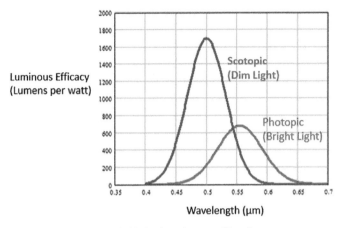

Figure 1.16.1 Luminous efficacies.

The lumen, the basic flux of photometry is the spectral power times the luminous efficacy curve. It is almost always the one for bright light shown in red which has a peak of 683 watts per lumen at about 0.55 μm. Scotopic vision. the vision of our rods in dim light, has a higher, more sensitive peak at about 507 μm.

A radiometric or photometric measurement is made with a radiometer or photometer, logically enough. Each has a responsivity, that is an output divided by the input that is a function of many things. The output is usually in some electronic form that has been calibrated by some standard and therefore gives a measurement in volts or amps that are equivalent to watts or lumens. The responsivity can be written as

$$R(\lambda,\ t,\ \theta,\ \varphi,\ x,\ y,\ s,\ p,\ RH,\ \dots)$$

It is a function of wavelength, time, angles, position, polarization, relative humidity and anything else you can think of. That is why the mantra in my classes on radiometry was Think of Everything.

It is not a fiction that the responsivity is a function of the moisture in the air. The radiometer we built for Venus measurements[26] had a different responsivity in the morning than it did in the afternoon. It was due to a difference in relative humidity. It is shown in the introductory photo.

Examples of radiometric calculations to determine if certain instruments are sensitive enough are in Sections A.4.3, A.4.5, A.4.28 and elsewhere.

---

[26]Tomasko, M. et al., "Pioneer Venus sounder probe solar flux radiometer," *IEEE Transactions on Geoscience and Remote Sensing*, GE-18 (1980).

Chapter 2
Optics in Nature

Light is all around us in nature: blue skies, red sunsets, mirages, and more. This chapter describes the concepts underlying the more familiar of them. But there are many more. They include why leaves turn color in the fall, how chameleons know to change their colors and how they do it, and how migratory birds find their way. These are described in the classic book *Rainbows, Halos, and Glories* by Robert Greenler[1] and elsewhere online.

As you will discover, blue skies are created by the spectrally differential scattering of small particles. Red sunsets are created in the same, but somewhat opposite, way and are enhanced by spectrally uniform scattering of the large particles in clouds. Sunsets are more vivid in the American Southwest, where the air is drier.

Hummingbirds turn their better sides to us and show us different colors by virtue of the interference properties of the layers of their feathers.

Rattlesnakes and pit vipers find kangaroo rats because the rats are warm blooded. Mosquitos find us because we exhale warm carbon dioxide.

2.1 Blue Skies

---

[1]Greenler, R. *Rainbows, Halos, and Glories.* SPIE Press (2020) and Cambridge University Press (1990).

Skies are blue because the blue part of the almost white light of the sun is scattered more than the other colors. As described in Section 1.7, the scattering of the molecules in the air is Rayleigh scattering. The degree of scatter, therefore, is proportional to the inverse fourth power of the wavelength. Red light has a wavelength of about 0.8 μm, whereas blue is about 0.4 μm, a factor of two. So blue light is 2^4 or 16 times more effective at scattering sunlight.

A nice, clear sky is mostly oxygen and nitrogen. Oxygen molecules have a diameter of 0.299 nm. Nitrogen has a diameter of 0.305 nm. Both of these are small enough to qualify as Rayleigh scatterers.

But the atmosphere consists of more than these two components. It also consists of water vapor (more so in the Eastern, Southern, and Northeastern United States; less over the western deserts). Water vapor is a collection of water droplets (H_2O, two hydrogens and one oxygen) that are about 10 to 15 μm in diameter, around 50 times the wavelength of yellow light. It is a Mie scatterer, as illustrated in Figure 2.1.1. Both the red and the blue waves encounter a surface that is large with respect to their wavelengths. They are therefore reflected according to the law of specular reflection. This type of reflection gives the atmosphere a whitish hue.

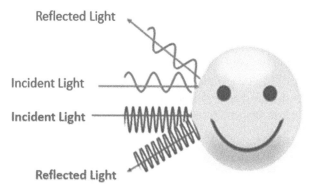

Figure 2.1.1   Mie scattering.

Thus, in desert climates with low humidity, the skies are bluer. [I like to think of them as less egotistical: there is less Mie in them! (Mie is pronounced "mee").]

That's a happy desert raindrop. All raindrops are happy ones in the desert.

2.2 Blue Water and Blue Ice

Why is water blue? Why is my pool (Figure 2.2.1) blue when it is pure water (with a little chlorine)? Why are the lakes in northern lower Michigan (such as Walloon Lake, Figure 2.2.2) blue? They do not even have chlorine, but they are pure. All such bodies of water that are nice and clean and contain no algae or similar organisms (e.g., oceans) look blue.

There are two reasons: reflection and transmission. On a clear day with a blue sky, the surface reflects specularly at about 2%, which makes it a little blue. But the main reason is that the water has maximum transmission in the blue part of the spectrum. Other colors are largely absorbed. These waters look blue even when the sky is overcast.

Figure 2.2.1 My pool.

Figure 2.2.2 Walloon Lake.

Figure 2.2.3 is a curve of the absorption of pure water in the visible[2] with a spectrum below it for reference. The point of lowest absorption is that of highest transmission. The horizontal axis is the wavelength in micrometers.

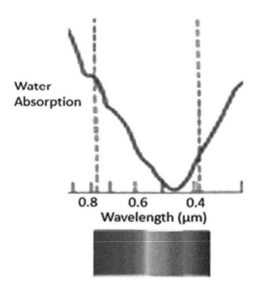

Figure 2.2.3 Water absorption.

Ice looks blue for the same reasons. Ice is solid water and has essentially the same transmission characteristics. Some ice floes look blue in part because they reflect the blue sky mostly near their top. But the main phenomenon is

---

[2]Wolfe, W. and Zissis, G. eds, *The Infrared Handbook*, US Government Printing Office (1978).

transmission. The deeper parts of the floes are even bluer because the deeper, denser parts squeeze out more air pockets that cause Mie scattering.

Icebergs are bluer at the bottom for the same reason that skies are bluer in the deserts. They are less egotistical (like the blue skies in Arizona)!

Figure 2.2.4   Icebergs with blue bottoms.

2.3 The Green Flash

The green flash, sometimes more yellow than green, is an atmospheric phenomenon that occurs at the horizon at sunset for just a few seconds. It is usually seen over the ocean or other smooth horizon. It is a result of atmospheric refraction.

The sun shines on the atmopshere that surrounds our planet. That atmosphere, because it is air rather than the vacuum of outer space, refracts the light. The refraction angle is different for the different colors. The reds are bent the least. The yellows in the middle of our sensitivity curve skim the surface. The blues are bent into the planet. This phenomenon is shown schematically in Figure 2.3.1.

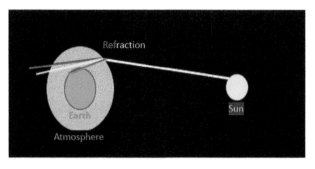

Figure 2.3.1   The green flash.

It may be useful to think of the atmosphere as a giant prism. The reds are bent the least, and so on, just like a prism.

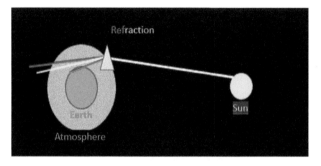

Figure 2.3.2   Prism representation.

I saw this once. I was on a bus with my fellow optinerds along the California coastline. Imagine the scene as we all saw this rare occurrence.

2.4 Hummingbirds

Hummingbirds (or "hummers" for short) whir their wings to hum at us, and they also flash their varying colors to please us—and maybe to entice or impress other hummers. As I am sure you have noticed, the colors seem to change as the birds change their positions with respect to the sun. This is the result of optical interference.[3] Hummers have layers of feathers on the outer portions of their bodies. These layers give rise to interference. The feathers themselves are brownish gray. From one viewpoint, the layer thickness is exactly a full wavelength of light, thereby generating constructive interference (e.g., for red light, as shown in Figure 2.4.1), but as the bird's position relative to the sun changes, the effective thickness can shorten and become exactly a full wavelength for blue light, as in Figure 2.4.2.

Figure 2.4.1 He's red.   **Figure 2.4.2** No, he's blue (or blue green).

---

[3]Greenewalt, C., Brandt, W., and Friel, D. "The iridescent colors of hummingbird feathers," *Proc. Amer. Philo. Soc.* **104**(3), 249–253 (1960).

The diagram in Figure 2.4.3 shows constructive interference for red light incident normally to the feather layer. There is constructive interference between the wave reflected from the front and the one reflected from the back. The bird gleams red.

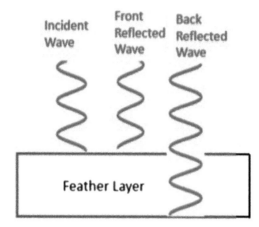

Figure 2.4.3   Hummingbird-feather constructive interference.

At the same time, in the same place, and in the same orientation, a wave with a different wavelength will interfere destructively. The hummer will not look blue.

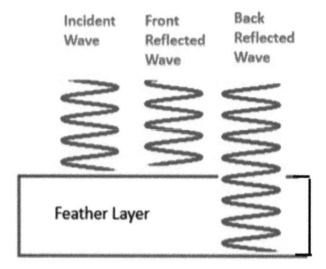

Figure 2.4.4   Feather destructive interference.

2.5 Lightning

We have all had it happen to us: One of our playful friends rubs his feet on the rug. He generates an electric charge and uses it to zap us. It is a small charge, but it is the same phenomenon as lightning and Saint Elmo's Fire on a different scale and in a different place.

It is a spark that is caused by a voltage difference between two bodies through an insulating medium. A voltage difference builds up until it exceeds the dielectric strength of the medium, and the spark occurs to relieve that difference. In air, the breakdown value is approximately 3 megavolts per meter, but it varies with humidity and pressure. That value is equivalent to 3 kilovolts per millimeter.

The high-altitude clouds rub on one another and generate a voltage difference between them and the ground (or between each other). When the difference exceeds the breakdown value of the dielectric strength of air, a spark occurs. The breakdown value is reduced markedly by the moisture in the air during a storm. The giant spark of lightning can be inside a cloud between two clouds or between the cloud and the ground. It is the latter that is dangerous.

There are approximately 45 lightning strokes on the surface of the Earth per second, and each one lasts for about a millisecond. A flash of lightning is a complicated phenomenon. It starts with a short step leader that prepares the path and consists of four or more individual strokes. It creates a plasma path in the air that is a region of free electrons. These electrons radiate as a blackbody at temperatures high enough to produce the white and somewhat bluish colors of lightning bolts. The spectrum even goes into the near infrared.

The zap from your playful friend has some awesome numbers. The voltage difference you generate by shuffling may be as much as 5000 volts. Then the zap goes over a few millimeters. That is an impressive voltage, but the current is small, and the power is on the order of millijoules (thousandths of watt seconds). No one gets hurt. But beware of the effect when dealing with electronic devices. Make sure you are grounded when you do something like replacing the inkjet on your printer.

Saint Elmo's Fire is usually observed on the tips of sailing vessels (St. Elmo is the patron saint of sailors). It is also caused by a voltage difference between clouds and ground. In this case, as opposed to lightning, it triggers at a lower voltage difference especially where there are pointed objects, such as the tops of masts. It radiates the colors of oxygen and nitrogen, the constituents of air.

2.6 Mirages

We have often seen mirages in the movies of camels and oases in the Sahara, but they also exist in the Southwest United States, where the sun is hot, and the roads are hot and dry. In front of us, there seems to be water on the road even though it has not rained in a month. It is a mirage caused by the hot road warming the air above it. The introductory figure is a mirage on the road between Hatch, CO and Deming, NM.

The air and the ray of light can be portrayed as in Figure 2.6.1, where the road is shown as black, and the air is increasingly dense and portrayed as darker with altitude. The air is rarest right next to the road where it is warmest, and it increases in density with altitude because it is less warm, and then decreases at higher altitudes, where it gradually disappears. A ray of light from the blue sky will be refracted by this density gradient, a refractive index gradient, as shown in the figure. A ray of light from the sky (shown entering on the upper right) looks like it is reflected from the road due to the gradual refraction of the ray by the refractive index gradient caused by the air density gradient caused by the temperature gradient.

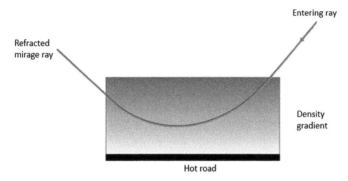

Figure 2.6.1 Heated-road model.

It may be easier to understand this phenomenon with a different model, one that is not quite true but closely resembles the situation. Imagine a layered structure rather than a continuous gradient, as shown in Figure 2.6.2. The ray from the sky enters at the upper right and is refracted toward the normal because it is rare to dense refraction. When the ray reaches the next layer, it refracts away from the normal since it is dense to rare refraction. This behavior repeats for each layer until it exits aimed at the observer.

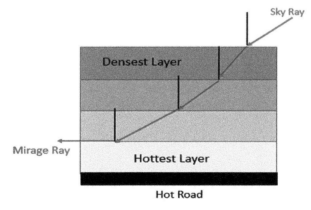

Figure 2.6.2 Layered model.

But why does the road look wet? It is all in your imagination: "There is a reflection of the sky from the road. The road would not make such a good reflection. It must be water." It ain't!

A closely related phenomenon is the Fata Morgana.[4] [I prefer the inaccurate translation, *the fate of Morgana*, not the literal one, *fairy Morgana*.] One example was published online by CBS News,[5] which I have recreated in Figure 2.6.3. A ship hovers in the air above the ocean. This optical phenomenon is named after the literary character Morgan le Fay, sorceress of Camelot and half-sister and antagonist of King Arthur. It was believed that these mirages, often of castles and islands, lured sailors to their deaths chasing them.

The phenomenon is caused by a temperature inversion that is opposite that of the road mirage described above. If you can imagine Figure 2.6.2 turned upside down, you can see how a ship on the surface of the ocean appears up in the air.

Figure 2.6.3   A Fata Morgana illustration.

The Fata Morgana may have been the inspiration for *The Flying Dutchman*. It may also have contributed to the sinking of the Titanic by hiding the iceberg and the Titanic from a nearby ship.[6]

[4]Wikipedia. "Fata Morgana (mirage)," https://en.wikipedia.org/wiki/Fata_Morgana_(mirage).
[5]CBS News. "Photos appear to show a giant ship hovering over the water off the English coast," https://www.cbsnews.com/news/floating-ship-optical-illusion-superior-mirage-cornwall-england (March 8, 2021).
[6]Smithsonian Magazine Online. "Did the Titanic Sink Because of an Optical Illusion?" https://www.smithsonianmag.com/science-nature/did-the-titanic-sink-because-of-an-optical-illusion-102040309 (March 1, 2012).

2.7 Mosquitoes

Who needs them? But they are here. Not so much in Arizona, where it is very dry and there are few places for them to breed. I think it is well known that the female of the species is the only one who bites. Not so well known is that it is the *pregnant* female who bites. She is after food for her new expected young and her own nourishment needs.

Mosquitoes home in on warm carbon dioxide from about 30 feet away. One countermeasure is to not exhale – ever. Inhaling is okay; exhaling is not! As mosquitoes get closer, they home in on various odors on the skin. One advertised repellant is the sound of an amorous male mosquito. I have no idea how they determined the sound or replicated it.[7] The online recommendations of how well it works are mixed.

2.8 Pit Vipers

These reptiles are a species of snakes that includes sidewinders and other rattlesnakes. They have small pits in their cheeks that act as pinhole cameras

---

[7]Amazon. "William and Joseph Mosquito Annoyer reviews," https://www.amazon.com/William-Joseph-Mosquito-Annoyer-Grey/dp/B000KNE47C.

to form an image on an array of heat sensitive nerve endings, as shown in Figure 2.8.1.

Figure 2.8.1 Pit-viper pit.

The sensors, situated a little to the side and below the eyes, are true pinhole cameras. They are little holes with about 2000 temperature-sensitive nerves behind them. The image is not very high resolution, but the nerves are very sensitive. Temperature differences of a small fraction of a degree (in Fahrenheit, Celsius, or Kelvin) are easily detected. Since there are two pits, one on each side of the head, they can also do a bit of ranging by triangulation (with a very short baseline).

Pit vipers usually wait in ambush for a warm body to come by, and then they strike. There is some evidence that they can follow the thermal trail left by their prey.[8]

There are about 150 different kinds of pit vipers, ranging from the hump-nosed viper, which is about half a meter long, to the Bushmaster, which is a little more than 3.5 meters long. They live in Eurasia and the Americas, notably the rattlesnakes of the southwest. Pit vipers in the United States include rattlesnakes but also water moccasins and copperheads. They are nocturnal hunters. There are 36 kinds of rattlesnakes, and 13 of them live in Arizona.[9] Please stay out of my yard!

---

[8]New York Times News Service. "Feared pit vipers yield their biological secrets of hunting, mating, killing." *The Baltimore Sun*, https://www.baltimoresun.com/news/bs-xpm-1991-12-08-1991342047-story.html (December 8, 1991).

[9]Brennan, T. and Holycross, A. *A Field Guide to Amphibians and Reptiles in Arizona*. Arizona Fish and Game Department. https://www.rosemonteis.us/files/references/016745.pdf (2006).

2.9 Rainbows

Rainbows are glorious, interesting, and once even a reported promise from God[10] that there will be no more floods. They are formed by the refraction and reflection of light by water droplets. A beautiful double rainbow over Hidden Valley, Arizona is shown here. It is a true Arizona rainbow, as attested by my granddaughter with the Big A on her shirt.

Figure 2.9.1 shows how this phenomenon can happen. A ray of sunlight, shown in white, impinges upon the front surface of the almost spherical drop, where it is refracted toward the rear surface. It is reflected from there to the front surface and then refracted out. The blue ray is refracted a little more than the red ray, as shown. This gives rise to the primary rainbow with the red ray on top because it is more vertical. The other colors fill in below it down to the blue ray. Figure 2.9.2 is an actual ray trace that shows the rays closer together and the transmitted ray.

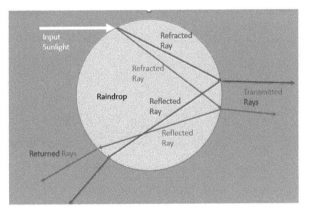

Figure 2.9.1   Rainbow ray schematic.

---

[10]Genesis 9:12-16.

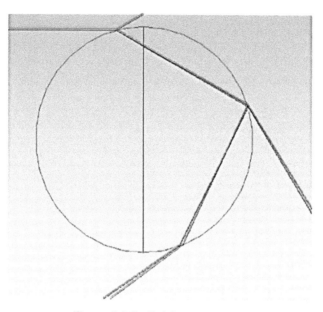

Figure 2.9.2   Rainbow ray trace.

Figure 2.9.3 shows a rainbow over Sabino Canyon in Tucson, AZ with a faint seconday bow in the upper left. Figure 2.9.4 shows one over Victoria Falls, on the border of Zambia and Zimbabwe. You can almost see the secondary bow.

Figure 2.9.3   Rainbow over Sabino Canyon.

Figure 2.9.4 Rainbow over Victoria Falls.

The secondary bow is formed when a ray is refracted into the droplet and makes two reflections off the back surface. The color arrangement is reversed as a result of the second reflection, and it is much dimmer as a result of the second reflection (shown schematically in Figure 2.9.5).

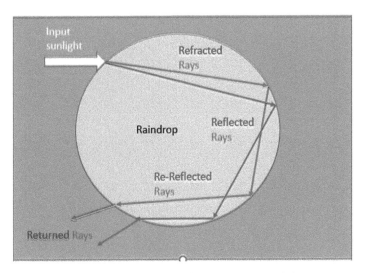

Figure 2.9.5 Formation of the secondary rainbow.

A tertiary rainbow is a rarely seen event. The color arrangement is again reversed so that it agrees with that of the primary bow, and it is much dimmer. The normal reflectivity of an air–water interface is about two percent (0.02), and the transmission, assuming no absorption, is 98%. Thus, the primary bow reduces the sunlight by a factor of two transmissions and one reflection: 0.98 × 0.98 × 0.02, or 0.0192. The secondary bow has an additional reflection at two percent and results in a total diminution factor of 0.000384 (about 0.004%). The tertiary bow is only 0.0000008 of the sun on it. Fortunately, the sun is very bright, but the tertiary bow is still **very** dim.

I have found no evidence of the existence of a quaternary bow, but there is such a thing as a supernumerary bow, wherein more pale blues and greens appear adjacent to the blues of the primary, and there may even be more bows.[11] This is caused by interference effects and is very seldom seen.

The rainbow always appears at about 42 degrees from the solar incidence angle. This is a result of Fermat's principle, which states that light follows the path of minimum time. In this case, the minimum time is the minimum total deviation angle. It takes the least time if it takes the smallest angle. As shown in Figure 2.9.6, there is an angle of incidence α upon entering, and its corresponding angle of refraction β, then an angle of incidence on the rear surface and its angle of reflection. Both are equal to β due to similar triangles and the law of reflection. Finally, there are angles of incidence and reflection upon exiting the raindrop. The sum of all these angles $2\alpha + 4\beta$ must be a minimum according to Fermat. This principle is described in Appendix A.2.9 using the required differential calculus.

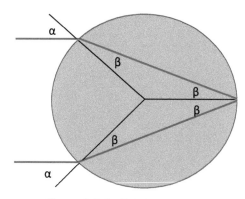

Figure 2.9.6   Rainbow angles.

I cannot close this section without a reference to a bow that you will probably never see anywhere else: a spiderbow. I saw the refraction from a

[11]Zhang, M. "Rainbowception: Photographer Snaps Rare Supernumerary Rainbow." Petapixel, https://petapixel.com/2018/09/24/rainbowception-photographer-snaps-rare-supernumerary-rainbow (September 24, 2018).

single spider strand outside my window. It is the same kind of spectral refraction seen in rainbows.

2.10 Sundogs and Halos

Sundogs are parts of full halos around the sun. They are formed in almost the same way as rainbows, but they arise from the interaction of sunlight with ice crystals.[12] They might be called icebows. Rather than refracting and reflecting from water droplets, as in rainbows, the light is refracted and reflected from tiny ice crystals in the upper atmosphere (where it is cold

---

[12]Greenler, R. *Rainbows, Halos, and Glories*, SPIE Press (2020) and Cambridge University Press (1990).

enough to form ice). The crystals are not familiar triangular prisms but hexagonal in form, although they still form spectra. As a result of their form, and as opposed to droplets, the arcs (bows) are at approximately 22 degrees. In theory, they form complete circles, but usually only sections are seen, called sundogs and pillars.

The introductory picture shows two sundogs over my daughter's house in Erie, Colorado. Figure 2.10.1 shows a more complete bow (halos). With a bit of imagination, you can see the entire circle.

Figure 2.10.1   Halos.

With even more imagination you can see the sun dogs very clearly!!!!!

2.11 Sunsets

Sunsets can be spectacular in Arizona (other places, too, but especially Arizona). They are at their best when there are clouds. The combination of Mie scattering from the clouds, which spreads the sunlight over the sky, and the Rayleigh scattering of the air, which filters the light, causes a spectacular sunset. As the sun descends below the horizon, its almost white light is scattered all over the western horizon by the clouds. The light then passes through the clear atmosphere of oxygen and nitrogen where mostly the red gets through due to the selective Rayleigh scattering of these particles, which are smaller than the wavelengths of the light.

Arizona and other sunsets in the Southwest are more vivid because there is less moisture in the air. The suspended droplets of water in more humid climes cause Mie scattering of the light that blurs the reds and yellows and induces a slightly whitish background.

There are certainly equally impressive sunrises for which all the same optical effects apply, but I have not seen them (because I sleep in).

Chapter 3
Optical Components

This chapter describes the individual optical elements that are used to make optical instruments. Everyone would agree that a light-emitting diode (LED) is an optical component, a single optical component. But it gets more arguable when you deal with prisms and lenses. A device called a prism is sometimes the combination of one or two individual prisms, but we still deal with it as a single prism. The same is true for lenses—a doublet lens is still considered a lens.

These are what I consider components: LEDs, baffles, blackbodies, and even lenses and prisms.

3.1 Baffles

Baffles are used to keep unwanted radiation from the image area of a telescope or other optical system. One form is the coronagraph, which is used to block the solar disc to view solar flares and similar phenomena. Another is the tubes that were designed for some of our anti-missile systems or the star tracker on the Apollo Lunar Excursion Module.

The **coronagraph** was invented by Bernard Lyot in 1931.[1] It operates in the image space of a telescope. (Section 3.10 explains that this is a field stop.) It is an occulting field stop, as shown in Figure 3.1.1. It blocks the central portion of the field of view. It is eponymously called a Lyot stop. The lens shown can be any type of optical system that creates an image of the sun, the image of which is shown in yellow with its solar halo of flares around it in faint yellow. The Lyot stop is the field stop shown in black. It blocks the central part of the sun and only lets its outer periphery be imaged by the second lens. It was used to study solar flares and the solar corona.

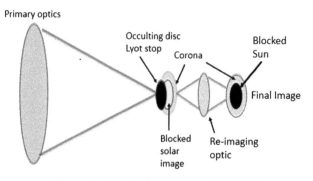

Figure 3.1.1   Lyot coronagraph field stop.

Baffle tubes operate in object space. They prevent radiation from out-of-field sources from getting to the image area. This is shown schematically in Figure 3.1.2, where it is assumed that there is some object to be imaged directly to the left, the arrow in blue. But there is a source off at an angle to the left and up (the circle in yellow). This is the case with some ballistic missile interceptors, e.g., the missile would be launched from a country to the east in the morning, and the rising sun is just out of the field of view. The simplest thing to do is make the tube long enough to shade the sun from the optical element, the primary mirror, as shown. The tube shown blocks the extreme ray from the sun from the mirror.

---

[1]Lyot, B. "A study of the solar corona and prominences without eclipses." *Monthly Notices of the Royal Astronomical Society* **99**, 580–594 (1939).

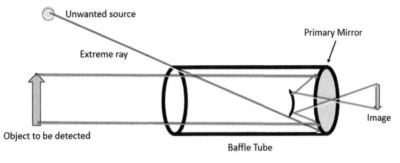

Figure 3.1.2 Basic baffle tube.

But the rays from that nasty source impinge upon the sides of the baffle tube and reflect to the primary mirror, as shown in Figure 3.1.3. They reflect either specularly or diffusely (or a bit of both), depending on the material of the baffle tube. Although important strides have been made in the fabrication of non-reflecting surfaces, so far nothing is perfect. Something else must be done. The procedure is to force as many reflections as possible before the unwanted radiation reaches the optical element.

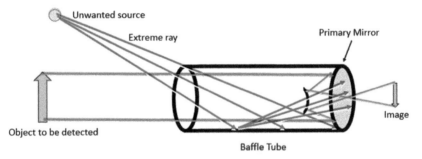

Figure 3.1.3 Baffle tube reflections.

This is where baffle vanes come in. One can determine where to place the first vane and how big it must be by constructing the second ray from the sun to the inside front edge of the tube. The vane shown in Figure 3.1.4 ensures that any ray from the sun at the assumed angle will reflect at least twice from the sides of the baffle.

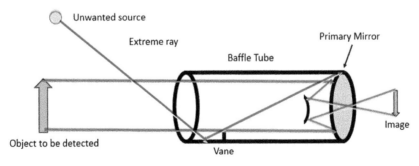

Figure 3.1.4 First vane design.

The process is to insert vanes to cause as many reflections as possible from the sides of the baffle tube to the mirror without restricting the light from the object to be sensed.

These are the concepts of baffle design.[2] The details can get much more complicated.

There are several computer programs available for the design of these baffles, notably GUERAP and ASAP.[3] One of my students applied ASAP to the improved performance of the Sidewinder missile optical guidance head.[4] She reduced the scatter to the detector by several orders of magnitude. The design of the Sidewinder infrared guidance is discussed further in Section 4.28.

Although baffle design seems straightforward and may not be very interesting, it was a considerable bone of contention in the design of the optical device that communicated between the first Apollo lander, the LEM, the Lunar Excursion Module, and the Apollo orbiter.[5] It has also been critical in the design of many of anti-ICBM detectors.[5]

3.2 Beam Splitters

There are beam splitters, and there are beam splitters. They all divide, split, an incident beam in some way. Some divide it spatially—they split light into two separate parts. Some divide it in amplitude. They send a portion of the intensity of the beam in one direction and the other portion in another direction. Some divide it into different polarizations.

---

[2]Fest, E. *Stray Light Analysis and Control*, SPIE Press (2013) [doi: 10.1117/3.1000980].

[3]*GUERAP II -Users Guide*, Defense Information Technical Center (1974), online; *About ASAP, The Breault Optical Design Program*, online.

[4]Fender, J. "An Investigation of Computer-Assisted Stray Radiation Analysis Programs," Dissertation, The University of Arizona (1981).

[5]Personal experience.

Spatial beam splitters are mostly reflecting solid prismatic forms, as shown in Figure 3.2.1. Light enters from the left and impinges on what is shown as a simple reflecting prism in the form of an axe blade. The input beam is divided into slightly less intense beams at right angles because this was assumed to be a 45-degree prism.

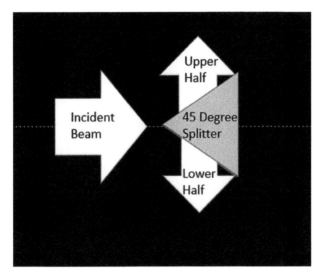

Figure 3.2.1 Spatial beam splitter.

Other configurations can divide the input beam into more subsidiary beams and aim them in other directions. For example, the spatial beam splitter shown in Figure 3.2.2 divides the input beam into four subsidiary beams and aims them about 10 degrees backward instead of 90 degrees up. The pyramid angles need to be five degrees more than 45 to do that.

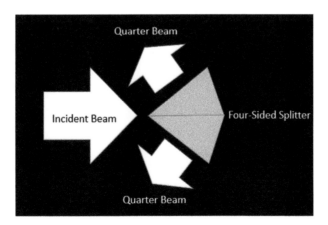

Figure 3.2.2 Four-sided spatial beam splitter.

The simplest form of an **amplitude beam splitter** is a transparent plane-parallel plate. In such a plate angled at 45-degrees an incident collimated beam will pass through the plate in the same direction displaced but not deviated nor changed in size. The reflected beam will be reflected 90 degrees and not changed in size. Both will be reduced in amplitude according to the reflectivity and transmissivity of the plate. This is shown schematically in Figure 3.2.3.

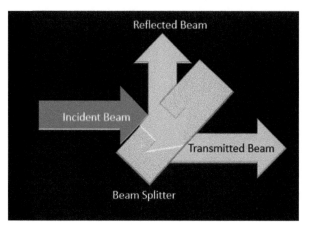

Figure 3.2.3   Simplified amplitude-beam-splitter geometry.

If it is just a plate of glass and the beam is visible light, the reflected beam will be about four percent of the amplitude of the incident beam and the transmitted one about 92%.

But it is not that simple. There will be multiple reflections. The incoming beam will be refracted down to the lower surface and reflected back to the upper surface, and it will join with the first reflected beam. This and other multiply reflected beams are shown in Figure 3.2.4. The multiply reflected beams are not coincident but displaced according to the thickness of the plate and its refractive index. This can be trouble.

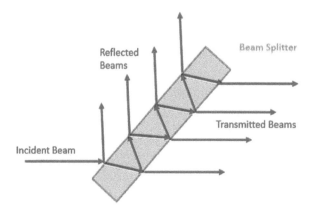

Figure 3.2.4   Multiple reflections.

The first reflected beam will have 4% the intensity of the incident beam. The secondary reflected beam will have losses induced by transmission (one minus the reflection) at the first surface, reflection at the second surface, and another loss in transmission at the top surface. So it is $0.96 \times 0.04 \times 0.96 = 0.036$, or 3.6% of the incident beam. Similarly, the top reflected beam will be 0.0006%. It can be ignored. The secondary transmitted beam has two reflections and two transmissions. It is 1.4% of the incident beam, compared to 93% for the primary transmitted beam. It is a small amount of about 1.5% and may be tolerable. But for some applications, these dim, displaced ghost beams are not acceptable.

One way to avoid this situation is to coat the first surface with a highly reflecting coating, say, 50%. Then the first reflected beam is 50%, and the first ghost beam is about 1%. Such a coating can be spectrally flat or spectrally varying. A second way to do this is to coat the second surface with an antireflection coating. Then there are no secondary reflections.

Another version of a beam splitter is a cube consisting of two triangular prisms that are attached at their hypoteni. Some are glued together, but others are held together by clamps. The simplest film is a metallic coating at about 50% reflectivity so that both beams are about equal in intensity. The multiple beam contributions are then negligible and congruent with the main beams.

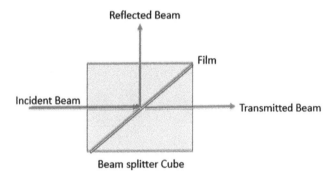

Figure 3.2.5 Beam splitter cube.

A cubic beam splitter has fewer and more benign ghost images. Figure 3.2.6 illustrates this. The primary and desired transmitted ray is shown in red. Part of it is reflected from the back surface as shown in yellow. It reflects directly back and directly down and back up and out. It is coincident with the primary beam. The upper beam, shown in blue is a bit more trouble. It is not coincident but is displaced by the refraction in the thin film. It is also reduced in amplitude by two reflections and two surface transmissions. For a glass cube with an epoxy film, there is virtually no reflection between the film and the glass. Both have a refractive index of about 1.5. Some things work right.

Exaggerated Film

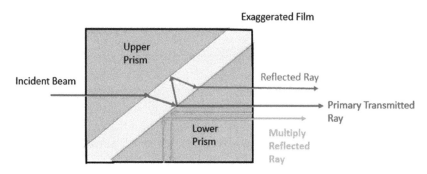

Figure 3.2.6   Detailed reflections in a beam splitter cube.

I consider polarizing beam splitters as polarizers. They are described in Section 3.13.

Interferometers use all sorts of beam splitters. Two important examples are the Michelson interferometer, which was used to prove the lack of a luminferous ether, and the Fizeau interferometer, which measured the speed of light in water and helped disprove Newton's corpuscle theory of light. A more modern example is the Sagnac interferometer, which was the basis of the fiber optic gyroscope.

3.3 Blackbodies

Blackbodies are theoretically perfect radiators. They are therefore fundamental to every type of radiation. In a crude sense, one can imagine that every radiating body has a blackbody inside it and an emissivity film around it that reduces the amount of radiation uniformly or in certain spectral regions. Of course, it is the properties of the material themselves that actually determine the emissivity, but this is a useful fiction. Nothing can radiate more at a given temperature either totally or in any part of the spectrum than a blackbody. The explanation of the spectrum of blackbody radiation by Planck led to the entire field of quantum mechanics.[6]

---

[6]Planck, M. "On an improvement of Wien's equation for the spectrum," *Verhandlungen der Deutschen physikalischen Gesellschaft* **2**, 202 (1900).

These theoretically perfect radiators are called blackbodies as a result of Kirchhoff's law, which holds that a good absorber is a good emitter.[7] His law states more precisely that the efficiency of emission is equal to the efficiency of absorption under identical conditions. Therefore, a perfectly black object, one that absorbs every single photon incident upon it, also emits every single photon that its temperature allows. Other bodies, gray bodies and colored bodies, are related to blackbodies by their efficiencies of emission, called emissivity. It has become customary to refer to perfect radiators as blackbodies. It seems simpler.

Blackbody radiation is a function of only temperature and wavelength. A representative blackbody spectrum for a body at 5778 K (solar surface temperature) is shown in Figure 3.3.1. There is a sharp rise from about 0.2 μm to a peak at about 0.5 μm and then an exponential decay to longer wavelengths. The peak is found by use of Wien's expression.[8] The product of the wavelength in μm and temperature in kelvin is 2898. Thus, the sun's surface at 5778 K peaks at 2898/5778, or 0.5 μm, in the yellow, as can be seen in the curve.

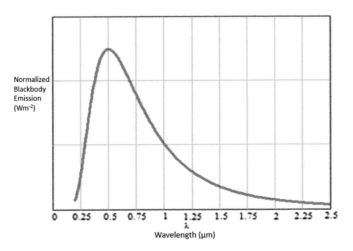

Figure 3.3.1   Representative blackbody curve.

A series of blackbody curves is shown in Figure 3.3.2 for decreasing temperatures: 6000 K, 5000 K, 4000 K, and 3000 K in purple, blue, green, and red, respectively. It illustrates the decrease in total radiation with decreasing temperature and the movement of the peak to longer

---

[7]Kirchhoff, G. "Über das Verhältnes zwischedn dem Emissionsvermögen und dem Absorptions Vermögen der Köper von Wärme und Licht," *Annalen der Physik und Chemie* **109**(2), 275 (1860).
[8]Wien, W. *Wiedemann Annalen* **53**, 132 (1894).

wavelengths with that same decrease. They decrease rapidly with decreasing temperature. The peak wavelengths are 0.5, 0.97, 0.72, and 0.97 μm. The blackbody curve for ambient temperature on Earth, about 300 K, is too small to show on this plot.

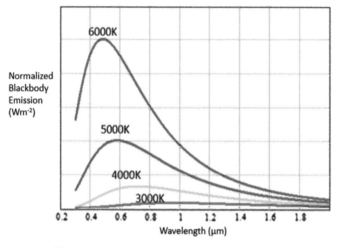

Figure 3.3.2   Normalized blackbody curves.

The locus of the peak wavelength as a function of temperature is shown in Figure 3.3.3. It is the wavelength of the peak in micrometers as a function of the temperature in kelvin. The peak for radiation at 300 K, approximately room temperature, is about 10 μm. And for the approximate temperature of the sun, it is just about 0.5 μm.

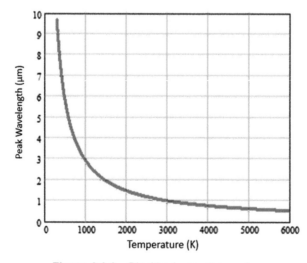

Figure 3.3.3   Blackbody spectra peaks.

The total radiation from a blackbody is a function only of the temperature and increases rapidly as the fourth power of temperature as shown in the next two figures. Figure 3.3.4 is an ordinary linear plot that shows how drastically blackbody radiation increases with temperature. The logarithmic plot of Figure 3.3.5 shows this in a perhaps more legible way.

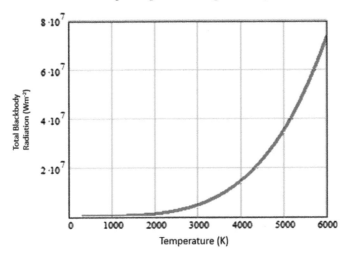

Figure 3.3.4 Total blackbody radiation.

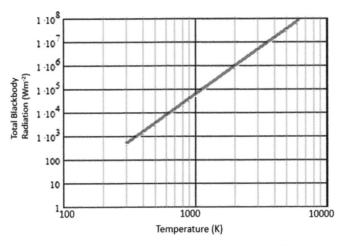

Figure 3.3.5 Total blackbody radiation, logarithmic scale.

The total blackbody radiation as described above is calculated based upon the Stefan–Boltzmann law. The value for the sun at 5778 K is 6.32×10^7 Wm^{-2}, which is very high in itself. But the sun is a big ball with a surface area of 6.07×10^{18} square meters. It radiates an astronomical amount of power: 3.83×10^{23} kilowatts!

The radiation from the Earth itself is not negligible. It is almost one-half kilowatt per square meter at room temperature (459 watts per square meter at 300 K). That is what keeps us all reasonably comfortable but gets us hot in the summer in Tucson when the temperature is 100°F, equal to 311 K and equivalent to 530 Wm^{-2}. An extra 70 watts of heat for every square meter! All of this is in the infrared. That is why it is so useful in driving, medicine and the military.

Blackbody simulators are instruments that attempt to radiate as a theoretical blackbody, and most come very close. They are typically cylindrical with a conical end. The secret of designing a blackbody is to have a large, uniform interior cavity with a small aperture from which the light radiates. These have been studied at length.[9]

The NBS, National Bureau of Standards, now NIST, the National Institute of Standards and Technology (NIST), has maintained two of these for decades, the platinum standard and the gold standard. They are rather complicated and expensive devices that only a government can afford.

Figures 3.3.6 and 3.3.7 are schematics of the two standards. The blackbody that was once the primary standard of light is shown in Figure 3.3.6. It consists of a cylindrical cavity with a conical end that is surrounded by platinum, followed by fused thorium oxide and then ordinary thorium oxide for insulation. The platinum is melted and then allowed to freeze. As it freezes, the temperature is constant at a known value. It is known as a freezing-point blackbody simulator.

It is no longer the standard of light; it has been replaced by a definition based on the luminous efficiency of the human eye.[10] It is not appropriate to

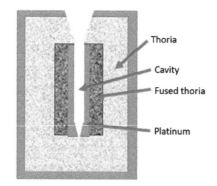

Figure 3.3.6 Standard of light schematic.

---

[9]Bartell, F. "Blackbody Cavity Radiation Simulator Theory," Dissertation, University of Arizona (1977); Wolfe, W. and Zissis, G. *The Infrared Handbook*, US Government Printing Office (1978).

[10]Convocation of the General Conference on Weights and Measures (26th meeting), Versailles: Bureau International des Poids et Mesures. 13 November 2018.

discuss its operation any further since it is no longer used. It is representative of how complicated and detailed these blackbody simulators can be.

The gold-point blackbody simulator is a cylinder with two conical ends, shown in the very middle of Figure 3.3.7. It is surrounded by the gold that is melted. That part is surrounded by graphite, which in turn is surrounded by quartz wool fibers and by cooling coils. It is operated in the same way as the platinum standard using the well-known temperature of freezing gold, the fusion temperature. It is another freezing point standard.

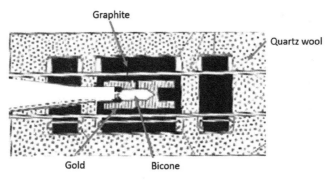

Figure 3.3.7   Gold-point blackbody simulator.

There are many commercial versions that are much cheaper and almost as good. They are typically about a foot in diameter and about a foot long or about a foot cubed. Most have the usual cylindrical cavity terminated by a cone and considerable insulation around the cavity. They range in temperature from ambient up to about 1400 °C. Their emissivities are about 0.999 due to the small aperture and long cylinder. They are controlled and heated electrically, and the higher the temperature is, the higher the price.

Figure 3.3.8   Typical cylindrical blackbody simulator.

I have a tip for the use of any of these blackbody simulators: Use it slightly off axis. The most nonuniform temperature of them is right at the conical tip. Don't look there.

There are also so-called extended-area blackbodies. They are about four inches square and have an emissivity of about 0.97. They do not have such long cavities. The relatively high emissivity is obtained by black paint and short, usually conical, cavities. They are not flat!

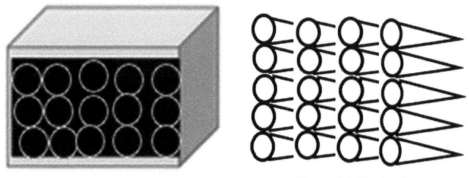

Figure 3.3.9 Extended-area blackbody.   **Figure 3.3.10** Interior.

Other versions are similar cavities designed into various instruments. They are usually cylinders with conical ends and may be in cylinders, cubes, or other configuration. Blackbody simulators are essential for the calibration of all sorts of radiometers and multispectral devices used in space and here on Earth for such things as monitoring the performance of electric generators and overhead transmission lines.

3.4 Cone Channel Condensers

Cone channel condensers are reflecting circular cones that condense and scramble light from a given field of view.[11] They are used in a number of infrared sensing instruments.

---

[11]Williamson, D. "Cone channel condenser optics." *JOSA* **42**, 712 (1952). Wolfe, W. and Zissis, G. *The Infrared Handbook*, US Government Printing Office (1978).

Figure 3.4.1 shows light rays entering two cones in cross-section: one in light blue, the other in darker blue inside it. Light enters both cones from the upper right. It reflects three times off the wider cone and exits, as shown by the black rays. It enters the narrow cone and reflects many times until it reaches the end, as shown by the red rays.

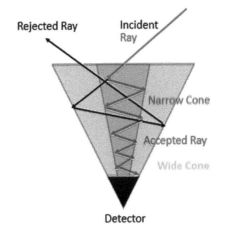

Figure 3.4.1 Two cone condensers.

The two cones are the same length. They differ only by the cone angle. The incident ray approaches each from the same angle and enters at the same place. The ray that enters the narrow cone reaches the detector; the one on the wider cone does not. This is analogous to the constancy of the angle-area product in afocal systems and a linear version of the area-solid-angle product in lens systems.

The limiting case is the cylindircal channel version, which is not a cone, not a condenser, but a hollow light pipe. The diagram shows that there is no restriction on the acceptance angle. It can accept almost the entire hemisphere. As long as the incident ray makes some small angle with the vertical side, it will continue down. But it does not condense.

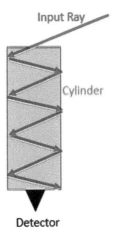

Figure 3.4.2 The limiting case: the cylinder,

A refractive cone with reflectorized sides is better than the open-air cone. The incident ray will be refracted further into the cone. Figure 3.4.3 shows the situation. The red ray is how it behaves in air. The blue ray shows its behavior with a material inside. A ray entering at a steeper angle will be refracted to coincide with what would have been the reflection trajectory of the air-filled cone condenser. The reflectorized material cone provides a wider field of view for the same cone angle but with a larger aperture for the same field and cone angle.

The advantage (i.e., the additional field angle for a given cone angle) is approximately proportional to the refractive index. Accordingly, the use of germanium in the infrared with its refractive index of about 4 is very attractive. It should be clear that a cone channel condenser that is filled with a material will be heavier than one without, and it will also be more expensive. But it will be sturdier.

Others have proposed the use of curved sides and rectangular or triangular shapes.[12] A circular cone is just a polygonal cone with an infinite number of infinitely short segments.Tracing rays in them is more complicated, but the concept is the same. One could also use a combination, say, a cone followed by a cylinder. That is too much detail to deal with here.

Cone channel condensers have been used in infrared devices to concentrate the light so that a smaller detector can be used and to scramble the light so that there are no "hot spots."

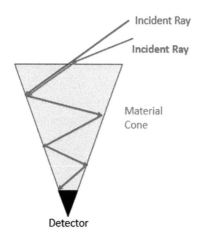

Figure 3.4.3 Reflectorized refractive cone condenser.

---

[12]Gush, H. "Hyperbolic cone-channel condenser," *Opt. Lett.* **2**(1), 22–24 (1978).

3.5 Detectors

Light detectors, or more generally, radiation detectors, are of two different types: thermal detectors and photon detectors. Thermal detectors convert the incident radiation into some form of heat and sense the change generated by electrical circuitry. Photon detectors operate by the direct conversion of photon energy to electrons that become free to participate in some form of electrical current or electron charge accumulation.[13]

Bolometers absorb the energy of the incoming radiation and heat up, thereby changing their temperature. This change in temperature results in a change in electrical resistance that can be measured in an appropriate circuit, such as a voltage divider or bridge or voltmeter.

Since they are sensitive only to the amount of heat they receive, bolometers have a spectral response that is independent of wavelength if they have a good black absorber on them. There are several kinds of bolometers: metals, semiconductors, and superconductors.

The bolometer was invented by Samuel Pierpont Langley, an American astronomer. In his memoirs, he notes that the device could detect the heat from a cow at a distance of a quarter mile.[14] This gave rise to a limerick[15] we all love:

Langley invented the bolometer,
A funny kind of thermometer.
It could detect the heat from a polar bear's seat
At a distance of half a kilometer.

---

[13]Dereniak, E. and Boreman, G. *Infrared Detectors and Systems*, Wiley (1996); Dereniak, E. and Crowe, D. *Optical Radiation Detectors*, Wiley (1984); Moss, T., Burrell, G., and Elis, B. *Semiconductor Opto-electronics*, Butterworth (1973).

[14]Langley, S. Biography Archived 2009-11-06 at the Wayback Machine High Altitude Observatory, University Corporation for Atmospheric Research.

[15]Wikipedia. "Bolometer," https://en.wikipedia.org/wiki/Talk%3ABolometer.

One of the criteria for creating a sensitive bolometer is a large change in the relative electrical resistance for a given temperature change, that is, the relative temperature coefficient of resistance $(1/R)$ (dR/dT). Semiconductors have higher coefficients than the metals that were first used. A typical value of the temperature coefficient of resistance for metals is about 0.0005 per degree, while for the semiconductors used in **thermistor bolometers** it is 0.02 per degree. Materials such as vanadium oxide and alpha silicon used in bolometer arrays have similar values.

The **superconducting bolometer**[16] also makes use of the temperature change in resistance. It does so at the superconducting transition temperature, where there is an enormous change in resistance from its finite value to zero, its superconducting value. That feature is both an advantage and a disadvantage. It is a large change in resistance, but considerable effort must be taken to ensure that the bolometer is in a stable environment. A very small change in local ambient temperature will trigger a false signal.

Thermocouples are based on the Seebeck effect discovered in 1841 by Johann Seebeck.[17] Whenever two dissimilar metals are joined, they generate a voltage difference if there is a temperature difference between the junction on the right and the ends of the wires connected to a voltmeter on the left. Radiation shines on the thermocouple; it heats up; it generates a voltage difference that is measured by an appropriate electrical circuit (maybe a voltmeter). Probably the most common pair is copper and constantan, an alloy of copper.

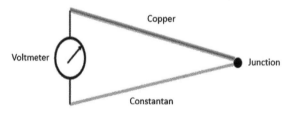

Figure 3.5.1   Schematic thermocouple.

Thermopiles are series-connected thermocouples; they increase the sensitivity.

In **photoconductive detectors**,[18] the incident photons give their energy to lightly bound electrons. The photons must have enough energy to free the electrons from their bound sites so only photons of short enough wavelength

[16]Andrews, H., Brucksch, W., Ziegler, W., and Blanchard, E. "A fast superconducting bolometer," *Rev. Sci. Instrum.* **13**, 281 (1942).

[17]Seebeck, T. "Magnetische Polarisation der Metalle und Erze durch Temperatur-Differenz," *Abhandlungen der Königlichen Akademie der Wissenschaften zu Berlin*, 265–373 (1822).

[18]Smith, W. "Effect of light on selenium during the passage of an electric current," *Nature* **7**(173), 303 (1873).

can do it. That wavelength depends upon the detector material. When this happens, the electrons are freed and increase the conductivity of the material. Incident photons change the conductivity.

In **photo emissive detectors**,[19] the photons have enough energy to kick the electrons out of the detector. They are said to overcome the work function. Such detectors also have voltage differences applied to them that accelerate the electrons from the cathode to the anode. The photo emissive material is in an evacuated cell (the cathode). The electrons are accelerated to the anode by a voltage difference. Incident photons cause electrons to be emitted.

Photomultiplier tubes are photo emissive detectors with many dynodes that multiply the number of electrons that the photons have generated.[20]

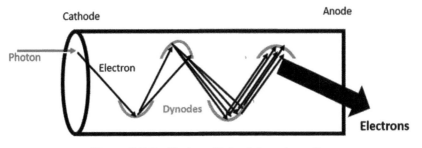

Figure 3.5.2 Photomultiplier tube schematic.

The **Golay cell** is named after Marcel Golay, who invented it.[21] It is a cell with a relatively black absorbing front surface that is filled with gas. The rear end has a flexible membrane that can expand as the gas is heated by the incoming radiation. The change is monitored by a light beam that reflects off it to a visible photodetector. It is relatively sensitive but slow and unwieldly... and obsolete.

Avalanche photodiodes are solid state photodiodes that have a large electrical bias across them. When an incident photon generates an electron, as in a normal photodiode, it is accelerated by that high voltage. The photon bangs into the molecules of the material of the next electrode and dislodges additional charges. They are accelerated the same way and generate even more charges, creating an avalanche of charges.

These various detectors are used in devices that vary from the expensive and complicated, such as the COBE DIRBE verification of the BIG BANG, to the simple and cheap motion detector on my garage door.

---

[19]Hertz, H. "Über einen Einfluss des ultravioletten Lichtes auf die electrische Entladung," *Annalen der Physik* **267**(8), 983–1000 (1887).

[20]Wikipedia. "Photomultiplier tube," https://en.wikipedia.org/wiki/Photomultiplier_Tube.

[21]Golay, M. "Theoretical consideration in heat and infra-red detection, with particular reference to the pneumatic detector," *Rev. Sci. Instruments* **18**, 347 (1947).

There are several ways to describe the performance of these detectors. One is the **signal-to-noise ratio (SNR)** that one gets from a given input. But this value requires a specification of the input. In order to normalize this, **detectivity** was defined. It is the SNR per unit input power with the somewhat unusual unit of reciprocal watts. It is the SNR per watt. But different-size detectors operating over different temporal bandwidths have different detectivities. Accordingly, Bob Jones defined **specific detectivity** as the detectivity per unit square root of the detector area and square root of the operating bandwidth.[22] The performance is a function of these square roots. The detector performance varies with the square root of its area because the power increases with area and the noise with the square root of area. The same is true for bandwidth. The value has been universally called specific detectivity and is designated as D^*, pronounced *Dee Star*.

Detector arrays were available back in the 1950s but were only linear arrays. Both photon and thermal detectors could be arranged in rows one next to the other with leads attached. Then, in 1970, Boyle and Smith invented the charge-coupled device (CCD), which could transfer charges from individual detectors along a single wire.[23] This gave rise to the two-dimensional detector array. Detectors could be arranged in rows and columns, and the output could be collected on single wires.

Both photon and thermal detectors take advantage of this advance, but the technologies are somewhat different.

The CCD itself is a two-dimensional arrangement of individual solid state capacitors that can store charge. The charge in one of them is caused to move to the adjacent cell by the application of a voltage on it. The charge moves to the next capacitor, and the process is repeated until the individual charge is sensed by an external circuit. It is analogous to a fire-fighting bucket brigade: a bucket of water is poured from pail to pail until it reaches the fire. The "bucket" of charges is moved from cell to cell until it gets out.

Infrared arrays—whether they are indium antimonide, mercury cadmium telluride, or microbolometers—need to couple to a silicon CCD to get the signal out. Each detector connects to one element of the CCD array and its output is processed by the CCD. Those connections can be tricky. It is not easy to align two planes (each about 25 μm in size) of a million pixels with an alignment uncertainty about 1 μm!

TV cameras now use arrays of millions of silicon and other semiconductor detectors. Infrared cameras now use photodetector or bolometer arrays with tens of thousands of elements.

---

[22]Jones, R., Goodwin, D., and Pullan, G. *Standard Procedures for Testing Infrared Detectors and Describing Their Performance,* AD 257597, Office of Defense and Development Engineering (1960).

[23]Boyle W. and Smith, G. "Charge coupled semiconductor devices," *Bell Syst. Tech. J.* **49**(4), 587 (1970).

3.6 Diffractions Gratings

Diffraction gratings are a means of generating spectra. They are a substitute for, and usually an improvement to, prisms. Their performance may be understood by first considering how a double slit produces an interference pattern because gratings combine diffraction and interference from multiple slits.

When monochromatic light impinges on two slits, each performs as a new source and propagates the light in spherical waves (shown as circles in Figure 3.6.1). These are sets of spherical crests and troughs shown as red and blue, respectively, in Figure 3.6.1. Where a crest meets a crest, there is constructive interference (red on red), and where a crest meets a trough (red on black), there is destructive interference; thus, there is an alternating bright and dark pattern of light, i.e., an interference pattern. And it is a function of wavelength.

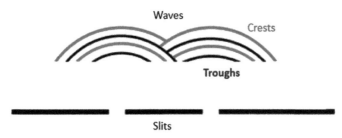

Figure 3.6.1 Double-slit interference.

If more slits are introduced, the waves add in the same way. Where the crests meet, there is constructive interference, and where the troughs meet the crests or other troughs, there is destructive interference. And a fourth, and a fifth, and so on, as illustrated in Figure 3.6.2 and indicated by the horizontal blue line.

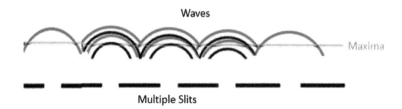

Figure 3.6.2 Multiple-slit interference.

When the light that is incident upon the grating is white (polychromatic), the pathlength to the points of coincidences of the maxima are different because the wavelengths are different. A spectrum is the result, as shown in Figure 3.6.3. There will be several places of constructive interference, called orders.

A diffraction grating may be thought of as a host of slits, thousands of them. The original grating (by David Rittenhouse in 1786[24]) was a hair wound on a screw in which the spacings between the windings acted as the slits. More modern gratings consist of wires drawn the same way (by Joseph von Fraunhofer[25]) or lines drawn on plates by very special engines that can accurately draw thousands of very narrow, parallel lines per inch (by John Strong and George Harrison). Most recently, gratings are produced by interferometric holographic techniques.

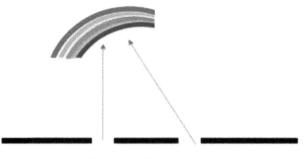

Figure 3.6.3 Spectrum.

Figure 3.6.4 shows a spectrum on my ceiling generated by the diffracting grating shown in the opening illustration. The source was the sun. On the right is the so-called zeroth order, in which all the colors constructively interfere to generate white light. It is a reflection like that of a mirror. To the left is the first-order spectrum, where there is constructive interference of the diffracted waves. To the left of it is a dark space, where there is destructive interference, and then the second-order spectrum appears dimly.

---

[24]Rittenhouse, D. "An optical problem, proposed by Mr. Hopkinson, and solved by Mr. Rittenhouse," *Trans. Am Phys. Soc.* **2**, 201 (1786).
[25]Fraunhofer, J. "Kurzer Bericht von den Resultaten neuerer Versuche über die Gesetze des Lichtes, und die Theorie derselben," *Ann. Physik* **74**, 337 (1823).

Figure 3.6.4 Diffraction spectrum.

Although the diagrams above show diffraction and interference in a transmission mode, most gratings (such as mine) are reflectors. The grooves can be altered to improve their performance, a practice called blazing. As you can imagine, when a grating is ruled by a mechanical engine with a tapered diamond tool, the grooves will be approximately rectangular with a triangular end. When the grooves are made interferometrically, they will be sinusoidal.

Blazing requires the rulings to be triangular in a special way. They need to reflect the light in the direction of a chosen order, thereby increasing its intensity. Blazed gratings have triangular grooves with their sides all facing in the same direction. The name probably came from the use of *blaze* to indicate a bright light.

Gratings are often preferred over prisms as spectrum generators. Obviously, they are flat rather than prismatic, so they need less room. They also have higher resolving power, although their transmission (or reflection) may not be quite as high.

Their resolving power, $\lambda/\Delta\lambda$, is equal to the order number times the total number of lines illuminated. An example is a modest 100 mm grating (like mine) with 250 lines per millimeter operated in the first order. The resolving power is then 100×250, or 25,000.

The same quantity for a prism is the change in wavelength over its spectral range divided by the wavelength range times the prism base, that is b $dn/d\lambda$. For a prism of borosilicate glass, a typical glass with a 10-centimeter base, the refractive index is 1.53004 at 400 nm and 1.51160 at 800 nm. That is a difference in refractive index of 0.01844 over a wavelength range of 400 nanometers, or 0.4 micrometers, or 0.0004 centimeters. The resolving power is then 461, a typical value for prisms.

Mechanical ruling of gratings reached its peak at MIT and Johns Hopkins University.[26] Blazed gratings with about 15,000 lines per inch (6000 lines per centimeter over about five centimeters) could be ruled by interferometrically controlled ruling engines.

Almost all diffraction gratings are now produced interferometrically.

---

[26]Strong, J. "New Johns Hopkins Ruling Engine," *JOSA* **41**, 3–15 (1951).

3.7 Fibers

The conduction of light in a fiber has been known for a long time. Its use in communication, medicine and other fields is a rather recent event.

The basic concept is that light exhibits total internal reflection when it travels from a dense to a rarer material at a sufficient angle (see Section 1.3). If light enters a glass fiber at a shallow enough angle, like the blue ray in Figure 3.7.1, it is totally reflected at the first glass-to-air interface. It reflects at the same angle by the law of reflection and is incident on the opposite side at the same angle, which is also large enough for total internal reflection. This process is repeated. The reflectivity at these surfaces is 100%; the loss in the fiber is due to absorption and scatter. If the ray angle is too steep, the ray is refracted out of the fiber, like the red ray.

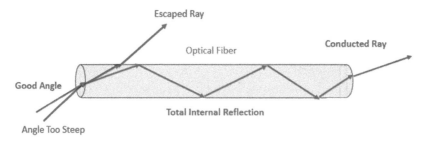

Figure 3.7.1 Reflection and refraction in a fiber.

Most fibers are clad. That is, they have another material on the outside. The external material both protects the transmitting fiber and ensures constancy of the refractive index ratio. Fibers that are used in unfriendly atmospheres, such as the ocean, have even more coatings for additional protection. Many, especially transatlantic fibers, have about eight to ten layers and enclose many individual transmission fibers, as represented in Figure 3.7.2.

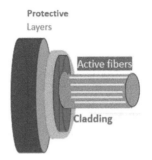

Figure 3.7.2 Representative optical fiber.

Decorative fibers are used on houses and Christmas trees to conduct the light of different colors from different sources to many different places. The lights can be red, green, blue, twinkling, or constant. The designs usually start with an LED source. Its light is fed by the fibers to a variety of places. In a flower, for example, the light would spread out into the shape of the blossom. In a Christmas tree, the light goes to little twigs or pinecones all over the tree. These fibers are often unclad or encased in one outer layer. [I have suggested to the manager of The Lodge at Chama that they use LEDs and fibers to put the light at the ends of the tines of their antlered chandeliers instead of at the bases.]

Fibers that are used for long-distance communication, for secure communication, for sensing bridge stability, and for medicine are discussed in corresponding sections in this book.

Figure 3.7.3 Chama chandelier.

3.8 Filters

In its broadest sense, a filter selects part of a whole. In a broad optical sense, it is a device that selects part of an optical beam from its total. Still in a large optical sense, this may be a temporal, spectral, or spatial selection. One example of temporal selection is optical ranging, in which a series of pulses is sent to a target, and the time for the round trip is measured to infer the distance. That is not the subject of this section (see Section 4.24 on lidar). A beam may also be divided spatially or in amplitude. That is the subject of beam splitters, covered in Section 3.2. Or the beam may be divided according to its state of polarization, which is covered in Section 3.12. This section is on spectral filters—devices that separate a beam into spectral regions, either narrow or broad.

There are mainly two types of spectral filters: absorption filters and interference filters.

Although **neutral density filters** are called filters; they are really attenuators not spectral filters. They reduce the transmission of light over a broad spectral range. Most of them are glass that is coated with aluminum or a similar metal. They can be graduated, that is, they have an attenuation that varies with position, as illustrated in Figure 3.8.1. That is usually accomplished by a gradation of the thickness of the metallic film. They can be linear or circular. Some circular ones are segmented as shown in Figure 3.8.2. Some consist of a pair of polarizing filters that adjust by varying the degree to which they are cross-polarized. Some are prisms that are partially transparent and move with respect to each other to vary the thickness and transmission, as in Figure 3.8.3. No neutral density filter covers the entire spectrum. It may be wide, but it is not forever.

Figure 3.8.1   Continuously variable linear attenuator.

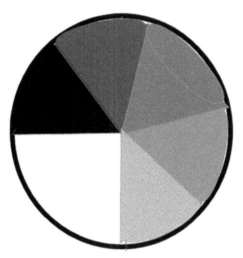

Figure 3.8.2   Segmented circular attenuator.

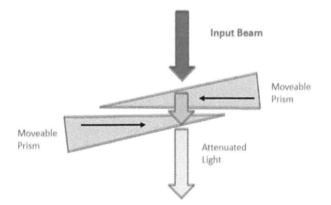

Figure 3.8.3   Partially transparent Risley prism attenuators.

Spectral filters are generally classified as long pass, short pass, and narrow band. The nomenclature is not always clear, but long and short pass usually refer to long and short wavelengths, not frequencies. It is better to indicate them as long-wave pass, short-wave pass, etc. Figure 3.8.4 shows these filters schematically.

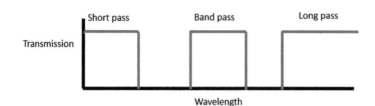

Figure 3.8.4 Filter specifications.

Absorption filters are simple in concept. They absorb in part of the spectral region and not in the so-called passband where they transmit. These bands are determined by the nature of the material. Most absorption filters have a relatively wide passband and, for the visible, consist of glass, plastic, or gels impregnated with appropriate dyes. All absorption filters are passband; none are either long- or short-pass filter. Two representative examples are shown in Figure 3.8.5. These examples show that absorption filters are not perfect. This is characteristic of all absorption filters. This figure from the internet shows typical absorption filter transmissions. They are Schott filters advertised by Edmund Optics. Their use in no way endorses or criticizes either company.

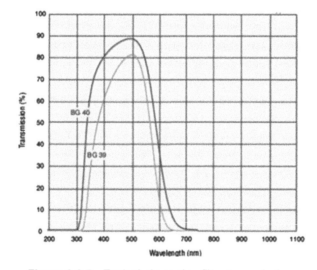

Figure 3.8.5 Typical absorption filter transmission.

Interference filters are more complicated and more precise. A wave that is incident upon a piece of material interferes with itself upon reflection. Figure 3.8.6 shows a wave on top that passes straight through a thin film, shown with exaggerated thickness. The reflected wave interferes with the straight-through wave, and it can be constructive or destructive, depending on the film thickness and the angle of incidence, which for now is assumed to be normal.

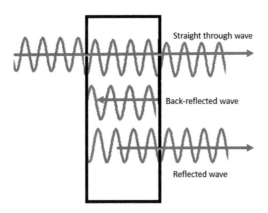

Figure 3.8.6 Constructive transmissive interference.

In the transmission filter shown in Figure 3.8.6, the incoming wave at the top passes right through; it is transmitted with a slight loss in intensity but with no change in frequency or phase. Part of it is reflected back to the first surface, as shown by the second, shorter wave. That portion is then reflected forward, passes through the second surface, and interferes with the first transmitted wave. As shown, this is constructive interference, but the thickness can be adjusted for either constructive or destructive interference. Of course, these waves are all superimposed, but I could not show them that way.

The design of interference filters centers around the use of quarter-wave optical thickness layers (QWOTs). Combined with a transmission filter, they generate destructive interference. The use of a double thickness does the opposite; i.e., it generates constructive interference. Interference filter design is the combination of these filters in stacks of different thicknesses to affect different wavelengths or colors.

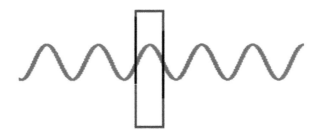

Figure 3.8.7 Quarter-wave optical thickness layer.

In the transmission filter, a wave that is reflected from the back surface to the front surface and back again suffers a half-wave phase delay—just enough for destructive interference. A double QWOT will do the opposite. Figure 3.8.8 shows a QWOT in action. The middle, reflected wave and its resultant forward-reflected wave is out of phase just enough to cause destructive interference.

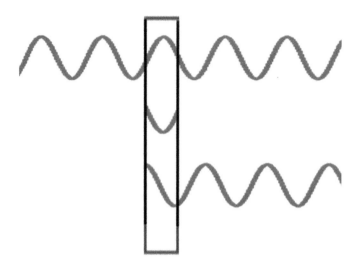

Figure 3.8.8   A QWOT interference illustration.

The design of interference filters is much like the design of lenses. There are computer programs that perform this task in much the same way that lenses are designed. An online site lists and provides links to several such design programs.[27] Some are free.

A special kind of filter is known as the **Christiansen filter**, named after Christian Christiansen.[28] It is based on the scattering of light and the lack thereof in the passband. The filter is a cell with particles suspended in a liquid. At most wavelengths, the particles scatter the light, but at the wavelength where the refractive indices of the particles and the liquid are equal, there is no scatter; the cell transmits. This is a rather cumbersome device and has given way to modern ones such as interference filters.

3.9 Lasers

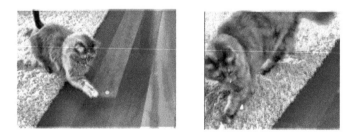

Lasers—now an accepted English word but once the acronym for **L**ight **A**mplification by **S**timulated **E**mission of **R**adiation—were invented in 1962 by

---

[27]Filter Design Software | *Nuts and Volts Magazine*, https://www.nutsvolts.com/magazine/article/filter-design-software.

[28]Christiansen, C. "Untersuchungen über die optischen Eigenschaften von fein verteilten Körpern," *Ann. Phys. Chem.* **23**, 298 (1884); **24**, 439 (1885).

Charles Townes and Arthur Schawlow.[29] Lasers emit light with very narrow linewidths in a collimated and phase coherent fashion (see Section 1.5). This makes them useful in a variety of applications, such as communication. It was Einstein who provided the thought in 1917 that radiation could be stimulated. Most radiation comes from spontaneous electronic transitions and molecular vibrations, but he suggested that it could also be the result of an incoming photon;[30] it could stimulate an electron, which could then stimulate a photon. That is, the photon could release its energy to an electron, which would move to a higher energy state. Then the electron could release its energy to a different photon and relax to a lower energy state. I do not think he envisioned the laser.

Although Charles Townes and Arthur Schawlow invented the concept of the laser, Theodore Maiman was the first to make one work.[31] He used a rod of ruby with reflectors on both ends with a flash tube wrapped around it. That is the essence of how a laser works.

As shown in Figure 3.9.1, a rod of ruby material, shown in red, is placed inside a region of intense light generated by a flash lamp wrapped around it, shown in yellow. The photons from the lamp transfer their energy to electrons in the ruby that move to a higher energy state. The electrons then drop to a lower energy state and emit photons. That is the laser light. The photons radiated at a wavelength of 693 nm in Maiman's case. The rod has a 100% reflector on one end, say, the left, in black, and a partial reflector on the other end, the right, in gray. The photons move back and forth in the rod from one reflecting end to the other thereby creating more photons that are in sync with them, since they are created right where the photons are. Some are transmitted through the partially reflecting end. That is the laser beam.

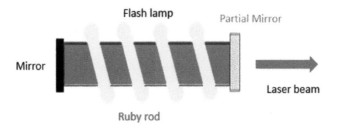

Figure 3.9.1 Schematic of Maiman's laser.

A laser beam is almost **monochromatic** for two reasons. All the electrons drop from one specific energy level to another, and the waves must have nulls at the ends. The energy difference is exact, and a photon has energy proportional to its frequency ($E = h\nu$). The full and partial mirrors are both

[29]Townes, C. and Schawlow, A. "Infrared and optical masers," *Phys. Rev.* **112**(4), 1940–1949 (1958).
[30]Einstein, A. "Zur Quantentheorie der Strahlung," *Physikalische Zeitschrift* **18**, 121 (1918).
[31]Maiman, T. "Stimulated optical radiation in ruby," *Nature* **187**, 493 (1960).

electrical conductors, so there cannot be an electric field in them. The optical electromagnetic field is shorted out. Thus, the waves must be zero at the two mirrors This means only waves of multiple half wavelengths that exactly fit in the cavity can exist. This is shown in Figure 3.9.2.

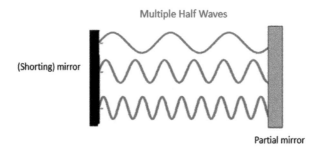

Figure 3.9.2   Multiple-wave restriction.

The waves are **in phase** because they generate new photons in synchronism, as noted above. They have a **collimated beam** since the photons have traveled back and forth in the narrow enclosure so many times, as if it were a very long, narrow tube.

Laser action in a gas follows the same in principle but is somewhat different in practice. Ali Javan, William Bennett, and Donald Herriot showed this could be done with a mixture of helium and neon,[32] the He-Ne (*heeny*) laser. It is a tube of about 90% helium and 10% neon gas. The ends of the tubes are anodes and cathodes with a voltage between them. The energy of this electric field is the stimulator in this case rather than a flash tube.

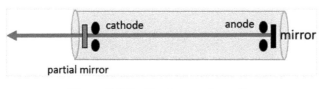

Figure 3.9.3   Gas laser schematic.

The He-Ne laser has three primary lines at 3.39, 1.15, and 0.633 μm. The line in the red, at 0.633 μm or 633 nm, has proven most useful. Other transitions in the visible are seldom used. Tuning to them can be accomplished with mirror adjustments or the use of spectral filters.

---

[32]Javan, A., Bennett, W.R., Herriott, D.R. "Population inversion and continuous optical maser oscillation in a gas discharge containing a He–Ne mixture," *Physical Review Letters* **6**(3), 106–110 (1961).

Kumar Patel showed how to make a gas laser with carbon dioxide.[33] It has the advantage of much higher output and efficiency, i.e., kilowatts instead of milliwatts, and as much as 20% efficiency. The CO_2 laser also consists of a mixed gas: carbon dioxide, nitrogen, helium, and xenon. A simplified discussion of the operation follows. The nitrogen molecules are excited by the electric field of the discharge tube. They in turn excite one of the resonant modes of the carbon dioxide (see Section 1.8). The carbon dioxide releases energy by emitting a photon and transferring the remaining energy to the helium atoms. They get excited but release their energy to the cooled walls of the tube. It was not the first time.

R obert N. Hall first established laser action with a solid-state diode in the infrared part of the spectrum.[34] In this case the excitation is current flow in the diode. The output wavelength is determined by the characteristics of the material. It may not be obvious that there is enough room in a diode for all this collimation and spectral definition action, but a wavelength of light is very small. A red photon has a wavelength of 0.8 μm. Thus, in a one-millimeter-thick laser diode pointer there are 1250 wavelengths, enough length for monochromatizating, phase synchronization, and collimation.

Soon other researchers with other materials generated diode lasers that emitted in the visible. The first one of these was a diode made of a mixture of gallium, arsenic, and phosphorus created by Nick Holonyak in 1962.[35] Now, they exist with peak wavelengths of just about any value one could desire.

In every case—a solid, a gas, or a diode laser—the basic concepts are the same. There is some mechanism for stimulation (excitation); there is a material with specific energetic transitions; and there is a cavity for synchronization, collimation, and monochromaticity.

Once it was done, everyone could do it! There are now probably more diode lasers than any other kind. Pointers for lecturers, in data centers for data transmission, CD players, and even cat teasers, as shown in the introductory image.

---

[33]Patel, K. "Continuous-wave laser action on vibrational-rotational transitions of CO_2," *Phys. Rev.* **136**(5A), A1187 (1964).

[34]R. Hall et al., "Coherent light emission from GaAs junctions," *Phys. Rev. Lett.* **9**, 366 (1962).

[35]Holonyak, N. "After glow," Illinois Alumni Magazine. May–June (2007).

3.10 Lenses

There are thousands of types of lenses, ways to design them, and ideas about them. The word "lens" is used in two different ways by lens designers and by me. A lens is a single element, but a lens is also a collection of two or more elements.

There are some concepts that need explanation before we delve into the many lenses that are used in cameras and elsewhere.

Object space is in front of the lens, where the object is. **Image space** is behind the lens, where the image is.

The **optical axis** of a lens is the line that goes through the centers of curvatures of both surfaces and is perpendicular to both surfaces. A completely spherical lens has many of these; any line that goes through the center is one, as shown in Figure 3.10.1.

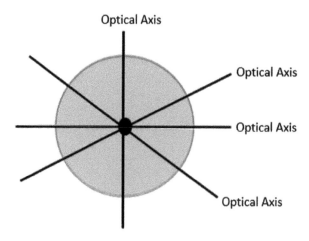

Figure 3.10.1   Optical axes of a sphere.

Most lenses have two different radii of curvature, one for each surface. Then there is a unique optical axis as shown in Figure 3.10.2. The lens is blue, and the circles that define its two curvatures are also shown.

The red line goes through the center of the lens but is not perpendicular to either lens surface; the blue line is perpendicular to one surface but does not go through the center. Only the black dashed-dot line is an optical axis. It is perpendicular to both surfaces and passes through both centers of curvature. Most optical diagrams show this as a dashed-dot line, as I did and usually do.

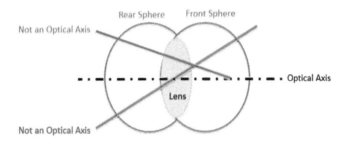

Figure 3.10.2 Optical axis of a lens.

The **focal point of a lens** is the spot on the optical axis where an incident ray parallel to the optical axis crosses the optical axis in image space. The focal surface is the region transverse to the optical axis at the focal point. It is often called the focal plane, but it is not always plane.

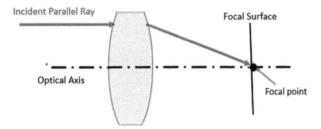

Figure 3.10.3 Focal point and surface.

A lens has two focal points, a front one and a rear one. A parallel beam can impinge on the lens from the left **or** from the right. Almost always we assume that the object is to the left and **the** focal point is to the right, but that arrangement is certainly not true with mirrors or even negative lenses.

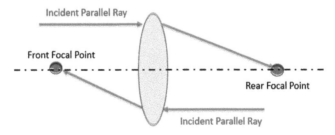

Figure 3.10.4 Front and back focal points.

The position and size of an image formed by a simple lens can be found by tracing two rays. The incoming ray that is parallel to the optical axis passes through the focal point as just described. The central ray that passes through the middle of the lens is not deviated. Any point on the object is imaged to a point on the image where these two rays intersect. For good measure, one can also draw a ray through the front focus that exits the lens parallel to the optical axis. These rays are shown in Figure 3.10.5.

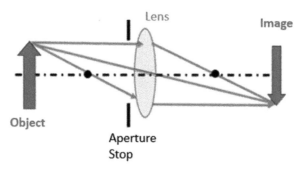

Figure 3.10.5   Optical diagram.

The **aperture stop** is a physical diaphragm that limits the size of the bundle of light that is accepted by the lens system. It is usually at the front of the lens system. A simple case is shown in Figure 3.10.6.

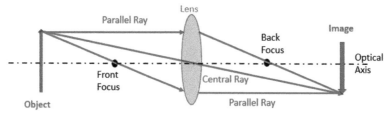

Figure 3.10.6   Aperture stop.

In some systems, the physical aperture stop is somewhere in the middle of the optical train. Then the image of that stop by all elements in front of it is the **entrance pupil**. It is the aperture stop as imaged by all the elements in front of it.

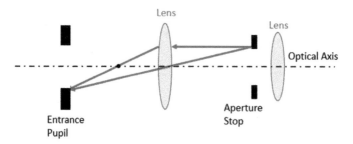

Figure 3.10.7   Entrance pupil.

The exit pupil is the aperture stop as imaged by all the elements behind it. The location of the exit pupil is critical in the design of binoculars to obtain proper eye relief.

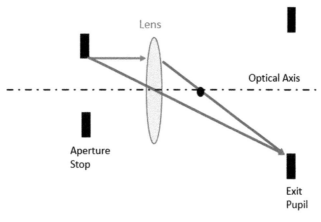

Figure 3.10.8   Exit pupil.

The **field stop** is the physical diaphragm that limits the extent of the field of view and is usually at or near the focal surface or an image of the focal surface. Figure 3.10.9 shows the imaging of an object by the parallel ray and the central ray. It shows a diaphragm in black next to the image. That component is a field stop that limits the image field of view. As shown, it will not allow the entire image through. The entrance pupil is the image of the field stop in object space.

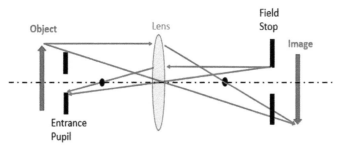

Figure 3.10.9   Field stop.

The **magnification** of the system is the ratio of the linear size of the image to that of the object. It is also equal to the ratio of the image distance to the object distance. This is shown by the similar triangles in light green in the object and image spaces of Figure 3.10.10. The ratio of the heights is the same as the ratio of the object distance to the image distance. The ratio of the image height to the object height is the magnification, and it is equal to the ratio of the image distance to the object distance

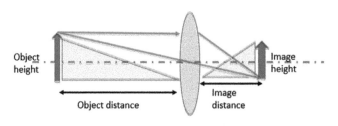

Figure 3.10.10   Magnification.

The reflective equivalent of a positive lens (that has two convex surfaces) is a concave mirror (with one concave surface). A spherical mirror is the easiest example to understand the concept. It is shown in Figure 3.10.11. Several rays parallel to the optical axis come to focus at several different positions on the optical axis, at different focal points. This is spherical aberration. The paraxial ray is chosen for approximate calculations. The focal point for it is the half radius.

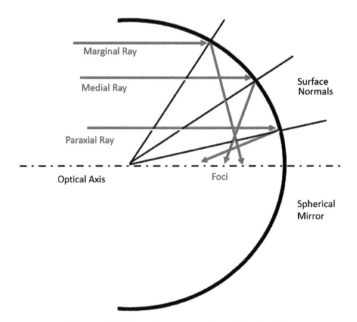

Figure 3.10.11   Concave-mirror focal point.

An equivalent paraxial ray trace can then be made for the concave mirror. It is shown in Figure 3.10.12. A ray from the object parallel to the optical axis reflects from the mirror and goes through the focal point. Another ray from the same spot on the object, the top of the arrow, is reflected at the center of the mirror. This ray may also be considered the equivalent of the ray that goes through the center of the lens undeviated. But it is reflected according to the law of specular reflection. A third ray could be used as a check. It goes through the focal point and reflects from the mirror parallel to the optical axis.

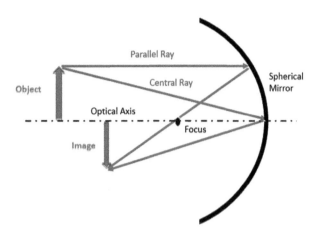

Figure 3.10.12   Concave mirror imaging.

This optical layout can also be done with a negative lens. The same two rays are drawn: one parallel to the optical axis, and one through the center of the lens. In this case, the refracted parallel ray needs to be extended backward since the lens diverges it. It goes back through the front focal point. The other ray goes through the center, as usual. The object is reduced and virtual.

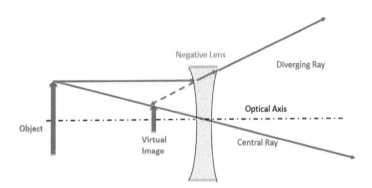

Figure 3.10.13   Negative lens imaging.

The reflective equivalent is the convex mirror. We can expect some virtual rays with this, as well. The parallel ray goes through the focus virtually. The other ray goes to the center of the mirror and reflects by the law of reflection to go back to meet the other ray virtually.

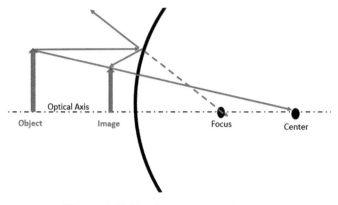

Figure 3.10.14   Convex mirror imaging.

These types of diagrams are used in other sections of this book to illustrate the concepts of magnifiers, microscopes, telescopes, and the like. Now, on to the description of the many lenses in existence. I will restrict this to some classic designs and basic ideas. I am **not** going to describe 10,000 different lenses!

We start with the **plane parallel plate**, which is not a lens but does extend a convergent beam. As shown in Figure 3.10.15, it does not change the geometry of a collimated beam, but it does move its focus.

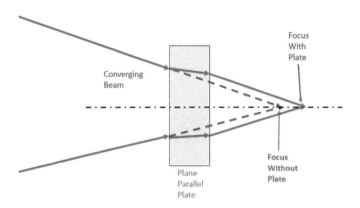

Figure 3.10.15   Plane-parallel plate refocusing.

The simplest lens is the **plano-convex lens**. It can be used with either surface in front (plano-convex or convex-plano). When collimated light enters the plane side first, it is not refracted until it exits. Then it refracts away from the normal, as shown and towards the optical axis, where it becomes the focal point.

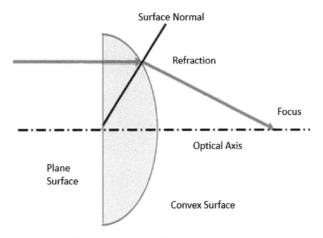

Figure 3.10.16 Plano-convex lens.

Collimated light that is incident upon the convex surface first, as with the **convex-plano lens**, is refracted at both surfaces. Both normals are shown in light blue in Figure 3.10.17, and the refractions at those points as red arrows. The first at the convex surface is toward the normal; the second at the plane surface is away from it.

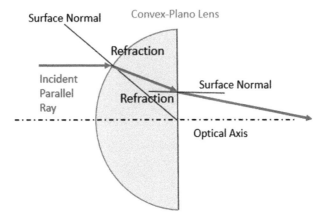

Figure 3.10.17 Convex-plano lens.

Probably the most ubiquitous singlet lens is the **biconvex lens**, shown schematically in Figure 3.10.18 with an exaggerated version to illustrate the refractive and focusing actions. The optical axis is perpendicular to both surfaces and goes through the middle of the lens. The incoming ray is parallel to the optical axis. It refracts at the front surface. Note that it bends inward toward the surface normal shown in light blue. It crosses the lens and refracts away from the surface normal because it is going from a dense medium to a rarer one, but the surface normal is different and causes a steeper exiting angle. It then intersects the optical axis, where it becomes the focal point. Singlet lenses like this are often used for magnifying glasses or the reading glasses you get at the drug store. They are also useful for some laser applications where chromatic aberration is not a concern.

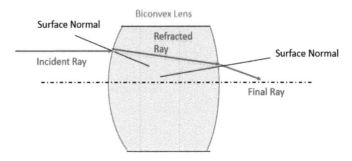

Figure 3.10.18 Biconvex lens.

Another useful singlet lens is the **meniscus lens**. It is a concave-convex lens as shown in Figure 3.10.19. It was used on the early Kodak Brownie cameras and on some camera obscuras. A meniscus is the upward curving shape of water in a cup, in which the attraction of the sidewalls pulls the outer edges upward.

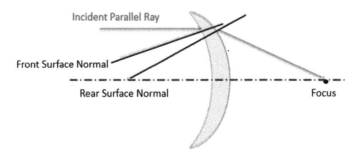

Figure 3.10.19 Meniscus lens.

Doublet lenses were introduced to correct for chromatic aberrations. They come in two versions and several designs. The versions are contact doublets and separated doublets, depending on whether the two lenses are in contact or not. The designer has a little more freedom with the separated doublet since there is one more surface and a separation that can be adjusted. They can be combinations of double-convex, convex-plano, convex-concave elements, and so on.

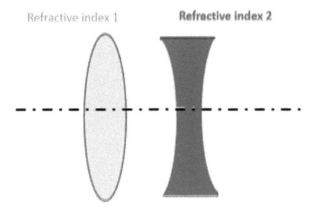

Figure 3.10.20   Generic doublet lens.

One useful doublet is a pair of identical meniscus surfaces with an aperture stop between them. It has the advantage of great symmetry, so that coma, lateral color, and distortion are cancelled.

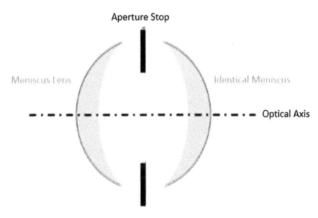

Figure 3.10.21   Meniscus doublet.

An **achromatic doublet** corrects for two colors and employs a biconvex and concave-plano duet of different glasses that have different refractive-index dispersions: one is relatively high, and the other is lower dispersion. "Flint" and "crown" are the corresponding terms for these types of glass. The flint lens refracts the shorter wavelengths more than the longer ones, and the negative crown glass lens compensates for this.

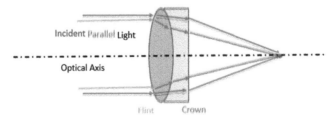

Figure 3.10.22   Achromatic doublet.

The **Cooke triplet** is the classic triplet lens.[36] It consists of two biconvex crown lenses with a double concave flint lens in between. The curvatures are adjusted so that they add up to zero and give a flat image plane. There are just enough degrees of freedom in this lens (six curvatures, five spacings, and three materials) to correct for all the third-order aberrations at a single wavelength.

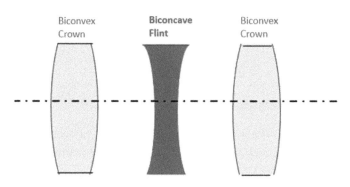

Figure 3.10.23   Cooke triplet.

The **Double Gauss** or **Biotar**[37] is one of the early and good camera lenses. It consists of two facing Gauss triplets and is the basis of many cameras. It uses six lenses and seven thicknesses for aberration corrections. There are many variations on the design; the Planar may have preceded it. It was designed before there were computer programs!

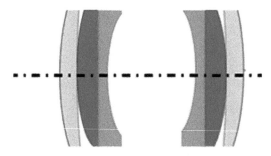

Figure 3.10.24   Zeiss Biotar lens.

The **Zeiss Tessar** is a second widely used camera lens. Although it has four (tessera) elements rather than the six of the Biotar, it has seven surfaces and three spacings to use for correction.

[36]Kidger, M. *Fundamental Optical Design*, SPIE Press (2002) [doi: 10.1117/3.397107].
[37]Kingslake, R. *A History of the Photographic Lens*, Academic Press (1989).

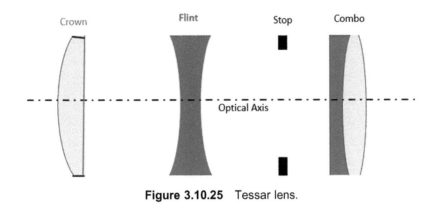

Figure 3.10.25 Tessar lens.

Wide angle lenses are all symmetrical. It is the symmetry that allows good imagery across a wide field. One example, the Roosinov lens from the Code V collection, cited by Shannon, is represented here.[38] It covers 113 degrees with acceptable resolution. It illustrates the use of symmetry in such lenses. The aperture stop is at the center, somewhat similar to the design of the Schmidt telescope. It thereby makes all rays on axis.

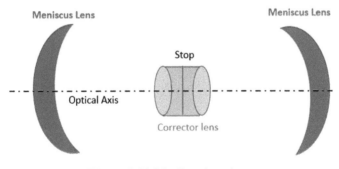

Figure 3.10.26 Roosinov lens.

An **afocal lens** can be a simple doublet, as shown in Figure 3.10.27. Two lenses are arranged so that their foci coincide. That means that a collimated beam that enters the first lens comes to a focus at the focal point of the second lens, which then sends it out as another collimated beam. This can be used as a relay in a periscope if the lenses are the same diameter, as in Figure 3.10.27. The lenses can be more complicated than shown to reduce any or all aberrations.

---

[38]Shannon, R. *The Art and Science of Optical Design,* Cambridge University Press (1997).

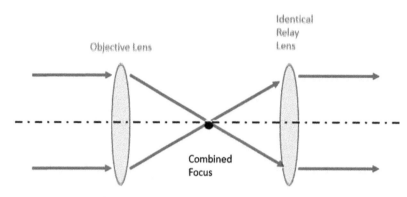

Figure 3.10.27   Equal-beam afocal relay.

An afocal system can also be a beam reducing or enlarging system if the lenses have different diameters. The beam divergence change is inversely proportional to the diameters of the lenses. This is the linear version of the constancy of throughput (see Section 1.9). If, for instance, the beam divergence of the incoming beam is one degree and the second lens is half the diameter of the first, then the outgoing beam divergence is two degrees, as shown in Figure 3.10.28.

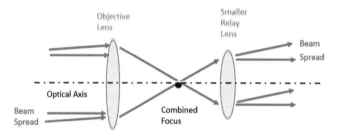

Figure 3.10.28   Unequal-beam afocal relay.

Mirrors can also be used as afocal relays, but this arrangement is more cumbersome. The use of a pair of mirrors will not permit equal-sized beam relays and there will be obscuration.

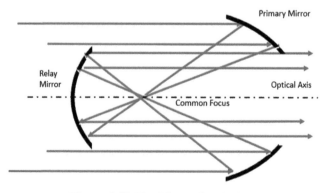

Figure 3.10.29   Mirror afocal relay.

Another pair of mirrors, that is, the mirror image of the pair of mirrors shown, can produce an equal-sized beam but with obscuration.

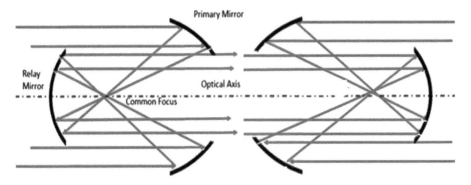

Figure 3.10.30 Complicated reflective, afocal equal-sized beam relay.

Zoom lenses allow us to change the magnification so we can see a distant object as if it were up close. It changes the object and image distances; it changes the magnification. It does this with a combination of afocal and imaging lenses. Figure 3.10.31 shows an afocal lens pair on the left combined with an imaging lens on the right. One afocal lens is moved back and forth to change the overall focal length of the zoom lens. The change in focal length accomplishes the change in magnification.

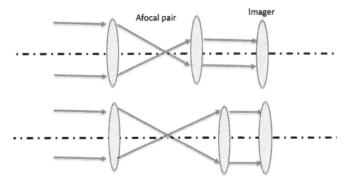

Figure 3.10.31 Zoom lens concept.

Gradient index (GRIN) lenses may have been first suggested by the fisheye lens of Maxwell described below, but so far, they seem to be of little use (except those in our own eyes). The basic idea is to design and manufacture a lens with a gradient in its refractive index rather than with a specific shape. Refer to the description of mirages, how the gradient in the density of the air produces a gradient in its refractive index that redirects an image of the sky to the viewer. It seems that manufacturing challenges have relegated them to the back burner.

The focusing property can be described in the same manner that is described in Section 2.6: replace the continuous gradient with several layers and leave it to you, the reader, to smear them out into continuity! (See Figure 3.10.32.) Such a GRIN lens can be in the form of a flat plate or combined with surface curvatures.

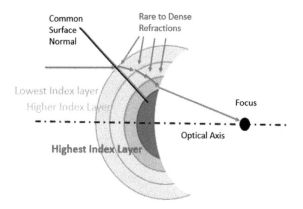

Figure 3.10.32   GRIN-lens layered model.

The **Fresnel lens** was invented by Augustin Fresnel in 1822. It is a way of making a thin lens do the work of a thick one by segmenting it, although it has its limitations. The diffraction limit of resolution applies to the dimensions of the individual segments. It is therefore not a high-resolution lens.

One way to understand the Fresnel lens is to imagine a thick lens being cut up into pieces and reassembled, although that is not the way it is made. Figure 3.10.33 shows a thick blue lens with imaginary borings in its left end that sort of hollow out the lens. A typical form is shown in Figure 3.10.34.

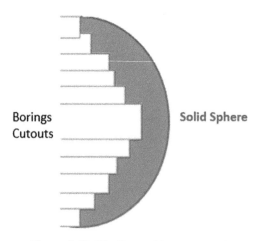

Figure 3.10.33   Fresnel lens concept.

Figure 3.10.34 Fresnel lens form.

The original use of the Fresnel lens was for lighthouses that had very large, very heavy lenses. It is now also used for recreational vehicles (RVs). This version uses a thin layer of plastic with the lens imprinted or molded into it. It can be attached to the rear window to provide a wide-angle view of everyone on the road who is behind you and wants to pass.

Immersion lenses are used in microscopes and with some infrared detectors. In microscopes, it is not a physical lens but a layer of oil between the microscope lens and the subject (see Section 4.27). A real lens is used in some infrared optics. It is called an immersion lens because the detector is brought into contact with the rear of the lens. Both hemispherical and hyperhemispherical versions have been used. The hemispherical version is shown in Figure 3.10.35. It forms the image at the center of the hemisphere and reduces the linear dimensions by a factor of the refractive index.[39] A hyperhemispherical lens, shown in Figure 3.10.36, forms the image a little past the center and reduces the linear size of the image by the square of the refractive index.[40] An example is shown in Figure 3.10.37.

---

[39]Smith, W. *Modern Optical Engineering*, McGraw-Hill (1990).
[40]Wolfe, W. *Proceedings of the Conference on Optical Instruments and Techniques,* Chapman and Hall (1962).

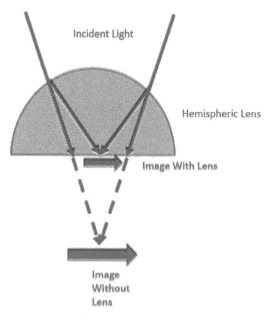

Figure 3.10.35   Hemispherical immersion lens.

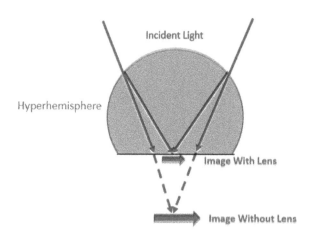

Figure 3.10.36   Hyperhemispherical immersion lens.

Figure 3.10.37   Hyperhemispherical immersion lens example.

The **Luneburg lens** is a sphere with a refractive index that varies with radial position (Figure 3.10.38). I know of no optical application for this lens. It has been used as an antenna and calibration method in microwave applications. The gradient of refractive index is $\sqrt{(2 - r^2)}$, where r is the relative radius. This requires that the refractive index be 1.41 at the center and fall off to 1.0 at the circumference.

The **Maxwell fisheye lens** is a form of Luneburg lens with a different refractive index profile. J. Clerk Maxwell (of electromagnetism fame) investigated it purely out of curiosity and compared it to what he thought a fish's eye might look like or what the fish might see. He did introduce the idea of gradient index lenses.[41] His formulation is a generalization of Luneburg's but actually preceded it. The expression for the refractive index as a function of radial position is $n_0/(1 + r^2)$. In this case, the refractive index n_0 at the center can be that of any glass. It decreases to one half its value at the circumference.

The Luneburg lens is difficult to make; the Maxwell fisheye is impossible in optics. For any reasonable glass, the refractive index is about 1.5 in the center and 0.75 at the edge (faster than the speed of light). The lens is possible with germanium in the infrared; it has an index of about 4.

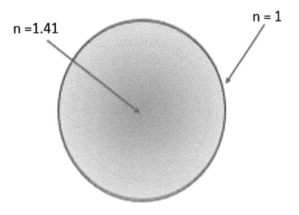

Figure 3.10.38 Luneburg lens.

The most recent development in lenses is the use of metamaterials. They consist of synthetic molecules. The lenses can then be made in any three-dimensional structure that is desired. GRIN lenses can be manufactured and refractive-index gradients can be generated both laterally and in-line. This is an area of ongoing research. I know of no commercial optical instruments that use them. Yet.

[41]Reed, J. S. webpage: James Clerk Maxwell: Concepts and Places (2016): https://homepages. abdn.ac.uk/j.s.reid/pages/Maxwell/Legacy/index.html.

3.11 Mirrors

Mirrors have a long history, starting with smooth lakes and placid pools. They may be divided into technical mirrors and common ones. The latter are those around the house and on cars. The former are those that are used in astronomical telescopes and other instruments. Common mirrors are almost always reflectorized on the rear, and technical mirrors on the front. If the reflective coating is on the rear, there is also a reflection from the front, a ghost reflection. It is about 4% with glass and is not evident in common usage. Such a mirror is better protected and longer lived than the front-surfaced mirror. It can cause troublesome ghost images in a technical instrument.

One-way mirrors are mirrors only in the sense that everything reflects. In their simplest form they are panes of glass. Recall that as you sit in your room at night with the lights on, you cannot see out through your picture window. The window is transmitting about 90% of the light from the night outside and reflecting about 10% of the light of your lamps. But it is much brighter inside than outside. Someone outside can clearly see inside. I have shown four pictures of and from my family room in Figures 3.11.1–3.11.4 during the day and at night. The leftmost image is from the inside out at night. It is ostensibly through the window, but what you see is the reflection of the inside of the room from the window. The other is the view from outside. You see the brightly lit inside from outside. The first two are at night. The next two were taken during the day. Figure 3.11.3 shows my garden from the inside of the house. Figure 3.11.4 shows my garden reflected off the window from the outside. They are slightly different because the angles had to be different, and the pairs are reversed by reflection.

Figure 3.11.1 Night inside.

Figure 3.11.2 Night outside.

Figure 3.11.3 Day inside.

Figure 3.11.4 Day outside.

One interesting application of an enhanced one-way mirror is the *mechitza*, a dividing screen used in orthodox Jewish temples. The screen consists of a collection of glass panes set at a 45-degree angle between horizontal screens, the tops of which are brightly colored and the bottoms are black. They are arranged as shown so that the male worshippers cannot see the women, but the women can see their men.

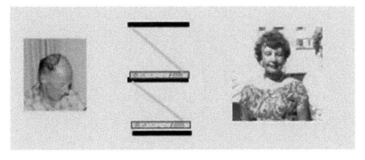

Figure 3.11.5 Mechitza.

One-way mirrors that are designed as part of an interrogation room generally have a partially reflecting coating to enhance the effect. They are also used for market research, psychology studies, teleprompters, and even in some modern trains (perhaps so that the conductor/engineer can see the passengers but not the other way around).

Wide-angle mirrors are used in cars as rear-view mirrors on the passenger's side. They always say "Objects in mirror are closer than they appear" because the mirrors are convex and thereby show a wider field of view. In order to get everything into that wider field, everything must be smaller. A smaller car seems to be farther away; it is our basic understanding. Some hallways have them at corners, and some ATMs have them so you can see who is waiting behind you. They are useful in banks and similar places as a part of security camera systems.

Figure 3.11.6   Side-view mirror.

Figure 3.10.14 in the section on lenses shows how the convex lens reduces the image size. It is also shown here in Figure 3.11.7.

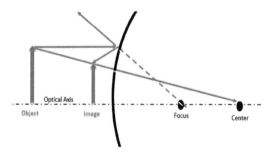

Figure 3.11.7   Long enough and at any distance.

It surprises most folks that a **vanity mirror** that provides a full image of your entire body only needs to be half your height. As shown in Figure 3.11.8 on the left, its top needs to be halfway between your eyes and the top of your head. The law of reflection sees to it that a ray from the top of your head reaches your eye. Similarly, the mirror needs only go halfway down to your feet for the same reason. Even fewer people believe that it is **not** a function of how far away you are from the mirror. It is not. See Figure 3.11.8 again. The law of reflection and the arguments above still hold, but the angles are smaller.

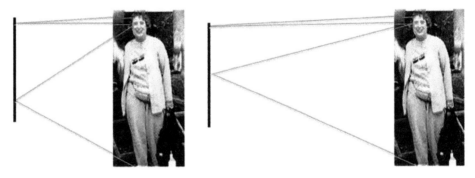

Figure 3.11.8 Vanity mirror

A common question asks, "Why do mirrors reverse things left to right but not up to down?" (MIT used it as a teaser for one of its courses on optics). They don't—they reverse things both ways. The law of reflection applies both horizontally and vertically and at all angles. However, we do not usually see it that way. We do see our left hand and arm looking like the right hand and arm of our image in the mirror. But we look up, and our head is up; we look down, and out feet are down. Consider the two images of the word REFLECTION shown in Figure 3.11.9. The left part shows reversal left to right; the right part shows reversal up to down. It is simply the law of reflection at work in all directions in my mirror.

Figure 3.11.9 Image reversal.

The **Mangin mirror**, the only rear-surfaced technical mirror of which I am aware, is a curved mirror with a curved refractive surface on its front. The two curvatures are different, and the combination of the two different curvatures and the thickness of the material allows for the correction of spherical and several other aberrations. The mirror was invented by Alphonse Mangin in 1876 and is used in searchlights, telescopes, and similar applications.

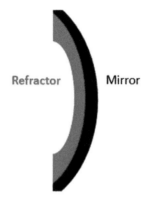

Figure 3.11.10 Mangin mirror.

Most **astronomical mirrors** are reflectorized with aluminum or gold. They are deposited on a blank made of a different material, such as glass or beryllium. Aluminum is often used because it is cheap and does not tarnish. Its oxide, the result of atmospheric oxidation, is aluminum oxide or sapphire. It is transparent in the visible and the near infrared. Gold is used on some infrared telescope mirrors because it has a higher reflectivity in the infrared part of the spectrum than aluminum. It is inert. Nevertheless, astronomical mirrors usually need to be recoated every several years because they get dirty.

The **Palomar mirror** was to be made of fused quartz, but that proved to be too difficult. Casting a 1.5-meter (200-inch) diameter mirror out of quartz is not easy! Instead, it was made of Pyrex, a borosilicate glass with a higher coefficient of thermal expansion. It is probably obvious that as the temperature changes, the shape of a mirror changes because of thermal expansion. The radius of curvature changes, and if it is not a sphere, it changes differently in different places. The observatory cools down at the end of the day. If it is not temperature controlled during the day, the mirror changes shape. Most observatories are open air. But there is another reason. During manufacture, the mirror blank is rubbed to attain its shape. The rubbing causes friction that causes heating and changes the figure of the mirror. During manufacture, the mirror had to rest after a certain time of grinding and polishing to come to equilibrium so that tests could be made on its figure.

A solid mirror blank is generally about one-sixth as thick as its diameter in order to stay rigid and maintain its figure. This factor can add weight quickly. The 1.5-meter-diameter Palomar mirror would be 25 centimeters thick and would weigh about 40 tons. Accordingly, it was formed as a **honeycomb**. Some 36 blocks were arranged in the mold into which the Pyrex glass was poured. This did not reduce the structural rigidity but reduced the weight by half. Most modern primary mirrors are now honeycombs. The process does not make astronomy any sweeter, just lighter.

Table 3.11.1 lists common materials used for mirror blanks. It is desirable to have a high strength-to-weight ratio to minimize the blank size, and it is also good to have a low coefficient of thermal expansion. Like most things in life, combinations of good things are hard to find.

Table 3.11.1 Mirror blank materials.

Material	Thermal Expansion	Elastic Strength	Density	Strength/Weight
	10^{-6} per K	GPa	gcm^{-3}	
Pyrex®	3.30	64	2.23	28.7
ZERODUR®	0.05	90	2.53	35.7
Cervit	0.05	89	2.53	35.2
Quartz	0.55	72	2.65	27.2
Beryllium	11.3	287	1.85	155.1
Silicon carbide	4.00	0.41	3.21	0.1

Adaptive mirrors or adaptive optics were invented by Horace Babcock in 1953.[42] They are intentionally thinner, but they have a series of pistons behind them that can shape the mirror. They are used to correct for atmospheric turbulence. They did not come into vogue until there were computers that could make enough calculations in real time. These adaptive mirrors use a star or fake star as a reference. The image of the star should be very close to a point. If the image is not a point, the pistons push and pull appropriately by trial and error until the image is minimized. Then the limitation on resolution is the telescope itself, not the atmosphere.

Zerodur and **Cervit** are mixtures of aluminum and silicon dioxides in glass and crystalline form. They have very low thermal expansion coefficients and have therefore been used in a variety of modern mirrors. Their properties are listed in the table and can be compared to Pyrex and quartz.

Beryllium was used as a blank for some ICBM interception systems. It has two characteristics that encouraged this. It has a low density and good strength. Its strength-to-weight ratio is high, so it can be a lightweight,

---

[42]Babcock, H. "The possibility of compensating astronomical seeing," *Publications of the Astronomical Society of the Pacific* **65**(386), 229–236 (1953).

high-performance mirror of the optical system of an intercepting missile. It is also impervious to nuclear radiation, so it would continue to work in the atmosphere of a nuclear detonation.

Silicon carbide is another material with a high strength-to-weight ratio, about four times that of the glasslike materials but somewhat less than beryllium. It was another consideration for a mirror blank for missile sensors.

The table above also shows the strength-to-weight ratios for some mirror blank materials. High values are important and are more important for those that fly. Zerodur, Cervit and quartz are best for terrestrial telescopes if they can be poured and figured well. Beryllium and silicon carbide are indicated for satellite applications. A GPa is a giga Pascal, but ignore the units; just do the comparison.

Thermal expansion properties of several low-expansion materials are shown in Figure 3.11.11. It shows the coefficients of thermal expansion (CTEs, relative change in length per degree) in parts per million. Note that the coefficients are a function of temperature, not a constant, and that fused silica, Zerodur, and ULE are so much better than the others.[43]

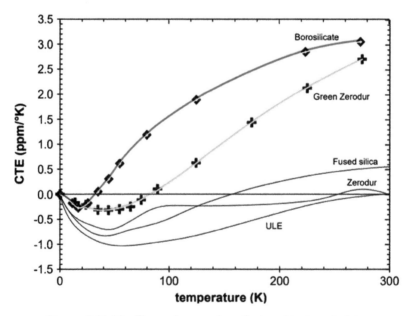

Figure 3.11.11   Thermal expansion of mirror blank materials.

One of the outstading developments of mirrors was accomplished by Roger Angel under the football stands at the University of Arizona. He

---

[43]Jacobs, S. "Thermal Expansivity and Temporal Stability Measurements at the University of Arizona," James C. Wyant College of Optical Sciences, https://wp.optics.arizona.edu/sfjacobs/wp-content/uploads/sites/47/2016/06/ThermalExpansivity2.pdf.

poured borosilicate glass into a mold that had an array of fire bricks in it to make a honeycomb. He arranged it so that the glass was at a molten but not liquid state. He then spun the entire structure thereby forming large convex parabolic mirrors of 8.4 meters in diameter, shown in Figure 3.11.12. He and his colleagues have since developed advanced grinding and polishing techniques for them. Six of them will be used in the Giant Magellan Telescope discssed later in Section 4.55.

Figure 3.11.12 Large mirror, little girl.

Mirror shapes vary from the simple plane mirror discussed above to specially shaped telescope mirrors called higher-order aspheres.

The spherical mirror is the next easiest after the plane one to understand, use, and make. Unfortunately, a sphere has spherical aberration. Incoming collimated light does not come to a focus. Figure 3.11.13 shows three parallel beams at three different heights. The reflections were carefully drawn using the law of reflection to show where the three beams intersect the axis, where they come to a focus. The beams are in red; the perpendiculars, the radii from the center of the mirror, are black. The nominal focus is half the radius, but the three beams intersect the axis at three different points somewhat near the half radius.

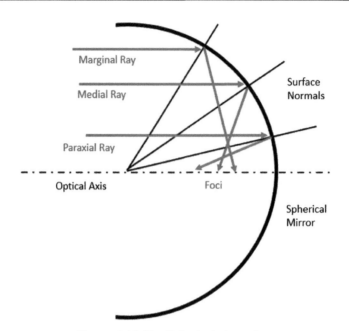

Figure 3.11.13 Spherical aberration.

It was known by most geometricians that rays or lines parallel to an axis all join perfectly to a point by a parabola. Just imagine reforming the sphere by pulling back the outer edges a little. You get a parabola and perfect on-axis imaging, as represented in Figure 3.11.14. The blue rays that reflect from the parabola will not be bent as much and come to a common focus. In one sense, a parabola is a laidback circle.

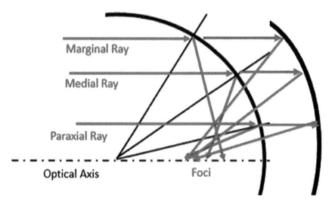

Figure 3.11.14 Spherical aberration correction using a parabola.

It was also known that rays emanating from one focus of an ellipse (and its solid cousin, an ellipsoid) converge to the other focus. These were properties of these special surfaces called conic sections, various slices through a cone: circle, ellipse, parabola, and hyperbola.

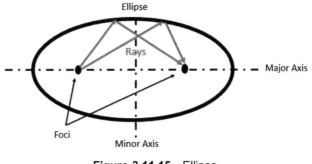

Figure 3.11.15 Ellipse.

The hyperbola is less well known. It is sort of an inside-out ellipse, as shown in Figure 3.11.16. Any ray aimed at the right-hand focal point will be reflected to the left-hand focus of the other, the virtual sheet of the hyperbola.

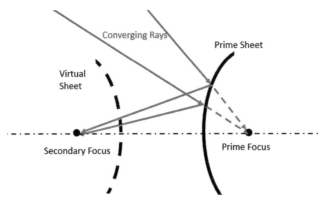

Figure 3.11.16 Hyperbola.

These conic sections—the plane, circle, ellipse, and hyperbola—are used as secondaries in many telescopes and other optical instruments. They are actually the three-dimensional equivalents: the plane, sphere, ellipsoid, and hyperboloid, as will be explained in Section A.6.3. Athough we often refer to them in terms of their 2D cousins, we do not believe in a flat Earth!

The spherical mirror is the easiest to make. One of my college projects required one two-meter-diameter spherical mirror and four higher-order aspheres of the same size. The spherical mirror took one month; the others each took six. We call these parabolas and ellipses aspheres. They are not spherical. Their rotational figure can be described by a quadratic, second-order equation. Some really good mirrors are tweaked aspheres. That is, the aspherical shape, parabolic for instance, is altered just a little to get better performance. We call these higher-order aspheres because the equation that descibes them has terms higher than the square.

3.12 Polarizers

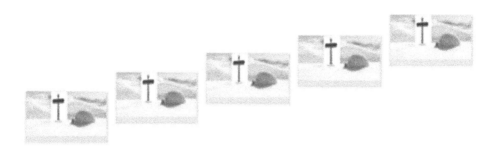

Polarizers[44] are devices that change the state of polarization of light. Some convert randomly polarized light into linearly polarized light. Some rotate the direction of linear polarized light. Others create circularly polarized light, and so on. They use different properties of the interaction of light with materials. These include parallel plates, parallel wires, polaroid material, and birefringent crystals.

Plane-parallel plate polarizers use the polarization property of reflection to create linearly polarized light. It was shown in Section 1.4 that, at an angle called Brewster's angle, the reflected light from a dielectric is linearly polarized. At exactly this angle, the light is completely polarized. But no beam is perfectly collimated, so no beam is at just this angle. There is a small spread, which leads to imperfect polarization.

Figure 3.12.1 illustrates this phenomenon. It is a plot of the perpendicularly polarized light and the parallel polarized light reflected from a dielectric. It is in terms of radian measure, but approximate degrees are also shown for familiarity. The green line is the ratio of the perpendicular reflection divided by the parallel reflection and is an indicator of the degree of polarization at that angle. The inverse is the degree of polarization, but that has an infinity and will not plot. There is a region near the polarizing angle where this ratio is very low, meaning its inverse is very high. There is a high degree of polarization, but it is not perfect. At one degree from the polarizing angle, the polarization factor is about 338, and at two degrees, it is about 655. There is good but not perfect polarizing action around the polarizing angle.

That is where a pile of polarizing plates come in. Figure 3.12.2 shows the enhancement by use of three plates. Unpolarized light enters from the left. Light polarized out of the plane of the paper (dots) is reflected at Brewster's angle; light of the other polarization is transmitted. The polarization is enhanced by the next two plates. Figure 3.12.2 shows three such plates. For a

---

[44]Shurcliff, W. *Polarized Light, Production and Use*, Harvard University Press (1962); Bennett, J. "Polarizers," in *Handbook of Optics*, Bass, M. ed., McGraw-Hill (1995).

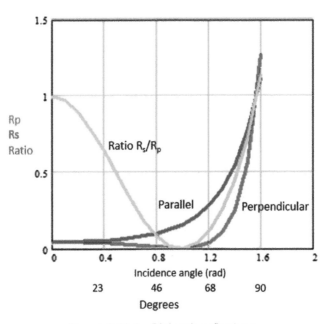

Figure 3.12.1 Dielectric reflections.

one-degree beam spread, the first plate introduces a polarization factor of 388; the second multiplies it; and the third does it again for a total polarization factor of almost 60 million. That should be good enough. The figure shows perpendicular light being attenuated. Parallel light will be reduced by the normal reflection factors.

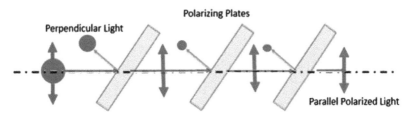

Figure 3.12.2 Plate-parallel plate polarizers.

Wire grid polarizers may have been the first type of artificial polarizers. They were a whole bunch of wires arranged parallel to each other in a plane. The light that passes through them is polarized, but not in the direction you might at first suspect. The polarization orientation that is transmitted is perpendicular to the direction of the wires. The mechanical analogy is wrong. The electric field is absorbed in the direction of the wires, not perpendicular to it, as indicated in Figure 3.12.3.

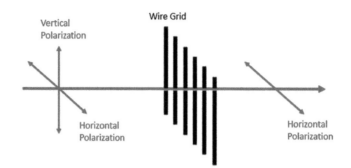

Figure 3.12.3   Wire grid polarization.

Early wire grid polarizers were just that: wires wound back and forth around closely spaced pegs. The theory was started by Fizeau and Rayleigh.[45] The grids typically have extinction ratios of about 100, that is, the transmitted component of polarization is 100 times that of the attenuated component.

Sheet polarizers were invented by Edwin Land.[46] They are a solid state version of wire grids. They are long chains of molecules that act very much like the wires. They are usually called Polaroids after Land's company. Or maybe he named his company after them.

Prism polarizers are based on birefringence, as discussed in Section 1.3. Light that travels in different directions in asymmetric crystals is polarized differently and has different refractive indices.

Although there are many variations of prism polarizers, the concept can be illustrated by just one. I chose the Glan–Thompson (G–T) prism, one of the first and most used. The G-T prism is shown schematically in Figure 3.12.4. It consists of two triangular pieces of birefringent calcite glued together. The thin film of glue may be used to enhance the total internal reflection, or it may be eliminated if the two prisms are clamped together. The

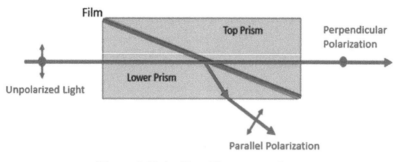

Figure 3.12.4   Glan–Thompson prism.

---

[45]Fizeau, M. *Compt. Rend. Acad. Sci.* **52**, 267 (1861); Lord Rayleigh, "On the passages of waves through apertures in Plane screens and allied problems," *Phil. Mag.* **43**, 259 (1897).
[46]Land, E. "Polarizing reflecting bodies," US patent 1918848A (1929).

calcite material splits the incoming beam into two with opposite polarizations. The parallel component is totally internally reflected at the interface because its refractive index is high enough. The perpendicular component passes through unaffected since its refractive index is low enough.

Circular polarizers turn unpolarized light or linearly polarized light into circularly polarized light, either clockwise or counterclockwise (viewed as it approaches). The secret of their operation is the quarter wave plate. A quarter wave plate is a birefringent material of a thickness such that the slower light is retarded by a quarter of a wavelength with respect to the faster polarization of opposite orientation. If they are adjusted to be of equal intensity, they will combine to give circular polarization.

3.13 Prisms

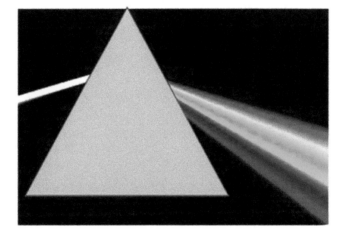

Prisms are of two major types: dispersive and non-dispersive. Polarizing prisms, which are a class of their own, are described in Section 3.12.

Dispersive prisms are used in spectrometers, multispectral imagers, projectors, television sets, and more. In every such case, they are used to disperse a spectrum, i.e., to spread the colors over space. The critical characteristic for a dispersive prism is its spectral resolving power, $\lambda/\Delta\lambda$. It is a measure of how well the prism spreads the light over space. The formula looks like it is upside down, but in this form, bigger is better. For a prism, it is the base length times the total refractive index change in the spectral band divided by that total spectral bandwidth. We might write that as $b\,\Delta n/\Delta\lambda$. It makes sense: spread the colors as much as you can with the total change in the refractive index Δn, and do it on a linear basis with a wider prism of base b. Then normalize with spectral spread $\Delta\lambda$.

As an example, the change in refractive index of borosilicate glass from 773 nm to 1014 nm is 5.4 times the change from 405 nm to 509 nm. Since the angular spread is approximately proportional to the refractive index spread, that is 2.33 times as much angular and linear spread of wavelengths in the violet than in the red and infrared. This is represented schematically in Figure 3.13.1.

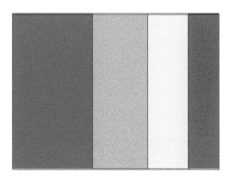

Figure 3.13.1   Nonlinear dispersion.

Nondispersive prisms are used to invert, revert, deviate and otherwise affect a beam of light in instruments like binoculars. There are even more of these than there are types of interferometers. The *Handbook of Optics* describes 30 of them,[47] and the Frankford Arsenal publication[48] even more. I will describe the design and use of only a few of the major ones to illustrate the concepts of their design. Smith discusses their uses in optical systems.[49]

These are all intended to deviate a beam in some way but **not** affect the spectrum. The primary way of doing that is to have perpendicular incidence on most outside surfaces and to use total internal reflection for most of the inside reflections.

The **right-angle prism** is the simplest. It is shown in Figure 3.13.2. It is, as it states, a prism with one right angle. Light is shone perpendicularly onto the front surface, a leg of the triangle, and is reflected off the angled one, the hypotenuse. The light is incident upon the hypotenuse at an angle of 45 degrees. This angle is sufficient for total internal reflection if the refractive index is 1.42 or greater, like almost all glasses and infrared materials. It is incident normally upon the other leg. It generates a 90-degree turn of the beam and no spectrum. There is no refraction.

---

[47]Wolfe, W. "Nondispersive Prisms," in *Handbook of Optics*, Bass, M., ed., McGraw-Hill (1995).
[48]*Design of Fire Control Optics*, US Government Printing Office (1952).
[49]Smith, W. *Modern Optical Engineering*, McGraw-Hill (1990).

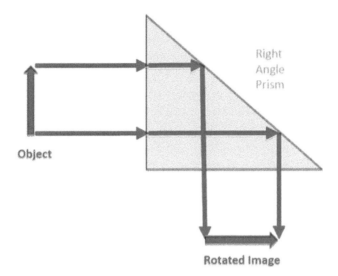

Figure 3.13.2   Right-angle prism.

The **Porro prism**, by Ignazio Porro, is also a right-angle prism used in a different orientation. The light enters the hypotenuse side and reflects off the two legs. It returns upon itself rather than making a 90-degree turn, and the image is inverted. The reflection angles are the same.

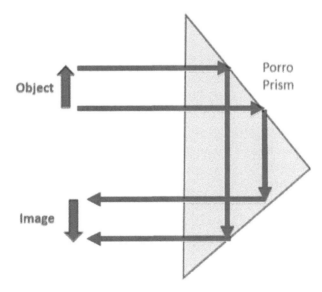

Figure 3.13.3   Porro prism.

The **Dove prism**, invented by Heinrich Wilhelm Dove, is used to invert an image without deviation. It does not revert or deviate, but it does invert. There is some dispersion upon entry at the angled front surface, but it is countered by the exit surface of opposite angle. The Dove is often used as an image rotator; the image rotates at twice the rate of the prism. A specific design is discussed in Appendix A.3.13.

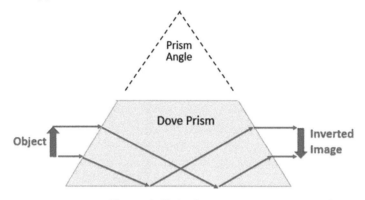

Figure 3.13.4 Dove prism.

The **roof prism** can be considered a variation of the right-angle prism in which the hypotenuse is replaced by an angled roof. It accomplishes both inversion and reversion, as shown. Inversion occurs because it is a right-angle prism; reversion occurs because it has a roof. These are often used in binoculars. A roof prism competes with a pair of Porro prisms, discussed above. I could not show all the rays.

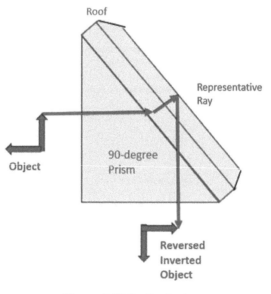

Figure 3.13.5 Roof prism.

Some prisms, which we do not usually think of as prisms, can divide a light beam by dividing its beam spatially rather than by its amplitude. The figure shows a beam coming from the left that is incident upon a pyramidal structure. It is divided into at least three beams (and three more I could not show). The beam is divided spatially; this type is also noted in the section on beam splitters (because it is one).

A prism is a solid geometric figure with similar rectilinear and parallel ends and parallelogram sides. It is not just the familiar triangular shapes. The one illustrated in Figure 3.13.6 has six sides. They can have any reasonable number of sides and divide the beam spatially in any reasonable number of resultant smaller beams.

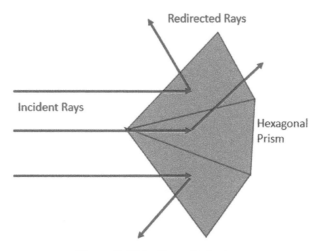

Figure 3.13.6   Pyramidal prism.

Risley prisms are used in two different ways. They can be used as a variable attenuator (discussed in Section 3.8). They act as an attenuating filter.

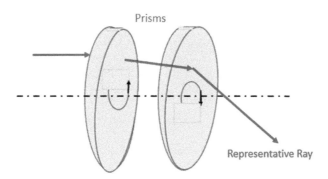

Figure 3.13.7   Risley beam-steering prisms.

They can also be used as beam steerers. Each prism bends the light in the usual way, and the combination can move it in both vertical and horizontal directions. The prisms can rotate at the same speed in the same direction with their apices aligned. Then they act as a rotating prism of their combined size. If they rotate in opposite directions at the same rate, one motion is cancelled out, and they generate a line the angle of which is determined by the angle at which the apices are aligned.

Some of the many patterns one can generate with rotating prisms are shown in Figure 3.13.8. The identifiers are as follows: **m** is the relative rotation rate, **k** is the relative prism angle, and **Φ** is the relative position of their apices when they start rotation.[50]

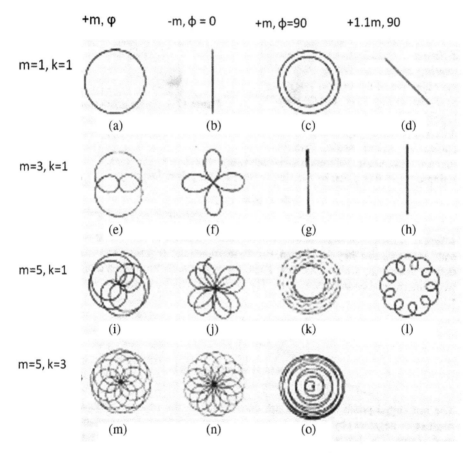

Figure 3.13.8 Risley prism patterns (reprinted from Ref. 49).

One application was a search-tracker in which a line was generated as in (b) of the chart and moved across a field by a mirror. When a target was identified, the relative rotation rates were changed to generate a rosette pattern, as in (m) or (n). Another was a horizon sensor in which the two prisms had different angles, as in (l).

---

[50]Wolfe, W. *Introduction to Infrared System Design*, SPIE Press (1996) [doi: 10.1117/3.226006].

3.14 Retroreflectors

Retroreflectors are devices that send a beam back in the exact same direction from which it came no matter from where it came. Any good mirror is a retroreflector if a beam is incident upon it perpendicularly, but these devices send the beam back upon itself no matter what the angle of incidence. There are two such devices: a cube corner and a cat's eye. I can show these in two dimensions and hope that you can visualize the third. They work best with collimated beams, although quite well with mostly collimated beams. Diverging and converging beams will spread out all over anyway.

The cube corner retroreflector, sometimes called the corner cube retroreflector, is just that: the corner of a cubical reflector. Purists insist that it is not a corner cube, but it **is** the corner of a cube, i.e., a cube corner. Figure 3.14.1 shows two beams (in red) that enter the corner at different angles from the upper left, the normal where they first strike the right side (in black), and where they then strike the lower side. The law of reflection is obeyed, and the beams both return upon themselves. It might help to realize that there are two right-angle reflections adding up to 180 degrees, a U turn. This is shown in two dimensions in Figure 3.14.1, but it holds true in three dimensions. A few Apollo missions placed some of them on the moon so that lasers on Earth could be used to measure the Earth–moon distance more accurately.

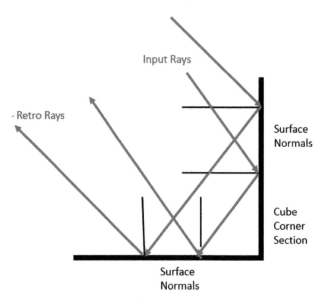

Figure 3.14.1   Cube corner retroreflector.

The cat's eye retroreflector is modeled after a real cat's eye. We have often seen the bright reflection from a cat in the beam of a flashlight, as shown in the introductory picture. The man-made version is used frequently in road signs and as bicycles reflectors. It is essentially a sphere with the appropriate refractive index and radius. A collimated beam enters from the left, as shown in Figure 3.14.2. It is refracted and focused on the rear of the sphere that has a reflective surface. The upper ray of that beam obeys the law of reflection and returns as the lower ray. It is the same with all the others. The secret is to choose the refractive material and the shape so that a parallel incoming beam is focused on the rear curved surface such that the central ray is perpendicular to it. The shape needs to be approximately spherical with a radius of a few millimeters and a refractive index of approximately 2.

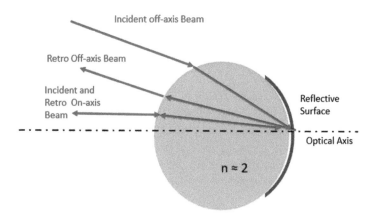

Figure 3.14.2   Cat's eye retroreflector.

A cat's eye is about 2 mm to 4 mm in diameter, and the refractive index is 1.74 to 1.75. My calculations in Appendix A.3.14 show that these are good values for a spherical retroreflector of that size (1.74 to 2.37 mm) and that they have a fairly wide margin. The devices on bicycles and road signs need not be precise. The headlight beams incident upon them are well collimated.

3.15 Sources

The sun is my doing, to paraphrase a title.[51] The **sun** is our basic source of light. All things come from the sun: growing things that are burned; fossil fuels that are burned; photons for solar panels and wind for turbines; even the sources that excite lasers. The sun is a giant ball of hydrogen and helium that undergoes continuous nuclear fusion. For our purposes, it consists of that gaseous interior surrounded by a photosphere that determines the optical properties that we see. The sun has a diameter about 100 times that of Earth (865,370 miles) and is about 330,000 times Earth's mass; it is about 98.86% of the mass of our entire solar system. Since it is 94.236 million miles from the Earth, the sun subtends just about one-half degree (865,370/94,236,000 equals 0.009 radians or 0.52 degrees).

The photosphere is essentially a blackbody ranging from 5777 K to about 5900 K (10,000°F). Too Darn Hot![52]

Figure 3.15.1 shows the spectral irradiance distribution from the sun at the Earth, above the atmosphere, at sea level, and with a 5900-K blackbody

---

[51]Steen, M. *The Sun is My Undoing*, Lume Books (1941).
[52]Porter, C. "Kiss Me Kate" (1948).

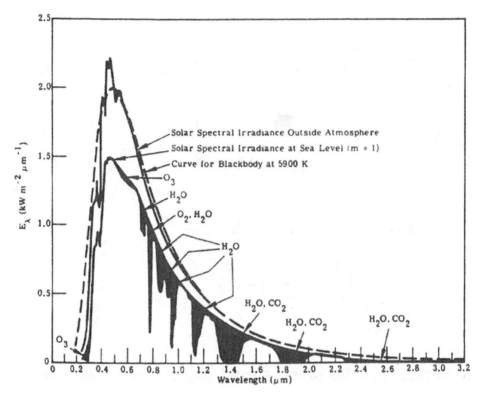

Figure 3.15.1   Solar irradiance (reprinted from Ref. 53).

reference curve.[53] The total irradiance above the atmosphere is 1368 Wm^{-2}. At the surface, it is approximately 1000 Wm^{-2}.

Although the **moon** is not a source but a reflector, the amount of moonlight is important in some applications like nighttime viewing. It becomes a source at night. Although calculating the irradiance from the moon can get very complicated,[54] a simplified treatment that shows the concepts is enough for our purposes. Although the reflectivity of the moon is not spectrally constant, it is almost so at a value of 12%. Since the moon also subtends the same angle at the Earth as the sun, we may take 12% of the solar irradiance for the full moon. And we can calculate a percentage decrease for the phases of the moon from full to dark. Obviously, when the moon is not visible, no light is visible from it.

Starlight can be important in nighttime viewing as well. How many lux are there on a moonless night? They are relatively sparse; contrary to Olber's paradox that the night sky should be filled with stars, it is not.

[53]Wolfe W. and Zissis, G. *The Infrared Handbook*, US Government Printing Office (1978).
[54]Willingham, D. *The lunar reflectivity model for Ranger Block III analysis*, JPL Technical Report 32-664 (1964).

Were the succession of stars endless, then the background of the sky would present us a uniform luminosity, like that displayed by the Galaxy—since there could be absolutely no point, in all that background, at which would not exist a star. The only mode, therefore, in which, under such a state of affairs, we could comprehend the voids which our telescopes find in innumerable directions, would be by supposing the distance of the invisible background so immense that no ray from it has yet been able to reach us at all.[55]

Poe was essentially right. Although there are about 10^{23} stars out there, some are so far away and moving so fast that their light has not yet reached us. Others are invisible because they are moving away fast enough that the red shift moves the light into the infrared (invisible) part of the spectrum.

A summary of these natural illuminance levels appears in Table 3.15.1.

Table 3.15.1   Illuminance levels.

Condition	Illuminance (lux)
Clear, sunny day	10,000
Overcast day	1000
Twilight	10
Full moon	0.01
Starlight	0.0001

Although it has been established that the human eye can detect a single photon,[56] a reasonable limit to seeing is approximately 0.000001 lux, quite a bit less than a moonless night. We can negotiate quite well in that level of illumination even without a flashlight. (I did it one entire summer at scout camp.)

Incandescent lights are the famous ones of Edison and the ones we have read by all our lives. Some are strips of tungsten arranged in a coiled coil. It is a tightly coiled wire that is formed in a less tight coil. Others are simple, tiny, fragile wires between tiny connectors. Although it is widely believed that they were invented by the great Edison, he was the one who made them practical. Many, many gradual improvements were made over the years: better filament materials and configurations, better vacuums, use of inert gas, and so on.

The modern incandescent bulb (which may soon become obsolete) uses a coiled coil of tungsten in an inert gas environment at a temperature of about 2700 K, just shy of melting.

---

[55]Poe, E. *Eureka: A Prose Poem*, Putnam (1848).

[56]Tinsley, J. et al. *"Direct detection of a single photon by humans,"* *Nat. Commun.* **7**, 12172 (2016); and my calculations.

Figure 3.15.2   Incandescent bulb.

Figure 3.15.3   Coiled coil.

Figure 3.15.4   Tiny wires.

The coil radiates with a high and constant emissivity, almost like a blackbody. Figure 3.15.5 shows a comparison of such a blackbody in red with the response of the eye in blue, both normalized to show the relationship. It is not a good match. Most of the output of the incandescent bulb is in the infrared and heats the room more than it lights it. Incandescent bulbs are cheap, inefficient, short lived, and almost obsolete. There is a comparison of them with others below.

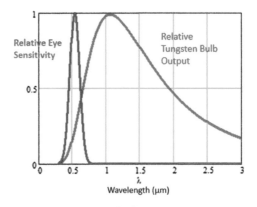

Figure 3.15.5   Tungsten bulb comparison with the eye.

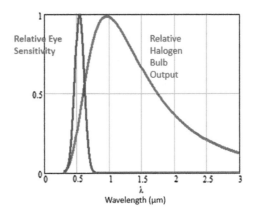

Figure 3.15.6   Halogen bulb comparison with the eye.

Halogen, **tungsten-halogen**, and **quartz-iodine lamps** are also incandescent lamps, but they are filled with an inert gas, a halogen gas (notably iodine and bromine but also sometimes xenon) that allows them to operate at a higher temperature. They use the same tungsten filament, but the inert gas has a chemical reaction with it to extend its life. The higher temperature also requires that the envelope be quartz rather than glass. Although this higher temperature increases the luminous efficacy and produces a whiter light, it requires the quartz and the halogen that make it more expensive. The match to the eyeball sensitivity is a bit better. The slightly higher temperature causes a small shift in the blackbody curve, as can barely be seen by comparing the peaks of the curves in Figures 3.15.5 and 3.15.6.

Fluorescent lamps are more efficient than incandescent ones and more expensive. Although Edison, Tesla, and others experimented with fluorescent lights, it was only in the 1970s that fluorescents came into their own as the familiar long tubes shown in Figures 3.15.7 and 3.15.8. A fluorescent lamp is a low-pressure mercury-vapor gas-discharge tube. It is a tube filled with mercury. An applied electric current, produced and controlled by the so-called ballast, excites the electrons in the mercury vapor. They produce ultraviolet light that excites the phosphors on the sides of the tube to generate visible light. Most fluorescents are quite white, but other phosphors can be used to make lights for other applications such as tanning, growing plants, inhibiting psoriasis, jaundice treatment, and as a germicide.

Figure 3.15.7   Fluorescent tube.

Figure 3.15.8 Fluorescent fixture.

More recently, manufacturers have twisted them into bulbs we now call **compact fluorescents** that are meant to be a more economical replacement for the incandescent bulb. They operate on the same principles; they just feature a different geometric configuration.

Figure 3.15.9 Compact fluorescent bulb.

Light-emitting diodes (LEDs) are the latest innovation in household lighting. They were invented in the early 1960s but did not become mature for almost 20 years. They consist of a small diode that has excess positive charges on one side (holes) and excess negative charges (electrons) on the other, with a barrier between them. When a voltage is applied, the charges flow toward each other, cross the barrier, and combine. (The positive charges could not be called positrons; that was taken.) But the electron left a vacancy when it left its stable place. The combination produces a photon. The wavelength of the photon depends on the nature of the material. The output is relatively monochromatic. The first LED radiated in the infrared at about 1 μm and was therefore not very useful for illumination.[57] Soon, other materials were used to get other wavelengths, many in the visible, and phosphors were used with them to get just about any color in the visible spectrum and into the infrared.

The LED itself is a very small device. Three of them in their capsules with leads attached are shown in Figure 3.15.10 along with a dime for comparison. They are combined with others to produce a great variety of illuminators, e.g., strip lamps under cabinets or even fireflies in a Mason jar.

---

[57]Biard, J. and Pittman, G. "Semiconductor radiant diode," US Patent 3293513.

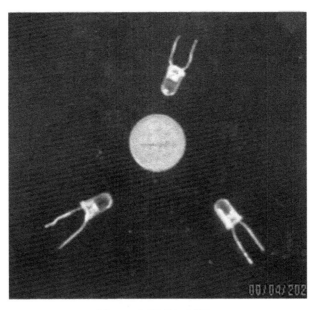

Figure 3.15.10 LEDs.

They are available in almost any color you desire, either by virtue of their inherent material properties or with phosphor coatings. A lamp can have more than one color so that it is blue by day and amber in the evening to not inhibit melatonin. They operate on milliamps and millivolts of dc current and voltage and have high luminous efficacy.

Table 3.15.2 is a comparison of the household sources of light. Efficacy is the luminous efficacy, the percent of the visible light output compared to the total radiant output. Efficency is an estimate of the average light output to the power presented from the electric wall outlet to the device, sometimes called the wall efficiency. Lifetimes are average and probably understated for LEDs because testing is not yet done. The total cost is an estimate of the cost for 20 years.

Table 3.15.2 Household sources of light.

Source	Unit cost ($)	Units (#)	Electricity ($)	Lifetime (h)	Total cost ($)
Incandescent	1	24	123	1200	147
Halogen	3	16	100	4000	148
CFL	2	8	31	8000	47
LED	–	1	12	50,000	7
Candle	0.25	96	0	5	24

The details, assumptions, and support for my calculations are in the next paragaph. It is fair to skip this one if you do not want to get into the weeds or if you believe me.

Cost is the best I could find by searching the internet, based on a 60-watt incandescent bulb and 20 years, operated 4 hours a day. The candle is there for the fun of it. CFL is the compact fluroescent, and 4 hours a day for 20 years is 29,200 hours. So one would have to buy $29200/1200 = 24$ incandescent bulbs, 8 CFLs, or only 1 LED. The 60-watt bulb takes 60 watts for 4 hours at 0.07$ per kilowatt hour. There are 29,200 hours times 0.060 kilowatts, or 1752 kWhr at $0.075 per, or $123. The LED takes 10% of that. The CFL takes 25% of the enrgy of an incandescent.

Sodium vapor lamps are used mostly in parking lots where they economically illuminate the area with yellow light, which is not necessarily pleasing but efficient and loved by astronomers. They emit at only the two sodium D lines at 589.0 and 589.6 nm. The tube is filled with an inert gas and solid sodium. After initiation, the sodium vaporizes. The electric voltage difference between the two ends of the lamp accelerates electrons and ions that recombine and thereby generate photons. In this case, it is the electron transition of the sodium electrons.

There are several kinds of **arc lamps**. They all work on the same principle as lightning and the winter rug-shuffling zap! When there is sufficient voltage difference between two points, a current will flow, even in air or another non-conductor. This flow of electrons in air causes ionization of the air molecules, and their relaxation causes the generation of photons. Lightning occurs when there is a sufficient voltage difference between clouds or the clouds and ground. Lightning is on the order of millions of volts. The winter zap occurs when you scrub your feet on the rug in a dry atmosphere and generate enough static electricity to shock your victim (about 1000 volts, but hardly any current or power). If you do it in the dark, you may be able to see the spark, the arc. All arc lamps operate on this principle. The differences are in the materials, gases, and configurations. The difference between the rug zap and lightning is only in the magnitude—AN ENORMOUS DIFFERENCE.

Carbon arc lamps are used in searchlights. They were perhaps made most famous during the Battle of Britain during WWII in spotting the Nazi Luftwaffe. They are a staple of county fairs. They use two carbon rods with a gap of about an inch between them and a voltage difference of about 80 volts, a current of 150 amps at a temperature of about 3000°F, thereby generating about 800 billion lumens per steradian (very bright). The ends of the carbon rods vaporize slowly. It is the carbon vapor in the arc that produces the light. Accordingly, there must be an accurate and consistent procedure for feeding the rods and for replacing them every hour or so.

Xenon arc lamps use the same electrical arrangement as the others. A small gap with a voltage difference across it. In this case, the xenon gas is excited by the current, and upon recombination, light is emitted. It resembles a blackbody at about 6000 K. The lamps are kept at about 30 atmospheres of pressure and in a quartz envelope as a result. So-called xenon arc headlights are really arcs of halides. Xenon is just the starter. They come in a variety of sizes and output power from as low as 75 watts to almost 2000 (and probably more).

Flames, good old campfires, were probably the first artificial sources of light (as well as heat for cooking and an atmosphere for eerie stories). The flames can be in the form of campfires, but also as torches and candles. A flame is generated when something sufficiently hot comes into contact with a combustible substance such as wood, kerosene, or a candle wick. Some of the material is decomposed into a gas, and the gas carries out the process of converting electron energy to photon energy. Wood, kerosene, and candle wax are all organic materials that are made up of chains of carbon, oxygen, and hydrogen (and sometimes other stuff). When the chains decompose, they separate into carbon dioxide, water vapor, and some incombustilbe materials that become the ashes. The colors—red, yellow, and sometimes blue—arise from different materials in the flame. The yellow is from that characteristic sodium yellow D-line doublet right in the middle of our visibility curve. Try throwing some salt on your next campfire to see it flame up yellow. The blue color, when it shows, comes from the breakdown of oxygen. It is usually near the bottom of the flame, where it is sucking in oxygen and is hottest. These colors are also dictated by the particle temperature. The bluer flame is hotter. Its blackbody curve peaks at about 400 nm, corresponding to a temperature of about 7250 K. The orangish flame is cooler, based on particles at a temperature of 5800 K.

Figure 3.15.11 Campfire.

Candles provide light in much the same way. They are just another flame. They burn oxygen and wax on a wick (Figure 3.15.12). The flame is better controlled than a fire and was long used as a household source of light. The

Figure 3.15.12   Candle.

wax at the bottom of the wick melts. It goes up the wick by capillary action and is vaporized by the flame. Since the wax is an organic material, the carbon, oxygen, and hydrogen become water vapor and carbon dioxide. Paraffin (from which most candles are made) is strictly carbon and hydrogen. Beeswax also has oxygen in its complicated organic chains, but that still gives only CO_2 and H_2O. The other major candle wax, stearin, has the same three molecules. Some candles contain esters for aroma.

Gas lamps were used for street lighting during the 1800s, before the era of electric lights. Gas lamps are flames, but they are controlled flames. They operate much like the candles described above, but rather than a wick, a small tube emits a controlled amount of gas that burns as a flame. At the time, there was a search for the gases that would provide the most light, that is, burn the brightest. Charles Clamond produced the eponymous Clamond basket by extruding a thread of magnesium oxide and "knitting" a basket with it. The flame would then cause the magnesium oxide to glow brightly, whitely. Later, Carl von Welsbach invented his mantle that is still in use today. It is a cotton bag that gets impregnated with soluble metal nitrates that burn away initially to metal oxides. You probably recognize it as the **propane lantern** you used on your last camping trip. They consist of mantles made of silk onto which various oxides are affixed. The silk burns away, and the oxides glow when the flame is on. They thus produce a nice white spectrum characteristic of the oxides. They are contained within a glass structure for safety and their own protection (they are fragile). They are also incandescent but use the oxides to get a nice spectrum. [That may also have been the last camping trip with it. LED versions are replacing them.]

Figure 3.15.13 Camp lantern.

No discussion of lighting would be complete without a discussion of **limelight.** These were cylinders of calcium oxide (lime) onto which an intense flame consisting of oxygen and hydrogen like that used for welding is directed, bringing it to about 2850 K. This setup produces a very white light. The limelight has been replaced by electric lights, although you can still be in it.

Globars and **Nernst lamps** are special sources used in spectrometers. Globars are rods of silicon carbide about one or two inches long that are heated to 1000–1650°C (1830–3000°F) and are a good approximation to a blackbody. Nernst lamps were essentially lightbulbs with ceramic rod filaments. They were first used for lighting and then in spectroscopy, e.g., Globars.

Chapter 4
Optical Instruments

This chapter includes more than sixty different optical instruments. They range from the very simple magnifying glass, which is just a lens, to ICBM detectors, which include telescopes, detector arrays, and electronics. They may have many moving parts, such as interferometers, or no moving parts, such as windows. The word "instrument" has several definitions and connotations, but I like the most general one for this chapter: "a means whereby something is achieved, performed, or furthered."[1] One would not normally think of a window as an instrument, but it sure does achieve keeping the heat out and letting the light in.

This chapter should be read a little at a time. Choose whichever application intrigues you and go there. The sections are arranged alphabetically and are independent but may depend on the material in chapter 1. There are a few cross-references, but that should not be a problem.

Enjoy the scope of endeavors the optical instruments support. Enjoy the fact that we have a good defense against ICBM attacks. Anticipate the advent of holographic TV, autonomous cars, better traffic control, better cancer detection, and maybe a peaceful world.

---

[1]Merriam Webster, Merriam-Webster.com.

4.1 Aeronautical Optics

Planes employ all sorts of optics, from the dials in the cockpit to altimeters, landing lights, and even CAT detectors, the last of which is the first subject of this section.

CAT detectors detect clear-air turbulence (not felines), as shown above. They are lidars, but they do not use time of flight or the red shift to measure the motion of the turbulence (see Section 4.24). Clear-air turbulence is the motion of invisible turbulent air. The motions that are most troublesome to aircraft are the vertical ones, as some say, the vertical "turbs." These are regions of air of different densities that move up and down very rapidly. Because it is their **vertical** motion that is troublesome, neither Doppler nor time of flight lidar is effective.

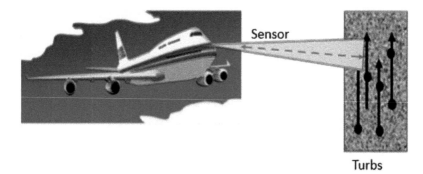

Figure 4.1.1   CAT detector.

A different technique is used. The backscatter from these turbs is measured. If the frequency in the variation of the amount of backscatter is high, it means that there is a considerable time variation in the density of the air; there is turbulence in that region. Since it is clear air, the scatter is of the Rayleigh variety and is spectrally dependent. CAT lidars therefore operate in the near-ultraviolet part of the spectrum to maximize the backscatter but where the air is still transparent (see Section 1.7).

Enhanced vison systems (EVSs) for aircraft landing are used during periods of limited visibility when there are mists or light fogs. Although some ordinary cameras and night-vision devices have been used, the primary imaging system is infrared. Both the 3–5-μm and 8-12-μm systems have been used. The latter are better because the longer waves are scattered less by the fogs. Depending upon the density of the fog, there is a decent image of the runway and its lights, but sometimes only the landing lights along the strip are visible. This will be a more significant problem as runway lights switch to LEDs.[2]

Contrary to the cited reference, there are four, not two, types of infrared devices possible for EVS. They are cooled indium antimonide (InSb) 3–5-μm cameras; cooled mercury cadmium telluride (MCT) systems in the same midrange band and also in the 8–12-μm long-wave band; and microbolometer imagers in the 8–12-μm band. I do not know of the commercial availability, but you can order almost anything for the right price.

Usually, an InSb system is chosen over a similar HCT system for the midrange infrared, but they both have about the same capabilities and cooling requirements. I think I have made it clear that a longer-wave system is preferable because there is less scattering. The choice then is between an MCT system that requires cooling and a less-sensitive microbolometer system that does not. Cooling is real trouble with expense, reliability, noise, and life problems. A typical microbolometer system provides a temperature difference of about 0.01 °C, which is fine for this application.

One solution described by the cited reference for the upcoming use of LEDs is the use of multispectral imagers to sense the visible light from the LEDs. I think a better solution is the use of some infrared LEDs to go with the visible ones to be used only when needed (or even Globars or incandescent bulbs). I will not go so far as to suggest candles or hot rocks!

When things get really bad, one can resort to a **synthetic vison system**. There is no detection of any surroundings. The GPS system knows where you are, where the airstrip is. and the surroundings. It presents all of this on the display without sensing it. This scenario at first seems scary, but it is not so bad. The GPS knows where you are and where all your surroundings are. A computer can then generate a synthetic image of your surroundings and you in it. You can use the instruments to come to a safe landing. I considered one of my trips guided by GPS. When I got to my destination, it showed me the building and the surroundings both at the right distances from my car. I was impressed, and I think it applies to this technique.

---

[2]Wikipedia. "Enhanced flight vision system," https://en.wikipedia.org/wiki/Enhanced_flight_vision_system.

4.2 Aerospace Navigation

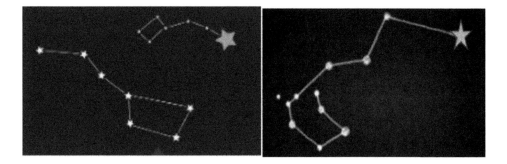

Two of the methods satellites use to determine their location and orientation are **star trackers** and **horizon sensors**. Star trackers, also known as star cameras, provide location information to satellites. They let the satellite know, for instance, that the North Star is at a certain pair of azimuth and elevation angles with respect to its position. A horizon sensor provides the information of orientation, of verticality, to the satellite. It gives the direction to the center of the Earth by measuring the horizon.

Star trackers usually use the North Star in the northern hemisphere and Canopus in the southern. The North Star is conspicuously located at the end of the Little Dipper. Not only is it very close to true north, it is also part of a well-defined constellation, very bright, and almost fixed in space. It has been used this way for centuries by ancient mariners. Canopus is one of the brightest stars in the southern hemisphere and is in the constellation Carina, which looks a lot like the Little Dipper, as shown in the introductory image. Star trackers locate these critical stars by imaging the surrounding star field and comparing it to a stored computer image.[3]

Horizon sensors use the 12-μm part of the spectrum of the atmosphere because it is stable. The Horizon Definition Measurement Program, of which I was a part, concluded, "The data produced during the study indicated that the carbon dioxide absorption band is an extremely stable radiation source."[4] The visible horizon is mostly variable clouds. The infrared band is dominated by the emission and absorption of carbon dioxide. Although it may be a villain in global warming, it is a hero in horizon sensing. It is a stable, high-altitude, reliable source of information.

The manner by which horizon sensors scan the infrared horizon varies a great deal. Some use fixed quadrant detectors and balance the outputs; some crisscross along diameters and equate the lengths; and some dither all the way around the edge, as illustrated in Figure 4.2.1. But they all use thermal

---

[3]Designs by my group at Honeywell.
[4]Bates, J. *Horizon Definition Study Summary*, NASA CR66432 (1967).

Quadrant Detector Images   Cross Scans   Dither Scan

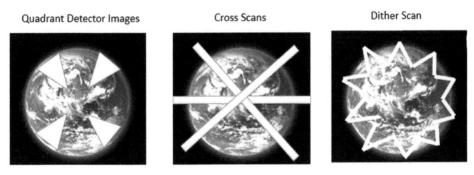

Figure 4.2.1 Horizon sensor patterns.

detectors in the 14–16-μm part of the infrared spectrum to sense the carbon-dioxide horizon.[5]

4.3 Automotive Optics

Our cars have a lot of optics in them: all the lights on the dash, headlights, taillights, rear- and side-view mirrors, and some stuff we do not even see. There are also the special glasses in the windshield and side windows. Night driving, lane changing, turn signals and even "look mom, no hands."

Interior lighting is now in the hands and minds of the imaginative. LEDs and fibers make all sorts of lighting possible. Some imaginary examples include a fake starlit roof instead of a T-top. That can be a host of small LEDs driven by tiny batteries or a bunch of fibers sticking their ends out all over the place. Cup holders can be accentuated with rings of color. Door handles can light up with the approach of a hand.

Dashboards are taking on an entirely different look. All the gauges are incorporated in a single display panel and can be selected at will. Regular gauges are now mostly LEDs, and the dashboard gauges can be on the windshield as heads-up displays.

---

[5]Wolfe, W. ed., *NASA Space Vehicle Design Criteria, Spacecraft Earth Horizon Sensors*. NASA SP 803 (1969).

Headlights are still mostly incandescent bulbs at the focus of a spherical or parabolic mirror with a lens in front. The lens can be rather complicated, incorporating small prisms to shape the beam. Halogen bulbs are used in some more expensive variations and in the obnoxious, **really bright** metal halide lamps. High and low beams are accomplished in at least two ways. Most use a pair of incandescent filaments, one at the focus of the mirror and one slightly above it. Other cars use two separate units. Because there is such a large market, there is considerable research going on to use LEDs, arrays of micromirrors, and free-form optics to improve brightness and beam shape and to reduce costs. There have been brief experiments with yellow lighting with alleged nighttime improvement in visibility. Not much, if any, gain was experienced.[6]

Other variations that were really not optical include the Tucker 48, which had a cyclops lamp in the center front grill that turned as the car turned to illuminate the road into which you were turning.[7] Others used hidden headlights that had "eyelids" that opened when the lights went on but retained the sleek style of the car otherwise.

Taillights have become almost exclusively LEDs because they are efficient and long lived. Some older cars still have incandescent bulbs behind crenellated covers that automotive buffs call lenses. These are shown in Figure 4.3.1. You can hardly tell the difference between them and the LEDs on the car in Figure 4.3.2, but the battery and the miles per gallon can.

Figure 4.3.1   Crenellated taillights.

Figure 4.3.2   LED taillights.

---

[6]Kneisel, K. *MedpageToday*, https://www.medpagetoday.com.
[7]Wikipedia. "Tucker 48," https://en.wikipedia.org/wiki/Tucker_48.

Flip mirrors are the ones inside the car at the top of the windshield, shown in Figure 4.3.3. They provide a direct view back during the day but a dimmed view of the oncoming headlights at night. They are a combination of a full mirror and a partial mirror. The little handle at the bottom moves the full mirror to a more inclined angle for night viewing. Newer cars do this automatically.

Figure 4.3.3   Rearview flip mirror.

Figure 4.3.4 shows the flip mirror in the daytime position. Figure 4.3.5 show the flipped mirror in the nighttime configuration.

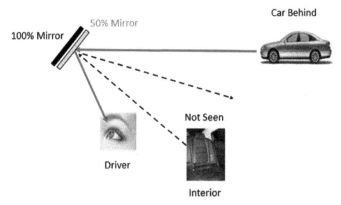

Figure 4.3.4   Flip mirror concept: daytime.

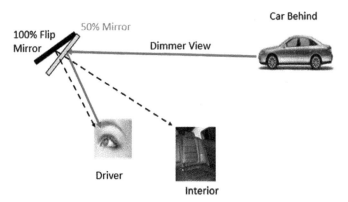

Figure 4.3.5   Flip mirror concept: nighttime.

Side-view mirrors on the passenger's side always say that cars are closer than they seem. They are moderately convex mirrors. Those on the driver's side are plane. An easy way to understand why they seem farther away is that they provide a wider field of view and therefore more objects in it. Since more objects are in the same space, they must be smaller. If they are smaller, you perceive them as farther away. Section 3.10 provides a ray trace that shows this as well. It is reproduced in Figure 4.3.7.

Figure 4.3.6   Passenger side mirror.

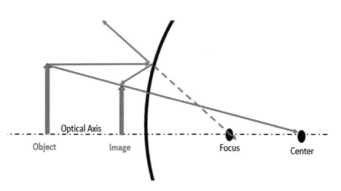

Figure 4.3.7   Convex mirror imaging.

Passenger window glass is treated so that it shatters into a gazillion small pieces rather than long shards that could be lethal. It is often called **tempered glass** after one of the techniques used to make it. A sheet of glass is exposed to a fairly high temperature and then quenched rapidly in a cooling bath. It is tempered. This process causes a contraction on the surface due to the cooling and a compressive strength as a result. The process can also be done chemically. When the glass surface is disrupted, dinged, or scratched, all that compression disintegrates the glass. You can usually see the result of compression with sunglasses. Striations of color will appear due to the changes in refractive index and density due to compression near the surface.

Windshields are formed of **laminated glass**. It is a piece of some form of polyvinyl, surrounded by two pieces of glass. In the event of an accident, such as a rock ding, the glass is held together by the plastic. The use of tempered glass in a windshield would be catastrophic. One ding and it is history. The invention of laminated glass was an accident by a French chemist. A glass flask that had contained cellulose nitrate and had not been washed fell on the floor. It held together. He recognized it for what it was. Thank goodness for alert chemists who procrastinate.

Backup and lane warnings are accomplished with cameras. The lane-warning camera monitors the lane markings and senses when your car is straying from the straight and narrow. The backup camera, of most use to me in the parking lot, has a field of view from 10 to 20 feet wide to the rear and senses when another car enters that field.

Head-up displays can now provide driver information such as mph, mpg, ETA, tire pressure, GPS location, and more on the lower portion of the windshield, where one can read it without taking one's eyes off the road, at least not very far. The alternative is an LED flat-panel display high on the dash.

Drowsiness detectors use several different technologies. One uses the pattern of wheel and pedal motions and frequency and precision of lane changes. A second uses physiological measurements such as heart rate and EKGs. The optical ones include cameras to measure the frequency of yawns and blinks, droops and nods. A droop is an extended blink that occurs upon drowsiness. I guess sleep is a very extended droop.

Adaptive cruise control slows a car that is in cruise control that gets too close to the car in front. It is an active optical system. A pulsed beam of light from an LED or diode laser is projected out front. It is filtered by a narrow-band filter that rejects most ambient light. The roundtrip time of flight is an indication of distance.

Night vision by means of infrared cameras was first demonstrated in a Cadillac over 20 years ago. The first design[8] used mercury cadmium telluride detectors, but later designs switched to microbolometer arrays.[9] The technology of that time was too expensive. Even though Cadillac charged about $2000 per unit, they lost money. Now there are many autos equipped with different designs of these night-vision devices. Figures 4.3.8 and 4.3.9 show what enhanced viewing is available. The left view is the visible one. The headlights are visible but not the people. The right view includes an inset from the night-vision system and shows the people in the middle on the road that you might have run into. Appendix A.4.3 shows how this can be done.

---

[8]My design.
[9]Hughes design.

Figure 4.3.8 Visible nighttime view.   **Figure 4.3.9** Infrared nighttime view.

4.4 Autonomous Vehicles

They are coming, and they are here. Optics is a large part of what makes autonomous vehicles autonomous. It includes GPS, lidar, cameras, and computers.

The **GPS system** is used to locate the vehicle. GPS is only marginally optical. It uses three or four of the 24–30 GPS satellites for triangulation (quadrangulation?). The lengths to the ground vehicle from the satellite are determined by the time it takes for the round trip of light. The onboard computer does the calculation and sends the information to the vehicle by radio signals. It keeps tabs on time with atomic clocks, and it corrects time variations due to relativistic effects. Time increases in the reduced gravity field of the orbital altitude, and it decreases with orbital speed. The uncertainty of this location has been improving with time from about 3 meters to 30 centimeters, about one foot.[10]

A **lidar** on top of the car spins a full 360 degrees to sense and range all objects forward, to the sides, and even the rear to see if something might be

---

[10]Wikipedia. GNSS (GPS) accuracy explained Jupiter Systems https://junipersys.com/support/article/6614.

catching up with you. Since light travels at one foot per nanosecond, it takes only 0.6 microseconds to detect and range on an object as far away as 100 yards. A lidar that rotates 360 degrees each second has plenty of time to detect and range on everything around it. Automotive lidars operate in the 1550-nm spectral range, the very near infrared. This gives them a little advantage in mists, but they do not work very well, if at all, in fog or heavy rain.

Lidar has difficulty in short-range detection as well, but it has a maximum range of about 200 meters and its angular uncertainty is 0.1 degree (1.7 milliradians). That means at the maximum range of 200 meters, the linear resolution is 0.34 meters, about a foot. In good weather and at a spin rate of once per second, the entire field is scanned while a 60-mph vehicle travels 88 feet. The total stopping distance, reaction time, and braking distance is 268 feet, or about 82 meters, and the range is 200 meters. You get three good looks at whatever is up there. Radar is used for shorter distances.[11]

Cameras in the front and in the rear of the vehicle provide images of the road ahead. Computers can process the images to extract reasonable range information. Certainly, edges of roads can be determined, and major obstacles in front can be detected.

There have been **trial runs** of both cars[12] and semis[13] that have been successful. One truck drove from Tulare, CA to Quakertown, PA.[14] That is almost across the entire United States: about 2420 miles.

I believe that the first commercial version of this technology will be convoys of several semis that travel our interstates. There will be only one human in them, a passenger in the front vehicle. Every eight hours or so, they will be platooned by another "driver." The trucks would not only have the lidars and GPS systems discussed above, they would also have systems to detect tire pressure and load shifts. This setup would reduce costs, speed delivery, and alleviate the shortage of qualified drivers (and the supply chain bottleneck that exists as I write this in October 2021).

[11]Khader, M. and Cherian, S. *"An Introduction to Automotive LIDAR,"* Texas Instruments, https://www.ti.com/lit/wp/slyy150a/slyy150a.pdf (2020).

[12]Hawkins, A. "Waymo's driverless car: ghost-riding in the back seat of a robot taxi," *The Verge,* https://www.theverge.com/2019/12/9/21000085/waymo-fully-driverless-car-self-driving-ride-hail-service-phoenix-arizona (December 9, 2019).

[13]Macdonald-Evoy, J. "Autonomous semi-trucks have been driving along I-10 for months and no one noticed," *AZ Mirror,* https://www.azmirror.com/2019/08/16/autonomous-semi-trucks-arizona-ups-tusimple (August 16, 2019).

[14]Blanco, S. "An Autonomous Semi-Truck Just Drove Across America to Deliver Butter," Forbes Online, https://www.forbes.com/sites/sebastianblanco/2019/12/12/an-autonomous-semi-truck-just-drove-across-america-to-deliver-butter (December 12, 2019).

4.5 Ballistic Missile Detection and Interception

The threat of an attack by ICBMs has been real ever since the USSR launched Sputnik in 1957. These attacks are countered with two technologies: the detection of the **launch**, and the interception of the warhead in **midcourse**. Interception during re-entry is very difficult; it is not yet an option.

The launch is accompanied by the emission of a great flame of carbon dioxide gas, water vapor, and even carbon particles that emits copiously in the infrared region of the spectrum, centered at about 2.5 μm. As the missile rises, the plume expands from about the size of the missile to several times its size. The introductory figure shows such a missile just after takeoff. The plume expands as it hits the ground to a diameter that is many times the size of the plume.

The first of these launch detectors was the Missile Defense Alarm System **(MIDAS)**, the king of missiles with a golden touch. It operated from 1960 to 1966 with limited success. We were not yet capable of orbiting a satellite in geostationary orbit at an altitude of 35,786 kilometers (22,236 miles). The MIDAS satellites, 10 to 12 of them, had to orbit in a polar orbit at about 1000 miles altitude. Twelve of these systems were launched with varying degrees of success. The program showed the way to a reliable ICBM launch detection system. Calculations of the detection capability are described in Appendix A.4.5.

The immediate successor, the **Defense Support Program**, had its first launch in 1970. The optical system was a Schmidt telescope with a lead-sulfide linear detector array that was swept over part of the Earth six times a minute by the rotation of the satellite. The final DSP satellite was launched in 2007 and has been replaced by the Space-Based Infrared System (SBIRS). (DSP was good enough to detect Scud missiles during Operation Desert Storm, the First Gulf War.) SBIRS consists of satellites in geostationary orbit and highly elliptical ones that can come much closer to the Earth for part of the orbit. Those in geosynchronous orbit detect missile launches, whether they are ICBMs or harmless satellites.

The powered vehicles can be tracked by radar or infrared until there is separation, and **midcourse** begins. The warhead then travels in an unpowered, predictable, parabolic arc toward its intended victim. It is about the size of an SUV and made of aluminum and ceramics. Because it sat on the launching pad for some time and flew through the air at a high speed, it has a temperature of about 300 K (80°F). Then the task is to detect an object this size radiating at this temperature with an emissivity of about 0.05 in a space background that is a blackbody at a temperature of about 60 K. My calculations in Appendix A.4.5 show that the missile radiates 120 watts per square meter (assuming a conservative emissivity of 0.05) in the 8–12-μm spectral band, while the background is 0.012 watts per square meter. That is a very nice signal-to-background ratio of 10,000!

The interceptor's guidance system uses a combination of radar and infrared, with the radar tracking during initial phases and the infrared doing the final homing. It is a hit-to-kill vehicle, also called an **exoatmospheric kill vehicle (EKV)**. The action is analogous to a big bullet striking another big bullet. The impact is enough to decimate the warhead without an explosion. The infrared system consists of a reflective telescope and an array of cooled mercury cadmium telluride detectors that operate in the 8–12-μm part of the spectrum.

Flight tests have shown an overall 82% success rate of interception including the early failures. Since the year 2000 with more mature vehicles, the rate is 89%.[15] The use of four interceptors increases these probabilities to 100%. We can live with that!

The **Terminal High-Altitude Advanced Defense (THAAD) missile** is a portable ground-based interceptor with similar infrared guidance. (It is not quite terminal, but that name was a good replacement for Theater Area, which was the original meaning of the acronym).[16] It is meant to intercept at close to the end of the midcourse flight or at the beginning of the terminal phase, before things get too furious. It is also a kinetic missile, but it is meant for IRBMs and hopefully ICBMs. It operates using an array of indium antimonide detectors in the 3–5-μm region. The THAAD success rate in tests is 100%.

The **Israeli Arrow** missile is part of the **Iron Dome Defense** system that protects the state of Israel and especially Jerusalem and Tel Aviv. It has a dual-mode guidance system: radar and infrared. The infrared system is a telescopic unit with an indium antimonide detector array cooled to 77 K that operates in the 3–5-μm spectrum. It has an explosive charge to destroy its target.[16]

---

[15]Wikipedia. "Ground Based Midcourse Defense," https://en.wikipedia.org/wiki/Ground-Based_Midcourse_Defense.

[16]Personal observation in project reviews.

4.6 Binoculars

Binoculars, as the name implies, consists of two oculars, two telescopes, one for each eye. Because they are used for terrestrial viewing such as bird watching or hunting, they must present a realistic, erect image to the eye. This is done with a Galilean-type telescope or with Porto or roof prisms. There are basically two types of binoculars, large and small. The larger ones are used for outdoor activities, and the smaller ones for the opera.

They consist of an objective lens system, an erector for some systems, and an eyepiece. The Galileans provide an erect image by the arrangement of the lenses.

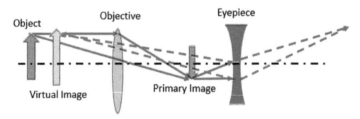

Figure 4.6.1 Galilean telescope schematic.

The Galilean design is described in Section 4.55 but is repeated here. It consists of a convex primary and concave secondary. The object, shown in green on the left, is imaged by the positive objective lens. The image is upside down and inside the focal point of the eyepiece, a negative lens. The eyepiece makes an enlarged virtual image, shown by the dashed lines. I do not know if Galileo used singlets or multiple elements for each and what their curvatures were, but they were positive and negative. And I do not know where he arranged the focal points.

The other design is shown **very** schematically in Figure 4.6.2. It shows the three main parts of a prismatic binocular: the objective lens assembly, the prismatic reinverter, and the eyepiece. It reverses the image side to side and up to down. It reverts and inverts.

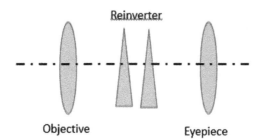

Figure 4.6.2 Prismatic binocular schematic.

The **objective lens** determines the light-collecting ability and the limits on resolution. The larger the diameter is, the better the resolution, the larger the area, and the more light it collects. If it is good enough, its diameter determines the diffraction-limited resolution. To make it good enough, these objectives are usually at least triplets of the Cooke variety (see Section 3.10).

There are two main designs of **prism erectors**: Porro prisms and roof prisms. Porro prisms are used in tandem. They are the reason for the kinky design of binocs, i.e., the two offset, cylindrical sections. A typical pair of prisms binoculars is shown in Figure 4.6.3.

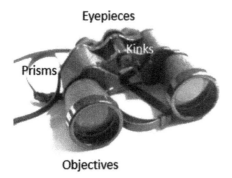

Figure 4.6.3 Typical prism binoculars.

The function of the prisms is illustrated in Figure 4.6.4. They revert and invert to give an image that is in the same orientation as the object.

The eyepiece provides an image with enough eye relief and size to provide comfortable viewing. Eye relief means how far from the eyepiece itself it is still comfortable to view the scene.

Figure 4.6.4   Prism binocular optics.

Opera glasses have a wider eyepiece distance than objective distance, and the objectives are not much larger than the eyepieces. Those shown in Figure 4.6.5 have the characteristic triangular kink indicating that they use prisms. Although you cannot see it in the figure, they are 6×15, meaning that they have a magnification of 6 and an aperture diameter of 15 mm. These are both relatively small numbers, but one does not need much magnification to see the diva better, and certainly you need no help in collecting light from the stage of an opera house. The sort of inverse arrangement of wider eyepiece distance is a result of compact design. There had to be a kink to accommodate the Porros. If the objective separation were larger, the glasses would get much larger.

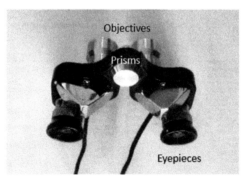

Figure 4.6.5   Opera glasses.

Most binocs, including opera glasses, have one adjustable eyepiece because our eyes are not identical. No two people are alike, and no two eyes are alike.

One of my colleagues predicts that binoculars will soon be a pair of digital cameras:[17] one for each eye that may include some telescopic optics but will not need erectoverters.

---

[17]Vukabratovich, D., private communication.

4.7 Borescopes

Borescopes are cousins to endoscopes. They are the industrial versions of tubes that look inside bodies. In this case, they are used on and in industrial and commercial bodies, not human ones. These include the sewer pipe from your home to the main, inspection of auto cylinders, air-conditioning ducts, jet engines, and even city sewer pipes. A major use involves the inspection of rifles.

Borescopes come in several varieties: rigid and flexible. They use lenses, fibers, and cameras.[18] As with endoscopes, a means of illumination and a means of imaging are the essential elements. For years, the illumination has been by a variety of incandescent lamps, but the trend now is to LEDs.

The essence of a rigid borescope is a mirror at the distal end and one or more lenses to relay an image of the object. The mirror usually has some sort of rotational capability, although not always 360 degrees, or it may just be the entire borescope that rotates. The illumination may come from a system that

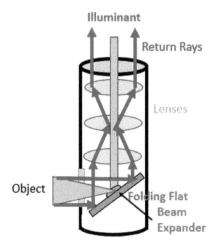

Figure 4.7.1 Rigid borescope schematic.

---

[18]Gradient Lens Corporation, "Our Technology," https://www.gradientlens.com/about-us/our-technology.

is in the middle of the imaging system, blocking it a little, or it may be around the outside.

A flexible system employing fiber optics can use a portion of the bundle for illumination and the rest for transferring the image. They incorporate many fibers to relay the image and usually lenses at both ends to form images.

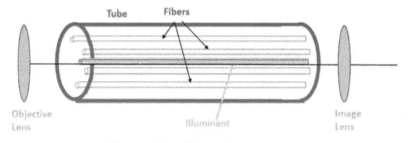

Figure 4.7.2   Fiber optic borescope.

The latest trend is the use of a digital camera at the distal end with LED lighting.

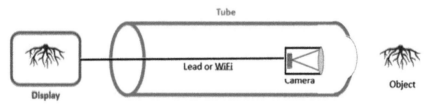

Figure 4.7.3   Digital camera borescope.

One borescope/endoscope designed at my college looks forward and backward but not directly to the side. The design is shown schematically in Figure 4.7.4. An annular mirror (shown in gray) is used to reflect part of the total field of view behind the sensor. That rearward part can be adjusted by adjusting the tilt of the mirror. The technique is not highly sophisticated, but it works. Presumably, there is a method for distinguishing the two images and for letting the light through the proper parts of the tube and rotating it.

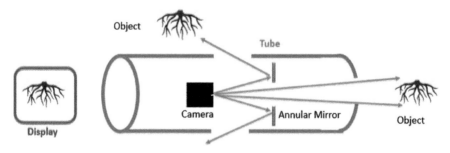

Figure 4.7.4   Rearview borescope schematic.

4.8 Cameras

The first cameras were **pinhole cameras**, a single hole in an enclosed box. No lens. No shutter. No film. Just a hole in a box. The hole limited the rays from a point on the object so that they came to a reasonable focus on the back of the box, as shown in Figure 4.8.1. The smaller the hole, the better the resolution but the dimmer the light. The image was upside down. These were quite popular during the Renaissance period, and Charlie Falco and Dave Hockney claimed they were used to outline paintings.[19] They were called camera obscuras or dark rooms. They were a source of entertainment, typically dark rooms with mirrors on the roof that reflected the light through the pinhole.

I used this once in a demonstration to sixth graders. It was a hit. They loved to see the upside-down images of their classmates on the wall cavorting on the patio outside. They took turns.

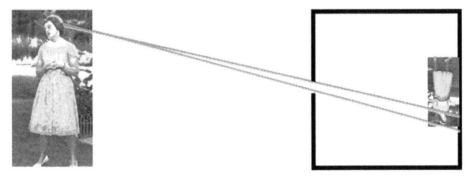

Figure 4.8.1 Pinhole camera.

---

[19]Wikipedia. "Hockney-Falco thesis," https://en.wikipedia.org/wiki/Hockney%E2%80%93Falco_thesis.

The essential components of a **modern** camera are the lens, the sensor, either film or a detector array, and a shutter. These components are shown in Figure 4.8.2 with both a front shutter and a focal plane shutter, as well as a flip mirror. Only one shutter is used in a single camera, and the flip mirror is only used in single lens reflex cameras. The shutter lets in the right amount of light and sets the timing. The lens forms the image. The sensor turns the image into something permanent, e.g., exposed film or digital memory.

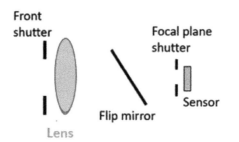

Figure 4.8.2 Camera essential components.

There are literally thousands of different camera **lenses**. Refer to Section 3.10 for that information.

There are basically two kinds of **shutter**, based on its location. The most common is at the aperture. You can see it by looking in at the front of the camera. The other is at the image plane, called a focal-plane shutter. It is faster because it can be smaller. There are several shutter geometries. Probably the most common, at least until the digital age, was the leaf shutter. It consists of about five to ten leaves. Different designs use different shapes. Most are some form of arc.

The first example of a leaf shutter, shown in Figure 4.8.3, consists of black triangles. A spring or other mechanism causes the leaves to move to the center, thereby blocking the light as shown in Figure 4.8.4. Opening the shutter, of course, is the opposite operation. The fastest leaf shutter is about 1 millisecond. Figure 4.8.5 depicts an arcuate shutter. The curved (arced) leaves rotate to open and shut the shutter. This is probably what you will see when you look into the front of your camera.

Figure 4.8.3 Almost open.   **Figure 4.8.4** Almost closed.   **Figure 4.8.5** Arcuate.

Most **focal plane shutters** are planes. They are thin, black, opaque rectangles that move across the image plane very quickly. The fastest such shutter takes about one tenth of a millisecond to open and shut.

The **single lens reflex (SLR)** camera uses a flip mirror to alternate between the viewer and the image plane. In this way the photographer can see the exact image. The flip mirror, shown in black in Figure 4.8.6, is in the optical train to reflect the image to the eye below, shown in blue. It flips out of the train to allow the light shown in red to go to the sensor.

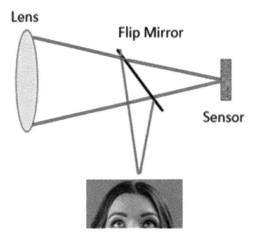

Figure 4.8.6   SLR camera schematic.

Another innovation was automatic adjustment of the light level. This can be done by adjusting either the shutter speed or the aperture stop, which also adjusts the aperture ratio.

The term **aperture ratio** can be a bit confusing. It is referred to as the F-number, the F/#, the F stop, the relative aperture, speed, and relative speed. It is the ratio of the focal length to the diameter of the aperture. The aperture is the area of the front lens that controls the amount of light that is accepted. It is usually controlled by a mechanical diaphragm. The overall speed of the camera is dictated by the cone of light that impinges on the film. The size of the aperture and the focal length determine that cone. The cone is two dimensional; the F-number is one dimensional. That is why the F-numbers have such an unusual sequence; square the ratios, and it makes sense. F/2 divided by F/1.4 is 1.43, but the square is 2.04. F/2.8 divided by F/2 is 1.4 but squared is 1.96. And so on. Each step change in F number doubles (or halves) the amount of light.

Those are the exposure adjustments that can be made, but there must be a measurement that dictates the amount. It is a **photometer**. For years, we used a separate photometer, a light meter. Only after circuit miniaturization was a fact could the photometer be incorporated into the camera body. It is a device that measures the amount of light that arrives at the aperture of the camera using a photo detector that is described in Section 3.5.

The next or perhaps concurrent step was **automatic focus**. Earlier systems used a variety of autofocus techniques. One was contrast detection. The lens or film was moved in and out a small amount, and the degree of contrast in part of the image was measured. The greater the contrast, the better the focus since the blur is decreased. Other types involved triangulation and even time-of-flight measurement with pulsed (non-laser) sources.

The **instant camera** was introduced by Edwin Land. Although he and his team made improvements in the camera design itself, it was the film that made it instant. Although there is considerable sophisticated chemistry involved, suffice it to say that Land essentially put the chemistry of the developing room in packets in the back of the camera and simplified the process.

Modern **digital cameras** have these features and more. They have better, more compact lens systems since all the lenses can be ground and polished to optimum (aspheric) shapes by computer-controlled methods. There is no need for a flip mirror since the exact image is portrayed on the LED screen. Simple infrared diodes or laser diodes are used as time-of-flight rangefinders. The light meter is integrated, and the output is instant on the LED screen.

Digital cameras focus the image on an array of detectors. In a sense, it is a triple array, one for each color. The arrays can contain as many as 25 million pixels. That would be about 8 million for each color. The portion of the object that is focused onto a pixel, one picture element, causes an electronic current or charge buildup. This is relayed electronically to the LED display. It has an equivalent number of pixels for each color.

The comparison of film and digital imaging is interesting. Film granularity, as I use it, is the diameter of the silver particles that result from exposure; it is about 1 μm. Pixel sizes of silicon photodiode arrays are about 5 to 10 μm, a bit larger. The ISO ratings (sensitivity ratings) of digital cameras are as high as 4,560,000, whereas the highest such rating for film is an ISO of less than 4,560.

The bottom line here is that digital detector arrays are much more sensitive but have poorer resolution, per se. That is, they have larger individual detection elements. Digital cameras can obtain better resolution by using longer image distances, and they can have smaller apertures because they are much more sensitive.

One specific comparison[20] is that a frame of 35-mm film with an ISO of 100 contains 20 megapixels. ISO 100 film is fine grained and relatively insensitive. That tracks with my simple Canon ELPH 180 with 20 megapixels. It has a lens about 7 mm in diameter, and it takes respectable pictures.

Many camera buffs prefer film, partly for artistic reasons, partly for dynamic range, and maybe for nostalgia. I prefer the convenience, the

---

[20]Langford, M. *Basic Photography, 7th Ed.*, Focal Press (2000).

compactness, the immediateness, and the ability to take as many pictures as I like and delete most of them. I am not a camera buff. I take casual pictures, and I wear clothes.

Figure 4.8.7 shows my ELPH 180, and Figure 4.8.8 shows a reasonably decent picture taken with it. It has 20 megapixels and an 8× optical zoom. These digital cameras range up to about 60 Mp.

Figure 4.8.7 My ELPH.

Figure 4.8.8 A respectable ELPH 180 picture.

Digital cameras can be made small. They have been advertised as spy cameras since they can be disguised as pendants, pens, sunglasses, and other very small things.

Consider a possible design as a representative concept. It uses a 1-megapixel array of silicon detectors, each 5 μm on a side. Each side is 1000 pixels long, that is, 5 millimeters, one half a centimeter, or less than 0.2

inches. Figure 4.8.9 shows such a concept with dimensions. Arrays have been made with pixels as small as 1.5 μm.[21]

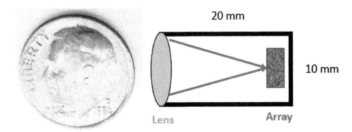

20 mm

10 mm

Lens   Array

Figure 4.8.9   Representative spy camera.

4.9 Camping Optics

Camping optics include three sources of light and perhaps binoculars and rifle sights. These latter two are camping activities, not camping tools, and they are covered in their respective sections. The sources are the flashlight you use to guide your way, the lantern you use to illuminate the tent and cooking area, and the campfire you use to warm yourself and where you tell scary stories.

Campfires are explained in Section 3.15. The heat causes the wood to give off gas. The heated electrons of the gas give up their energy to photons. There is light in the visible and warmth in the infrared.

Camp lanterns were discussed in the same section. Almost all the old propane-fueled Welsbach mantle lanterns have now been replaced by LEDs fueled by batteries. They are also more versatile. They can illuminate just a half sphere or the full one. The brightness can be adjusted. They can charge other devices via a USB port and can be charged via a similar connection to a

---

[21]Zhang, Y. et al. "Sub-wavelength-pitch silicon-photonic phased array for large field of regard coherent beam steering," *Opt. Exp.* **27**(3), 1929 (2019).

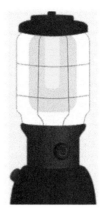

Figure 4.9.1 Propane lantern.   **Figure 4.9.2** LED lantern.

solar panel or even be hand-cranked when all else fails. They will last for about an hour when everything is draining the battery but several hours in normal campground use.

Flashlights consist of a source placed at the focus of a concave reflector. A glass plate is placed in front for protection and for additional focusing on occasion. The reflector has an amount of curvature that depends on the desired beamwidth. Real parabolas are in only the most expensive models. There are many varieties, from small pen lights and units for a visor to large handheld units used to illuminate a work area.

Two examples are shown in Figure 4.9.3. On the left is a standard version with several LEDs and a switch for bright and brighter. Brightness is controlled by the number of LEDs that are activated. On the right is a device meant to be clipped to the visor of a cap. In the middle is a dime to show the sizes. Remember that the LEDs themselves are very small. They can be arranged in all sorts of geometries and colors, e.g., for visors or at the end of flexible arms or long, extended ones.

Figure 4.9.3 LED flashlights.

4.10 Colorimeters

Colorimeters are of two types. One type of colorimeter measures the tone quality of colors. The other is a type of spectrometer. The latter is described here. It measures the transmission of a substance at a given color, a given wavelength. Every substance has a unique spectrum. Each has individual, specific frequencies and wavelengths where it absorbs and transmits. Medical instruments take advantage of this fact. Every medical laboratory is equipped with them.

One blood test is for your level of hemoglobin. Hemoglobin has a nice, simple spectrum. Figure 4.10.1 shows a simplified optical spectrum of it. If your levels are too low, there is a deficiency of your red blood cells that may indicate anemia. The colorimetric test for it is a spectrally narrow beam of light centered at either 0.55 or 0.63 μm. As the blood passes this beam, the amount of absorption is measured and compared to a standard. Similar tests are made for the concentrations of oxygen, carbon dioxide, bilirubin, and the other constituents of blood and urine. About 75% of the medical laboratory tests are done this way.[22]

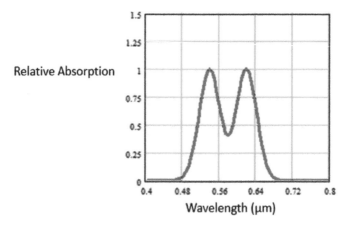

Figure 4.10.1 Simplified hemoglobin spectrum.

[22]Nahm, K. private communication.

The tube of the colorimeter holds a series of blood drops taken from a series of patients. They are separated by globs of air, as shown in Figure 4.10.2. The drop from each patient passes by individual beams of light of different wavelengths (colors). Each wavelength is related to a specific constituent of the blood. The blood sample (drop) of each patient is stepped along in a process called peristalsis, like that of twitching muscles. Modern instruments make several hundred measurements per hour. Even my wife, who was a terrific medical technologist, was not that fast!

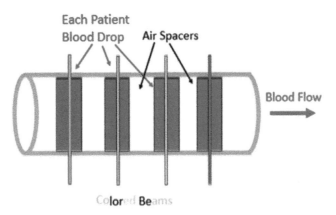

Figure 4.10.2 Schematic of blood testing.

The med techs have been replaced by automatic machines called Sequential Multiple Analyzer Computers (SMACs). They test many samples a day using colorimetry and peristaltic pumps. The SMAC 20 tests 20 different components of your blood.[23]

After your favorite medical vampire (phlebotomist) has extracted your blood, it is tested mostly by optical means. I was married to my favorite blood sucker for 60 years!

---

[23] *SMAC 20 Blood Test | Walk-in Lab*, http//www:Walkinlab.com.

4.11 Colposcopes

Colposcopes are used for optically examining the cervix, either in conjunction with a Pap smear or independently. Clinical devices are usually mounted on a variety of stands with arms that maneuver the scope into position. All that auxiliary equipment, as important as it may be, is not relevant to an optics discussion. The essence of the optical system is shown in Figure 4.11.1 in a more versatile version than the illustration above. The light from a source is reflected from a folding mirror to the patient. The light is reflected from the patient to a beam splitter that sends a portion of the light from the patient to a lens below it, where it is imaged. Light that passes the first beam splitter is divided by the second. A portion goes to the next lens below it and lets the rest go to the final one that is in line with the patient. This allows three simultaneous looks at the cervix. These could be the doctor, a student, and a camera. It is possible to introduce all sorts of innovations and adaptations: zoom lenses, color enhancement, differential imaging, and so on. A typical, modern system uses an LED as the source and an array of about 1000×2000 solid state pixels for imaging.

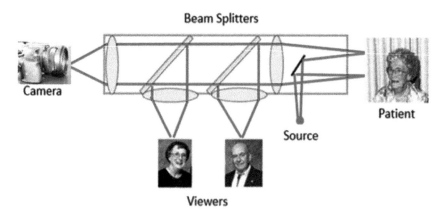

Figure 4.11.1 Colposcope optics schematic.

4.12 Communications

Optics has provided incredible advancements in communications. These advancements include hundreds of television channels, high-speed worldwide fiber optics links, satellite relays, internet searches, and more. This section is devoted to our increased capability in communicating phone and email messages around the world. This has been accomplished by advances in free-space communication and with the aid of laser diodes and fiber optics.

Some of the earliest forms of free space communication were smoke signals and semaphore. We now use electronic means that entails modulation of the signal spectrum with a carrier wave. The carrier must be at least ten times the signal frequency or else the message becomes jumbled, as shown in Figure 4.12.1, where the carrier is only five times the signal frequency. In the figure, the top red line is the assumed signal frequency. The blue line is the assumed carrier at five times the signal frequency; the bottom green line shows the result when they are multiplied together, when the carrier modulates the signal. You can see that the modulated green wave is just a jumble. A proper carrier wave at many times the signal frequency is shown in Figure 4.12.2. The envelope of the green wave is a good representation of the signal, both positive and negative maxima.

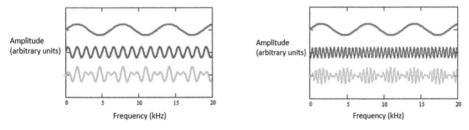

Figure 4.12.1 Inadequate modulation.   **Figure 4.12.2** Adequate modulation.

The range of our hearing and speech tops out at about 20,000 cycles per second, 20 kHz. AM radio ranges from 550 to 1720 kHz, kilocycles per second. Figure 4.12.2 shows a 20-kHz wave of speech superimposed on a 1000-kHZ AM carrier. AM radio works.

Optical communication takes advantage of this relationship between signal and carrier frequencies by providing a very-high-frequency carrier. FM radio carriers are in the range of 88–108 MHz, 88–108 × 10^6 Hz. Even high-definition TV is only 54–216 MHz, 216 × 10^6 Hz. Yellow light, in the middle of the visible spectrum, is 5 × 10^{14} Hz. Light has a frequency of more than a million times either of these values. I say if you can get an advantage of more than a million, take it.

Free-space optical communication is limited mostly to short-range terrestrial links and space relays. The terrestrial links are limited by atmospheric conditions such as fog, rain, and snow. The ground-based spatial links are relays that are usually microwaves.

Optical communication in fibers is useful in several ways. It is efficient for sending lots of information long distances and for more secure and lighter systems for short distances. An example of the latter is communication on naval ships. Wire-based, insulated cables have what are called evanescent waves. A little of the electric energy radiates outside the wire or outlet. You can demonstrate this for yourself with a handy electrical gadget shown in Figure 4.12.3. It detects whether your 110-volt ac house line is active or not by measuring near it. It senses the electric field outside the wire or outlet. Such evanescent waves have measurable amounts for about a few wavelengths outside the cable.

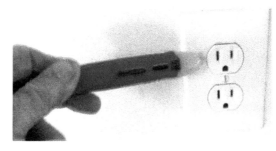

Figure 4.12.3   Electrical tester.

Optical fibers for long-range communication are all over the world, literally. They are used to send tremendous amounts of information from one place to another in this age of information. One connection covers 28,000 kilometers, linking the UK, India, and Japan. It sends 120,000 voice channels at a rate of 10 Gbaud (gigabits per second). Approximately every 50 kilometers, there is an optical amplifier that boosts the signal at a wavelength of 1480 nm.[24]

---

[24]Wikipedia. "Fibre-optic Link around the globe," https://en.wikipedia.org/wiki/Fibre-optic_Link_Around_the_Globe.

These fiber optic cables are capable of sending terabytes of data in seconds. They are based on the high frequency of light. Their construction is straightforward: the essential fiber is surrounded by another with a lower refractive index that ensures a constant total internal reflection. That is covered by one or more layers of protection depending upon the environment (see Section 3.7).

Similar but less bulky fibers are used in computer clouds. These facilities were once called central computers. They have massive computing capabilities because many units are interconnected by fibers and laser diodes that transit information back and forth.

4.13 Computer Optics

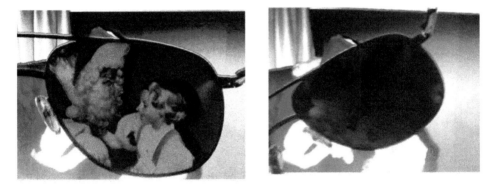

Currently, there are three areas in which optics plays a part in the operation of our home computers: the electronics, the display, and the mouse. Laser diodes and fiber optics have a big role in the giant computers that are now called clouds and used to be called central computers.

The very impressive **computer electronics** of today are a result of a couple Nobel-worthy inventions and some important developments. The invention of the transistor in 1947 freed us from the space-consuming vacuums of electron tubes. The invention of the integrated circuit in 1959 put all of them together. The impressive optics involved with making integrated circuits with more and more components in smaller and smaller spaces is described in Section 4.33. We can now make integrated circuits with transistors and connects as small as about 10 nm.

Computer displays are said to be either LED or LCD, light-emitting diode or liquid crystal display. But they are both LCD; it is just the source of light that is different.

The LCD display uses a source of light that can be an LED or other light that shines on a combination of polarizers and liquid crystals. As shown in Figure 4.13.1, the light is shone onto a polarizer that for illustration purposes I will assume is vertically linearly polarized. It then passes through the liquid crystal that has electrodes attached and then to another polarizer. The liquid crystal is a liquid with long, linear crystals in it. When an electrical voltage is applied, those crystals can be made to come into alignment and act like a polarization rotator because they are twisted crystals. That also explains why you should not wear polarized sunglasses when you are using a computer. The display is polarized. The introductory pair of images show the polarization effect. My (onetime) little girl on Santa's lap is dimly visible through a colored polarized sunglass lens, as shown on the left, when it is in the right orientation. But she disappears when the polarized sunglass is rotated.

Figure 4.13.1 shows the LCD in the off position; no voltage is applied to the liquid crystal; the crystals are oriented randomly. Unpolarized light comes from the left. It is polarized by a grid polarizer. It impinges upon the liquid crystal, and nothing happens because they are in random order. It emerges with the same polarization and impinges upon the polarizer oriented in the opposite direction. Nothing (nada) gets through. Recall that polarization in the plane of the paper is shown by arrows and out of the plane by dots. Also recall that light polarized perpendicular to the grid lines is the component that is transmitted.

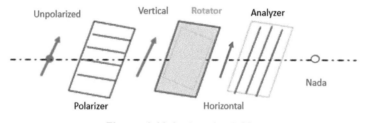

Figure 4.13.1   Inactive LCD.

If the liquid crystal is activated, the situation is as shown in Figure 4.13.2. Unpolarized light is polarized in the plane of the paper by the first polarizer, rotated by the activated twisted crystals in the liquid to be out of the plane of the paper, and then transmitted by the second polarizer.

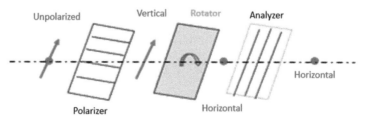

Figure 4.13.2   Activated LCD.

A **computer mouse** has eyes. It shines either an LED or a diode laser onto a surface and takes an image of it with a tiny camera. As the mouse moves, the consecutive pictures are compared to determine how much one is offset compared to the previous. This is the same technique used in map-matching missiles (Section 4.28). It happens thousands of times a second. Early devices required the mouse to stay on a pad that had a grid on it. To my surprise, modern mice can operate on surfaces that seem to have no texture, as shown in Figure 4.13.3. (The slight shading is due to the illumination gradient.) That is one smart rodent that can image about a million pixels and compare them to another million pixels all in the blink of an eye or the twitch of a hand.

Figure 4.13.3 Mouse on a uniform pad.

The first central computer of my experience was the Michigan Digital Analog Computer (MIDAC). It consisted of two rooms, each about the size of my garage, full of vacuum tubes. It had to be air conditioned. I was one of its first users in 1954; it was built in 1953 in the Willow Run Laboratories. It was preceded by the Electronic Numeric Integrator and Computer (ENIAC) at the University of Pennsylvania, built in 1945. ENIAC's capabilities can now be programmed on a single chip about 10×5 mm.

The central computers of today, the clouds, are massive and impressive. IBM's Watson can process 800 gigabytes in one second. Most central computers use about 100,000 individual processors. A typical unit looks something like that in Figure 4.13.4. Each vertical rack is a processor.

Figure 4.13.4 Processors in the cloud.

Each processor is connected to another by a bunch of optical cables that look something like those in Figure 4.13.5. I do not need to multiply the number of processors times the number of racks times the number of cables to convince you of the billions of optical cables and diodes at their ends that are used in these clouds.

Figure 4.13.5   Optical cable connectors.

4.14 Emissometers

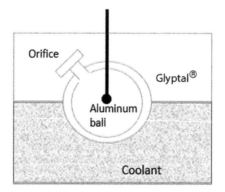

Emissometers are technical devices that measure the emissivity of materials.

There are two types of emissivity: total hemispherical and spectral directional. Total means the entire spectrum and hemispherical means half a sphere, the overlying hemisphere of the object. Directional in this case is normal incidence.

Total hemispherical emissivity was measured by Lou Drummeter and Manny Goldstein in an evacuated chamber.[25] It is shown schematically in the introductory figure. An aluminum ball is hung by a fine wire in a spherical chamber. Detailed calculations of this determination appear in Appendix A.4.14.

Spectral directional emissivity was measured by Joe Richmond at what was then the National Bureau of Standards.[26] He did it with a spectrometer and a blackbody. He brought the blackbody to the same temperature as the sample, which was a secondary standard strip lamp, and compared the outputs. One possible arrangement is shown in Figure 4.14.1.

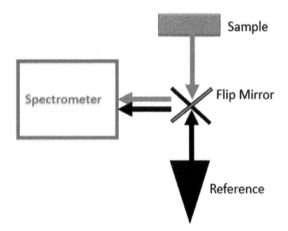

Figure 4.14.1   Possible spectral directional emissivity measurement arrangement.

Many methods for evaluating the emissivity of opaque materials are based on reflectivity measurements. They are described in Section 4.43. If a material is opaque, the transmission is zero. Any incident light is either reflected or absorbed. The complement of the reflectivity is the absorptivity, which is equal to emissivity.

---

[25]Drummeter, L. and Goldstein, E. "Vanguard Emittance Studies at NRL," in *Surface Effects on Spacecraft Materials*, F. J. Claus, ed., Wiley (1960).

[26]Richmond, J. "Some Methods Used at the Bureau of Standards for Measuring Thermal Emittance at High Temperatures," in *Precision Measurement and Calibration: Radiometry and Photometry*, Hammond, H. and Mason, H., eds., *NBS Special Publication* **300**(7), 182 (1971).

4.15 Endoscopes

Endoscopes, inside scopes, are used for peering at our insides, our endos—from the top or the bottom. The early versions consisted of an array of lenses in rigid tubes. That was obviously not comfortable or effective. It is said that some of the early applications were done only on sword swallowers. That makes sense but stretches credulity. Improvements included the design of relatively flexible tubes with lenses (sort of like miniature periscopes). The first fiber optic endoscopes were developed by H. H. Hopkins and Narinder S. Kapany in the UK, A. C. S. van Heel in the Netherlands, and Brian O'Brien, Sr. in the US simultaneously and independently.[27] They consisted of many individual fiber optics formed in a coherent bundle. Coherent in this case means that the spatial arrangement of fibers at both ends is the same.

The principle of operation is relatively easy to understand. A source at the receiver end sends light down a few central fibers to the object at the distal end (colored yellow in Figure 4.15.1). Each fiber at the distal end accepts light focused on it by a lens. It sends that light by multiple internal reflections to the eyepiece at the proximal end. With enough small fibers, a decent image can be obtained. A typical device contains several hundred thousand fibers, each about 10 μm in diameter. That might be 1000 ×1000 pixels, a pretty good image. (HDTV is about 1000 × 2000). The overall diameter is then about 14000 μm, about 14 mm—not too uncomfortable. There have been other designs that eliminate the lens at the distal end. One is to curve the end of each fiber so that it is a convex-plano lens. Another is to just not use anything and count on the individual fibers to collect light from the individual pixels of the object.

---

[27]Hopkins, H. and Kapany, N. "A flexible fibrescope, using static scanning," *Nature* **173**, 39–41 (1954); van Heel, A. C. S. "A new method of transporting optical images without aberrations," *Nature* **173**, 39 (1954).

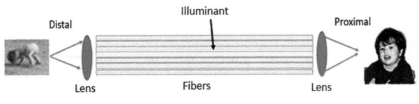

Figure 4.15.1   Concept of a fiber optic endoscope.

The next development was the use of a small camera in the distal end that sends an image back to a display device to the proximal end with a wire. As you can see, this makes for an even-smaller-diameter, more flexible tube. Advances in WiFi transmission techniques with smaller and smaller transmitters permit the elimination of the wire and the receiver in the tube.

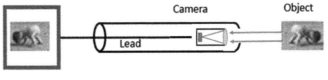

Figure 4.15.2   Camera-based endoscope.

The final improvement (so far) is a small capsule containing a small camera that can be swallowed. It travels down the intestinal tract and can even be guided by magnetic means. A schematic is shown in Figure 4.15.3. It continually transmits images to an external receiver, memory, and display so suspicious areas can be explored more intensely. The capsule is about the size of what one might call a horse pill, 25×11 mm $= 2.5 \times 1.1$ cm, about 1.0×0.4 inches, which is approximately the size shown in the introductory picture. One version of the capsule is a so-called tethered endoscopic capsule with a cord on the end of it to pull it back out.[28] No, thank you; I know how to get rid of it!

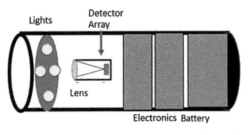

Figure 4.15.3   Endoscopy capsule schematic.

---

[28]Several are shown on the Internet, Medicaldesignbriefs.com, for instance.

4.16 Gyroscopes

We are probably all familiar with the old-fashioned gyroscopes, the little iron toys that have several different spinning parts. Their professional cousins were used for navigation. Optics has replaced them for that use with optical gyroscopes and fiber optic gyroscopes (FOGs).

The essential operation of any gyroscope is to sense a change in direction based on the stability of its spinning parts. This change can then be used by an airplane, missile, or drone to compensate for any change in its heading. Iron gyros use the inertia of spinning wheels; optical gyros use the phase of light (or, if you will, the spinning of photons).

Figure 4.16.1 The old iron gyro.

The optical gyroscope, or ring laser gyroscope, was invented in 1963 based on the operation of the Sagnac interferometer.[29] The Sagnac has two

---

[29]Macek W. and Davis, Jr., T. "Rotation rate sensing with traveling-wave ring lasers," *Appl. Phys. Lett.* **2**, 67 (1963).

beams propagating around a rectangle in opposite directions, as shown in Figure 4.16.2. One is shown in red, the other in blue. They overlap, but I could not show it that way. I also did not show the unimportant refractions in the beam splitter. The two beams combine and interfere (see Section 4.21 for more information). Assume that the interference is constructive with the interferometer at rest. When it rotates clockwise, the light that is also moving clockwise has a slightly longer path than its counterclockwise counterpart. The two beams get just a little out of phase. If the source is a red laser beam, the distance, from complete constructive interference to complete destructive interference is a half wavelength, or 0.4 μm. If the interferometer is 10 centimeters on a side, that is a rotation of only about 4 microradians, or about 1 arc second. That is sensitive!

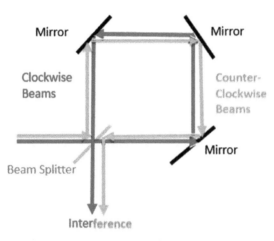

Figure 4.16.2   Sagnac interferometer schematic.

Then along came the fiber optic gyro. The arms of the interferometer were replaced by fibers. The beams are injected the same way by a beam splitter into a fiber. They travel in opposite directions for the length of the fiber rather than just from mirror to mirror. This extended length results in considerably more sensitivity and simplicity and reduction in weight. The first FOG was demonstrated by Victor Vali and Richard Shorthill in 1976.[30]

At least one high-performance FOG that has 250 meters of fiber has a drift of 1 degree per hour. Typical high-end iron gyros drift at about six times that rate. Maybe at the other end of the size spectrum are devices one inch in

---

[30]Udd, E. "Optical Gyros," *SPIE Professional*, https://spie.org/news/spie-professional-magazine-archive/2016-october/optical-gyros (October 1, 2016).

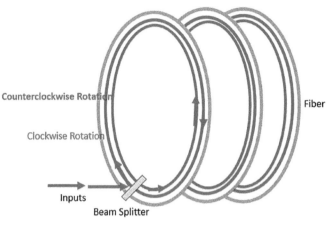

Figure 4.16.3 Fiber optic gyro.

diameter by about two inches long for use in drones. They do not need the extremely low drift rates.

4.17 Flyfishing Optics

Flyfishing is my favorite sport. It gets me outdoors and into nature. It challenges my dexterity (or lack thereof) and my patience. And it involves optics: finding the fish, fooling the fish, landing the fish, and putting them back.

Flyfishers should always wear **polarized sunglasses**. Perhaps the primary reason is to keep the hook out of your eye, but the light reflected from water is polarized. If you wear a polarizer, it blocks one component of the light so that there is less glare from the stream, and the fish are easier to see.

Section 1.4 showed the reflectivities of the two polarizations of light on glass. That is repeated here for water, which has a slightly lower refractive index. The figure shows that sunglasses that block the perpendicular polarization will decrease the reflectivity significantly.

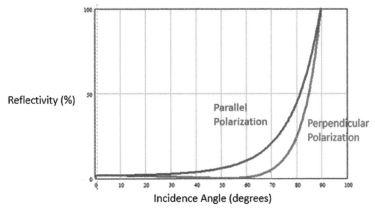

Figure 4.17.1   Reflectivity of water.

The **refractive effect of the water** might lure you into a fruitless cast behind the fish. The caster who is unaware of the phenomenon of refraction thinks the fish is where the top, straight yellow line of Figure 4.17.2 indicates. He casts a bit in front of him as shown by the green line. But because of the bending of light by refraction, he has actually placed the fly right behind him. Unless he is lucky and the fish turns around, that cast is futile.

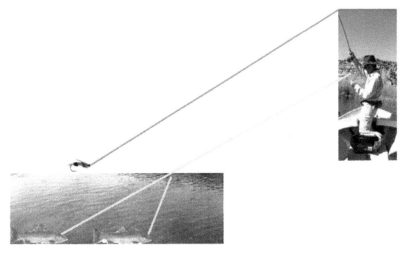

Figure 4.17.2   Get it right.

Fish have a **window**, a limited area on the surface where they can see above the surface. They will see your lower extremities if you are wading or the bottom of the boat if you are floating. But the significance of this window is to get the fly into the window so that the fish sees it all: the entire attractive fly, not just a few bottom feathers.

Figure 4.17.3   The fish's window.

The flyfishing outfit consists of a rod and reel, a line, a leader, a tippet, and a fly. I love to fish on top. That is, I float the fly and hope the fish will come up to it. Then I want my **fly line** to have minimum contrast with the background but still be visible to me. In this photo, the green fly line is just visible.

Figure 4.17.4   The fly line, rod, and reel.

The leader is attached to the end of the line. It is about nine feet of nylon monofilament that is reasonably transparent. It tapers down to a diameter of about 1 millimeter.

Figure 4.17.5   Leader attached to the line.

A tippet is attached to the end of the leader. It has an even smaller diameter than the leader and is the part of the overall rig that comes closest to the fish (except for the fly). It is where another bit of optics arises in the flyfishing world. Fluorocarbon has a higher strength-to-weight ratio than monofilament. It is better, just like with astronomical and anti-ICBM mirrors. Thinner, less visible sections will still hold that fish without breaking. And it is less visible because it has a better refractive-index match to water, which is 1.333. Mono has an index of 1.525, and fluoro has an index of 1.420. You can do the approximate calculations, but it is 34% versus 40%. Fluorocarbon tippets are much less visible, but they cost much more. If you are a fisher, I recommend fluoro for tippets 5× and smaller, but not for leaders or larger tippets.

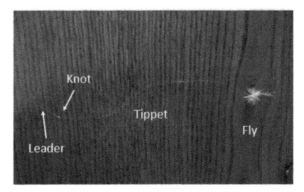

Figure 4.17.6   Tippet on the end of the leader and the fly.

And remember, the worst day on the stream is better than the best day at the office.

And the best time to go fishing is when you can.

4.18 Holography

Holos and *gramma*, Greek for "whole writing," are the origins of the words "hologram" and "holography." It is a way of presenting true three-dimensional images.

The idea was conceived by Dennis Gabor in 1948.[31]

The first realization was by Emmett Leith and Juris Upatnieks at my old stomping grounds, the Willow Run Laboratories of the University of Michigan, in 1962.[32] That is a bit like Maiman, who did what Townes and Shawlow predicted theoretically (see Section 3.9). Emmett did what Gabor predicted theoretically.

As an interesting aside, the Ann Arbor papers completely missed the significance of holography. The headlines and articles were all about how Emmett had accomplished imaging **without lenses**, not that they were 3D.[33] I guess that is not the first nor will it be the last time the media has gotten or will get it wrong.

The process is one of optical interference. A laser beam is shone onto the desired scene through a beam splitter. One portion of the beam is used for a reference. The laser light reflected from the scene is combined with the laser light from that reference. It is recorded as an interference pattern and called a hologram. Recall that the interference of two light fields is three-dimensional. Therefore, a hologram is a recording of a three-dimensional interference pattern.

---

[31]Gabor, D. "A new microscopic principle," *Nature* **161**, 777 (1948).
[32]Leith, E. and Upatnieks, J. "Reconstructed wavefronts and communication theory," *JOSA* **52**, 1123 (1962).
[33]My observations—I lived there then.

Figure 4.18.1 shows how a hologram is produced. A laser beam is divided by a beam splitter into the signal beam and the reference beam. The signal beam is transmitted as a bunch of individually transmitted beams reflected or transmitted from the various parts of the object to a film. The reference beam is routed to eventually join the signal beam on the film where they interfere according to their relative phases.

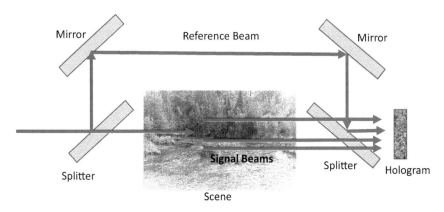

Figure 4.18.1 Hologram construction.

The holographic image is then produced by shining a different laser beam on the film, as shown in the diagram of Figure 4.18.2.

One simple but very abstract way to understand this is the principle of reciprocity. The interference pattern formed by the signal and reference beams produces the original object when a laser is shone back through it. A more

Figure 4.18.2 Image reconstruction.

physical way to understand it is that each of the waves generated by the hologram combines with the others to generate the reproduced image.

In the early days, the images were monochromatic. They were the color of the reconstruction laser. But much has been done since 1962, and we now have colored holograms. We almost have holographic TV.

When Emmett first accomplished his hologram, there was much hullabaloo. All sorts of things were predicted for this new modality. Many have come to pass.

The military can now generate and use holographic maps of the battlefield or the potential battlefield. The holograms must first be obtained by local imagery, but they can be rolled up, carried, and unfurled to show soldiers the lay of the land in three dimensions.

Vast amounts of data can now be stored due to the three-dimensionality of holograms. One report quotes 500 megabytes per cubic millimeter.[34] That translates into a more understandable 500 gigabytes per cubic centimeter. A cube 10 centimeters by 10 centimeters by 20 centimeters can store the entire contents of the Library of Congress, as indicated in the introductory picture.

Our credit cards are protected by small holograms for security. It is tough to forge a hologram. In the original, it looks 3D and changes color as it is viewed from different angles.

White light or rainbow holograms were developed in 1969.[35] The process is to illuminate the object with red, green, and blue lasers and coat the interference pattern on the film with a conductor to place free electrons on the surface. The film, the hologram, is then illuminated with white light, and the free electrons produce the colored photons by giving them their energies.

Progress is being made on holographic TV. Both MIT and the Wyant College of Optical Sciences have been working on holographic TV. Just imagine watching *Avatar* on our TV sets without glasses (in fact, watching everything in three dimensions). The issue is recording materials.

Even if the signal is sent to our TV set as a hologram, our TV has to interpret it. It has to shine another laser on it to generate the three-dimensional reconstruction. The television needs to do it at about 30 times a second. The material must record the hologram, erase it, and record the next image in $1/30^{th}$ of a second. Neither institute has found that material at this date.

---

[34]Wikipedia. "Holographic Data Storage," https://en.wikipedia.org/wiki/Holographic_data_storage.

[35]Benton S. "Hologram reconstructions with extended incoherent sources," *JOSA* **59**, 1545–1546 (1969).

4.19 The Human Eye

The human eye is one of the most impressive of all optical instruments. It was, of course, designed by the most impressive One of all.

There are many very good descriptions of the human eye in the literature and online. They are mostly from an anatomical or biological point of view. I have taken a different approach. This is a description of the human eye as if it were a digital camera or digital camcorder.

The eye is in a convenient carrying case, the head. It goes everyplace we do and is always available for use. We never lose it, but we may get lost. It

Figure 4.19.1　The carrying case.

also contains the recorder, processor, and memory. It is highly portable and secure, able to go wherever we go and not get lost.

The eye has a lens cap, an adjustable aperture, an adjustable lens, and a very complicated detector array. All of these are encased in protective fluids.

The **lens cap** is the **eyelid**. It automatically covers the eye when it is not in use, overnight, during sleep. It also senses when some alien object might hit the eye. Something like a golf ball, raindrop, fishing fly, or even excessive photons. It is better than the normal camera lens cap that has no sensor.

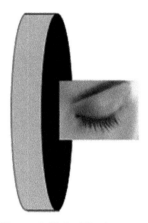

Figure 4.19.2   The lens cap.

The **cornea** is similar to the dome of a missile guidance system or the protective plate used in some underwater cameras. Its dual function is protection and focusing. The cornea, as illustrated in Figure 4.19.3, performs much but not all of the focusing. It is what LASIK reforms in that operation (see Section 4.23).

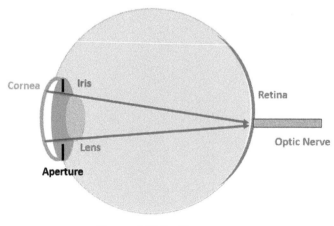

Figure 4.19.3   The cornea.

The **iris** is the colored portion of the eye that contains the **aperture stop**, which is a circular hole that varies from about 4 mm to 8 mm in diameter, depending on the incident light. Iris was a messenger of the Greek gods, represented by the rainbow of many colors. Perhaps that is why it is called that. The many different colors of eyes. The change in the size of the iris is caused by the illumination level sensed by the retina, the photometer of the eye. That is the first and immediate adjustment your eye makes when you go from a dim to a bright room, or the other way around.

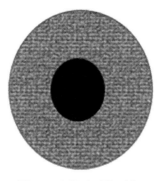

Figure 4.19.4 The iris.

The **lens** is the critical focusing element. It adds to the focusing of the cornea, as shown. It is a biconvex, gradient index, almost circular, aspheric element about 10 mm in diameter by 4 mm thick, diagrammed in Figure 4.19.5. The gradient in the refractive index is from about 1.406 in the center to about 1.386 at the periphery. It consists of fluidic elements called **cristallins**. The lens changes its shape and therefore its imaging properties when the ciliary muscles act on it. The lens performs about one fourth of the imaging; the cornea most of it. It is the most important, in my view, because it is the dynamic element that allows us to view at infinity, as in Figure 4.19.5, and then read this book, as in Figure 4.19.6, even though the cornea does most of the work. If the object is close to the eye, the muscles make the lens bulge for increased focusing action.

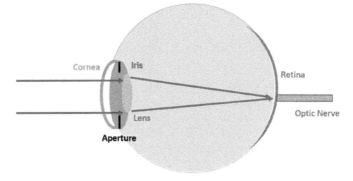

Figure 4.19.5 Viewing infinity.

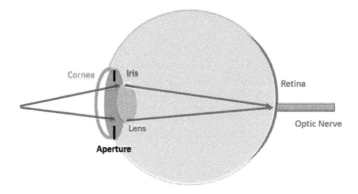

Figure 4.19.6   Viewing a book.

The **detector array** is surely the most complicated. It is called the **retina** by the doctors and most others. It is about three quarters of a sphere with a 32 mm diameter. It consists of 7 million cones and about 100 million rods in a non-uniform and interdigitated pattern. Cones are called that because they are cone shaped, about 50 μm long and from 0.5–4.0 μm in diameter where they sense the light. They are located almost entirely in the center of the retina but are thinly disbursed over its entire surface. They exist in three varieties to detect the three colors. The rods are shaped like (you got it) rods that are typically 2 μm in diameter and 100 μm long.

There is a dense collection of cones, shown in blue at the center of the retina, 160,000 per square millimeter, and no rods (in red). There is a sparse distribution of cones in the periphery, not shown. There is a denser collection of rods in the periphery with none at the center. That is a very sophisticated detector array.

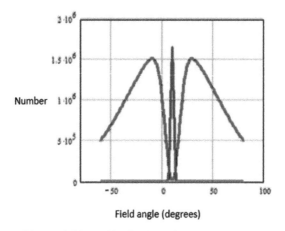

Figure 4.19.7   Distribution of rods and cones.

Some[36] have equated this format to 576 megapixels! Compare that to your digital camera. You can also compare it to an eagle's eye. The eagle has about 100 million cones; we have less than 10 million.

The **sensitivity** of the human eye has been calculated to be a single photon in its most sensitive condition by me and others. My calculations are shown in Appendix A.4.19. It has also been measured with reasonable probability.[37] Not with certainty because human subjects were involved; they are incredibly uncertain.

The **dynamic range** is an astounding factor of 10^8, 100,000,000 times. This is accomplished in two ways. The first is the change in the aperture. That was shown above to be a factor of four. The rest is a switch from cones to rods and a change in the chemistry of the rods. That switch results in a change in their sensitivity.

The **spatial resolution** of the eye is also astounding and variable. You can calculate it in at least two ways: by the Rayleigh resolution criterion and by the angular subtense of a cone. The Rayleigh calculation is $1.22\lambda/d$, where λ is the wavelength of light, and d is the diameter of the aperture. In this case, they would be 0.5 μm for yellow light in the middle of the visible spectrum and 4 mm (4000 μm) for the diameter of the eye in daylight. The calculation is $1.22 \times 0.5/4000$ and equals 0.000153 radians, which equals 0.008738 degrees, or 0.525 arc minutes.

A second way to calculate acuity is the angular resolution of a cone at the entrance of the eye. The end of a cone is 0.5 μm in diameter. The distance from the back of the lens to the retina is about 24 mm. The angular subtense of a cone is therefore from 20 to 167 microradians, 0.001 to 0.01 degrees, or 0.06 to 0.6 arc minutes.

Both calculations agree with most commonly accepted values[38] of about one-half arc minute, but the eye is hard to pin down. It varies with people and conditions. You can just about read a license plate a quarter mile away. A cop can too.

Although it seems we can see this well over our entire field of view, this is not so. We have that acuity of about one-half arc minute over about two degrees in the center of our field of view. We see the rest of the field by rapid scanning of this foveal field. This leads to another improvement of our spatial resolution capability, i.e., partial pixel resolution. By this motion, we can have resolution somewhat better than that defined by the diameter of a cone.

The **spectral sensitivity** is different for different levels of illumination, the so-called photopic and scotopic regions. Cones (in red) sense in the photopic region when there is sufficient illumination, and rods do it when it is relatively

---

[36]Clarkvision Photography - Resolution of the Human Eye, Clarkvison.com.

[37]Tinsley, J. et al. "Direct detection of a single photon by humans," *Nature Comm* **7**, 12172 (2016).

[38]Myron. Y. and Duker, J. *Ophthalmology, 3rd Ed.*, Elsevier (2009).

dark. The two curves are shown in Figure 4.19.8. They are my approximations based on Gaussian curves with centers at 555 nm and 507 nm, respectively, and standard deviations of 0.002 nm and peaks at 683 and 1700 lumens per watt. They are within a few percent of the measured values.

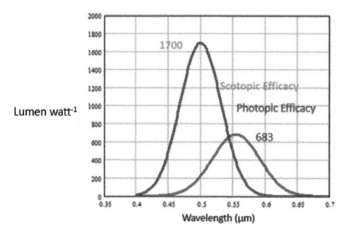

Figure 4.19.8   Eye sensitivity.

We see different **colors** because there are three different types of cones, one for each color. They are not, as I once thought, clearly and distinctly three separate responses in the blue, green, and red, but they overlap considerably. They look approximately like the curves in Figure 4.19.9. We are most sensitive to yellow at about 555 nm, shown in green—almost as sensitive to the reds and least to the blues.

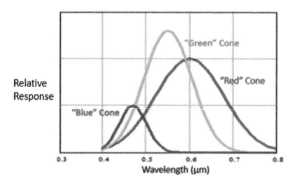

Figure 4.19.9   Spectral response of cones.

The **temporal response** time of the human eye is about 0.1 second. It is usually measured and specified as the **flicker fusion frequency**.[39] That is the

---

[39]Hecht, S. and Shlaer, S. "Intermittent stimulation by light: V. The relation between intensity and critical frequency for different parts of the spectrum," *J. Gen Physiol.* **19**(6), 965 (1936).

frequency of flashing light pulses that no longer appears intermittent. The flickering has fused. As with most things with the eye, it depends upon many factors. In this case, it is the color of the pulses, their intensity, the background, the pulse shape, the condition of the observer, and who he or she is. That is, the physical prowess and condition of the observer. For most applications, it is all right to assume 0.1 second.

Stereopsis (solid appearance) or three-dimensional viewing is accomplished in two ways, both with one eye and with two. This is something a regular camera cannot do.

Binocular depth perception is accomplished because our two eyes are separated. This is done in part by triangulation and in part by what I call side viewing. I think others call it parallax. It is seeing the sides of an object from two different angles, one for each eye. Binocular depth perception is only effective for a limited distance, about 50 to 100 yards. You can determine your own by blinking your eyes: sight on a distant object and close one eye. Then open it and look with the other eye. If it appears to move side to side, you still have depth perception at that distance. If so, try a more distant object.

Monocular depth perception is obtained in several ways: size, occlusion, relative motion, and muscle relaxation. We sense, for instance, that one cup is farther away than the other because it appears to be smaller, and we know its size. The cup in Figure 4.19.10 appears farther away than the one in Figure 4.19.11, but they are equally far from your eye, on the same page. We can also infer its approximate distance because we know how big it is and how big it appears.

Figure 4.19.10   Distant cup.

Figure 4.19.11   Close cup.

We also think that one cup is farther from us than the other if it is occluded in part by the first one, like the two in Figure 4.19.12. We can also tell something about the distance of things based on their motion with respect to each other. A very subtle clue is from our muscles. We have a sense of the degree to which our muscles contract or relax to adjust our lens for the appropriate object distance.

Figure 4.19.12   Occluding cups.

In summary, the human eye is the most remarkable optical instrument I know. It has adaptive, gradient index, sensitivity-adjustable, three-dimensional, multicolor, rapid-response, giga-pixel optics, a sensitive lens cap, and remarkable carrying case with incredible memory.

4.20 Infrared Cameras

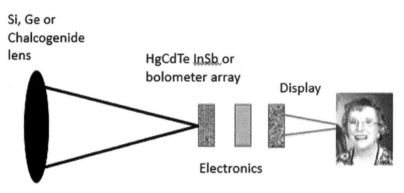

Infrared cameras, also called thermal cameras, thermographs, or thermal imagers are devices that form images of the thermal radiation from objects in the infrared part of the spectrum.

The first infrared imaging device was the Evaporagraph, invented by Max Czerny in 1929.[40] It was a thin membrane covered with a volatile liquid. White light was shown on it, and the film generated interference colors according to its thickness. The thickness was determined by the amount of evaporation, which in turn was determined by the intensity of the infrared radiation on it. It was refined and developed further by Bruce Billings and his colleagues at Baird Atomic, Inc.,[41] but it never really succeeded.

---

[40]Czerny, M. "Über Photographie im Ultrarot," *Z. Physik* **53**, 1–12 (1929).

[41]Arnquist, W. "A survey of early infrared developments," *Proc. IRE* (September 1958).

The first reasonably successful devices used a single detector and a scanning mirror. One example of this is the Thermograph built by R. Bowling Barnes. It consisted of an imaging mirror, a scanning mirror, and a single thermistor bolometer detector.[42] The scanning mirror moved an image of the detector over the scene. Barnes was interested in the medical applications, so the object would be a patient. One special application was breast cancer screening. In that case, an area of about 500 millimeters by 300 millimeters would need to be scanned with a spot size of about one millimeter. That comes to 150,000 pixels. The time constant of a thermistor bolometer is about 5 milliseconds. So, the time for a full scan turned out to be 750 seconds, or almost 15 minutes.

The technology improved with the improvement of detectors. These included indium antimonide (InSb) and mercury cadmium telluride (HCT), both photon detectors. InSb operated in the mid-infrared range from about 3–5 μm, while HCT could be used there or in the longer-wave band from 8–12 μm. Both had time constants much faster than a bolometer, 1000 times faster, at about 1 microsecond. Then the image was almost instantaneous, i.e., less than a second.[43]

Modern infrared imagers use infrared detector arrays thanks to the invention of the charge-coupled device by Boyle and Smith in 1972. It allows the extraction of signals from otherwise unreachable portions of a two-dimensional array. The arrays may consist of InSb, HCT, or microbolometers.

Cooling is a complication (I think anathema), but it is accomplished in one of two ways. Both indium antimonide that operates in the 5-μm region and mercury cadmium telluride that operates in the 12-μm region need to be cooled to about 77 K ($-320°$F). This is accomplished either by a Joule–Thompson refrigerator or the expansion of nitrogen gas from a pressurized container. Obviously, the latter works for only a brief time, i.e., until the gas runs out. The former generates unwelcome noise and vibration, but they are getting better. Microbolometers do not need to be cooled, but they are about ten times less sensitive and ten times slower than the other two. It all depends upon the application.

Whereas visible cameras sense the light that is reflected from objects, infrared cameras sense the radiation that is emitted from them. The design depends upon the emission of the object, required sensitivity, cost, duration of the application and even, sometimes, the personal preference of the designer.

An infrared camera has the same components as a digital camera, but they are of different materials. The introductory figure shows them. There is a lens that is made of a material such as silicon or a special chalcogenide glass

---

[42]My observation.
[43]Some of my designs.

because regular glass is not transparent in the infrared. The optics are either refractive lenses of the same material or reflective mirrors, such as those in telescopes. The radiation is focused onto a detector array. In the visible, this would be silicon. In the infrared, it is HCT or bolometers. Then the electronic processing and the displays are identical. The major difference is in the material of the lens and the detector array. Those used for infrared cameras are more expensive and not quite as sensitive.

The applications are described in the appropriate sections; they include the military, industrial, medical, agricultural, and forensic sectors of our society. Infrared cameras have a wide range of applications. It is useful to think of them as television sets that sense temperature differences of a small fraction of a degree.

4.21 Interferometers

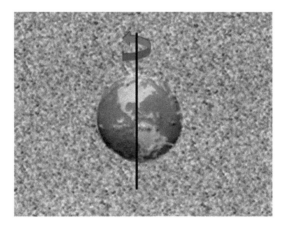

There are literally dozens and dozens of interferometers that are used for dozens and dozens of purposes. They are generally classified as double-beam or multiple-beam and common-path or double-path devices. Multiple-beam interferometers generate sharper fringes; common-path devices are more stable. They also separate into those that divide and recombine the amplitude (amplitude division) and those that separate and recombine sections of the wavefront (wavefront division).

Surely the most famous and useful is the **Michelson** interferometer, invented by A. A. Michelson, the first American Nobel Laureate in physics. He was interested in determining the existence of a thing called the luminiferous ether, the stuff that was thought to support the propagation of light. It didn't, and it wasn't (see Section 5.7). The interferometer has a source of light, not shown on the left in Figure 4.21.1. It is split into two beams by a beam splitter. They travel in directions perpendicular to each other and are reflected by the mirrors, then recombined by the same beam splitter that is then a beam combiner. Their combination produces an interference pattern.

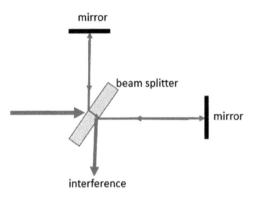

Figure 4.21.1 Michelson interferometer.

Michelson and Morley carried out their experiment in 1887 at Case Western Reserve on a very stable optical table, a giant slab of concrete mounted on a pool of mercury.[44] It was assumed that the ether wind flowed in a certain direction. If so, it would affect the interference differently with different orientations of the interferometer. It did not, after many trials. It was one of the events that led Einstein to relativity.

It is the basis of the LIGO experiments on the existence of gravity waves and the modern Fourier transform interferometer spectrometer that was used in the COBE-DIRBE experiment (see the corresponding entries in the Appendix). The Michelson is a double-path, double-beam instrument.

An aside on the **luminiferous ether** (light-bearing air) seems in order. It was well known at the time, the mid 1800s, that water carried water waves and air supported sound waves. Something had to support optical waves. But that was a very weird something. It had to be everywhere. It was even in vacuums. It had to be stable; the Earth moved through it. It had to be solid enough to support optical waves, but it had to be ethereal enough for everything else to move through it. In one sense, it was the "dark matter" of that time. It was there, but no one really understood it. But it wasn't there. Michelson and Morley were the first to prove its absence while trying to prove its existence, and many others supported those findings.

It was all around the Earth. In position. Unmoving. But the Earth rotated through it, thereby generating a relative motion. That caused the luminiferous wind that Michelson and Morley sought to measure.

---

[44]Michelson A. and Morley, E. "On the relative motion of the Earth and the luminiferous ether," *Am. J. Sci.* **34**(203), 333345 (1887).

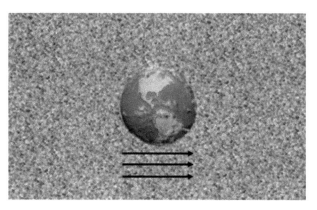

Figure 4.21.2 Luminiferous wind.

The **Twyman–Green** interferometer is a close relative to the Michelson. It is the Michelson with collimated light. It was invented by Frank Twyman and Arthur Green in 1916. It is used mostly for optical testing[45] but is sometimes described as a Michelson. It **is** a form of a Michelson.

The **Sagnac** interferometer is a double-beam, common-path interferometer in which the beams travel in opposite directions around a rectangle.[46] It was invented by George Sagnac in 1913 to further investigate the existence of ether. The two beams that travel clockwise (blue) and counterclockwise (red) are actually superimposed and will interfere differently if the instrument

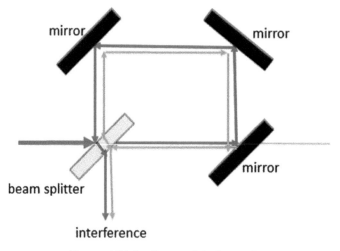

Figure 4.21.3 Sagnac interferometer.

[45]Malacara, D. *Optical Shop Testing*, Wiley (2007).
[46]Sagnac, G. "L'éther lumineux démontré par l'effet du vent relatif d'éther dans un interféromètre en rotation uniforme," *Comptes Rendus* **157**, 708–710 (1913).

rotates since one path will be just a little longer than the other during a measurement time. It is the basis for the fiber optic gyroscope.

The **Mach–Zehnder** interferometer looks much like the Sagnac, but there is a significant difference: one of the Sagnac mirrors is replaced by a beam splitter. It was invented by the two Ludwigs, Mach and Zehnder, in 1913.[47] The distance between each beam splitter and each mirror in which the sample and reference lie can be adjusted considerably. This behavior makes the device quite versatile. It is a two-beam, separate-path, amplitude-division interferometer.

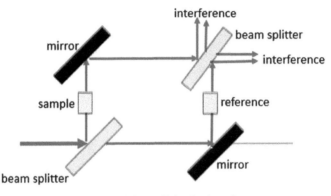

Figure 4.21.4 Mach–Zehnder interferometer.

The **Jamin interferometer** consists of two thick refracting plates that are reflectorized on the back. Four beams are generated from the combination of front and back reflections, but two of them are blocked. The others interfere.

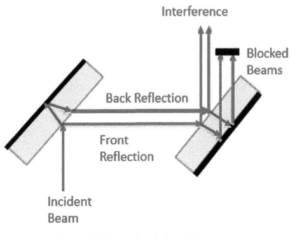

Figure 4.21.5 Jamin interferometer.

[47]Zehnder, L. "Ein neuer Interferenzrefraktor," *Zeitscrift für Instrumentkunde* **11**, 275 (1891); Mach, L. "Über einen Interferenzrefraktor," *Zeitscrift für Instrumentkunde* **12**, 89 (1892).

There are two different **Fizeau interferometers**. One is used for testing optical components and the other for testing the luminiferous ether. The one for testing optics is depicted in Figure 4.21.6. It shines collimated light onto a sample in front of a flat. The interference between them is focused by another lens.

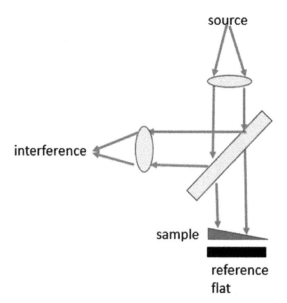

Figure 4.21.6   Fizeau optical test interferometer.

The other, and first, **Fizeau interferometer** was designed to test that pesky ether before Michelson, but it was inconclusive. According to the ether theory, the speed of light would be increased or decreased by the speed of the medium

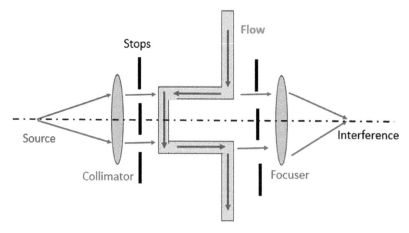

Figure 4.21.7   Fizeau ether test interferometer.

in which it moved. It was called the ether drag. Fizeau arranged his instrument as shown to interfere the two beams that would be of different speeds and therefore different phases. He found the drag, but it was too small and therefore inconclusive. We needed Michelson. This is a wavefront-division, double-path interferometer.

The **Rayleigh interferometer**[48] is based on the same idea as the Fizeau, but it has two distinct tubes for the comparison. And it is a material rather than a directional comparison. It was one of the instruments that helped prove the wave nature of light by showing that light travels slower in a medium than in air.

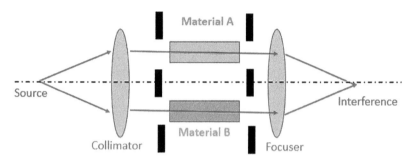

Figure 4.21.8 Rayleigh interferometer.

I am not sure that you can call this a **Newton interferometer**, but it is an arrangement that produces Newton's rings. It is much like the simple Fizeau. The source at top is collimated and passes through a beam splitter to the

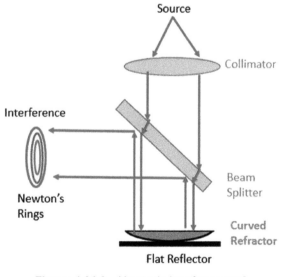

Figure 4.21.9 Newton's interferometer?

---

[48]Hariharan, P. *Basics of Interferometry*, 2nd Ed., Academic Press (2007).

curved sample and then a reference flat. The interference is between curved waves from the sample and plane ones from the flat. Newton was the first to observe and discuss these interference rings, but he could not describe them in terms of his corpuscular theory.[49]

The **Lummer–Gherke** plate is a multiple-beam, separate-path interferometer that uses multiple reflections from a long thin plate, as shown in Figure 4.21.10. There are many more interreflections in a typical plate than I have shown. There are several ways to introduce the light into the plate. Two are used to angle the front end or to put a prism at the top of the front end. The device was invented by Otto Lummer and Ernst Gehrke in 1903, but it has been replaced by the Fabry Perot interferometer and is not worth further description.

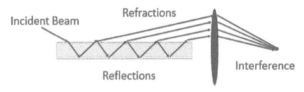

Figure 4.21.10   Lummer–Gehrke plate.

The **Shearing interferometer** is extremely simple, consisting only of a well-made glass plate. The two beams from the front (reflected) and back surfaces (refracted, reflected refracted) interfere. The interference occurs in the small region where they overlap, indicated by the tan area. The other region can be blocked as well as the ghost images, which are very dim anyway.

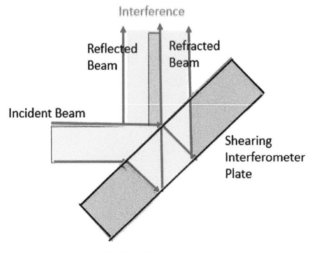

Figure 4.21.11   Shearing interferometer.

---

[49]Newton, I. *Opticks*, London (1704).

The **Fabry–Perot interferometer** uses a pair of plane-parallel plates in collimated light to accomplish multiple-beam interference. This is therefore a multiple-beam interferometer. The multiple beams overlap, but I could not show them that way. And I could not show an infinite number of them with decreasing amplitude, but that is what they are. This device was invented in 1899 by Charles Fabry and Alfred Perot.[50]

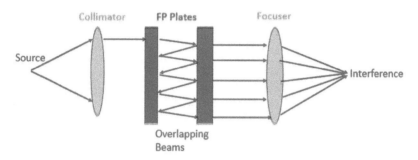

Figure 4.21.12   Fabry–Perot interferometer.

The multiple beams enhance the sharpness of the peaks. This is shown in Figure 4.21.13. The red line represents a single interference. The blue line is four; green represents eight; and magenta represents 30.

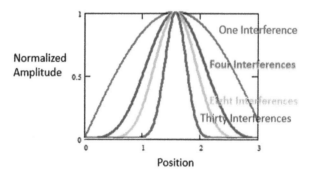

Figure 4.21.13   Fringe sharpening.

The **Michelson stellar interferometer** is used to measure the diameter of stars. It uses the fringe contrast as a measure. The larger the star is, the more places from which the light is emitted and so the more individual beams with random phases (the less coherent the light). The fringe contrast is a measure of the degree of coherence and therefore of the area over which all these individual beams originate. The two outrigger mirrors shown in Figure 4.21.14 reflect the light to the interior mirrors, and they send it to optics that brings together the two separate beams for interference. The

---

[50]Fabry, C. and Perot, A. "A new form of interferometer," *Ann. Chim. Phys.* **16**(7) (1899).

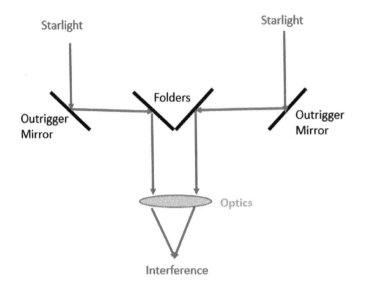

Figure 4.21.14 Michelson stellar interferometer.

device was proposed by Michelson in 1890 and realized at the Mount Wilson Observatory in 1920. Betelgeuse was found to be 380 million kilometers (240 miles) in diameter.[51]

The **phase-shifting interferometer**[52] takes us from the astronomical to microscopic. It is used to measure the minute roughness of surfaces, less than 1 nm. Note that an aluminum atom is about 0.3 nm in diameter.

The process is to form a two-dimensional interferogram of the surface with an instrument like a Michelson or Twyman–Green interferometer. This is done by shining a coherent source such as a laser into the interferometer. It reflects off the reference surface and the test surface. The reflections are combined and interfere. The reference surface is very flat; the test surface has the structure to be measured. There is a different amount of interference between each point on each surface so that a two-dimensional interference pattern is generated. But it is only relative. It takes two more independent measurements to quantify the interference pattern, i.e., to actually measure the height distribution as a function of position on the test surface. The three positions of the reference are shown at the top of Figure 4.21.15.

---

[51]Michelson, A. and Pease, F. "On the application of interference methods to astronomical measurements," *Astrophys. J.* **53**, 249 (1921).

[52]Wyant, J. and Creath, K. "Two-wavelength phase-shifting interferometer and method," US patent 832,489 (1989).

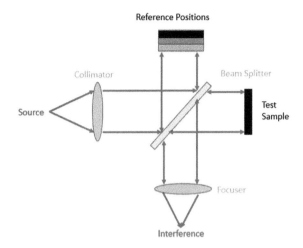

Figure 4.21.15   Phase-shifting interferometer.

Optical coherence tomography[53] is a technique for obtaining three-dimensional imagery of a patient using interferometric techniques.[54] It is essentially a Michelson interferometer with a scanning mirror and imaging system. The source is a low-coherence source so that interference occurs over only a limited range. That range is set by the position of the reference mirror to produce constructive interference at a limited vertical position. The scanning mirror covers the entire surface of the sample by an x-y scan. One vertical slice is taken for each position of the reference mirror. A three-dimensional interference pattern is thereby obtained. Maybe this should be called optical incoherence tomography (OCT).

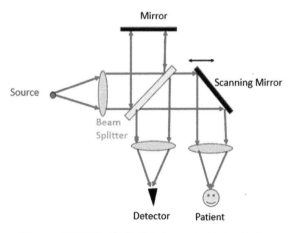

Figure 4.21.16   Optical coherence tomography.

---

[53]Fercher, A. "Optical coherence tomography," *J. Biomed. Opt.* **1**(2), 157–173 (1996) [doi: 10.1117/12.231361].

[54]Cote, J. "New Optic-Disc Topographic Technique," Dissertation, University of Arizona College of Optical Sciences (1996).

4.22 Laser Damage

Laser damage can be either intentional or accidental. Intentional damage includes the use of a laser as a weapon. The most significant accidental damage is that to the eye. Damage during industrial operations is just a mistake.

Almost from the very invention and demonstration of the laser, they were considered as a potential **weapon**. They can have considerable energy in a concentrated beam. There is no windage or gravitational attraction; it is true line-of-sight aiming and hitting. But it was not simple. There were several deterrents. One was efficiency. Lasers are 10–20% efficient. A very powerful supply is necessary to get a sufficiently powerful beam. The atmosphere has a tendency to spread the beam, and many targets reflect much of the beam. In particular, ICBMs and aircraft can be metallic with reflections greater than 90%. One early worker in the field cited a suit of armor as the best defense against a laser weapon, as noted above.

The energy profile of a laser is approximately Gaussian. This means that there is more heat in the center than nearer to the edges. There is therefore a gradient in the density of the air and a resultant gradient in the refractive index. It is warmer towards the center, less dense toward the center, and has a smaller refractive index toward the center. It makes a biconcave gradient index lens and tends to spread the beam.

Appendix A.4.22 shows that it takes a finite amount of time to penetrate any material with a laser (at least several seconds). It therefore requires some very accurate tracking at least for a short time to keep the beam on the same spot long enough.

Lasers, and especially laser pointers, are a potential source of **eye damage**. I felt the heat of a 1-kilowatt carbon-dioxide laser beam in our laboratory on the side of my head. I ducked right away. I got away. Most of you will not encounter high-energy lasers like that, but you will encounter laser pointers. Laser pointers are red and green and only about a milliwatt. Eye damage by laser pointers is unlikely but possible. It will not happen if the laser is properly certified and if you do not let it shine in your eye for more than a blink.

The maximum permissible exposure in watts per square centimeter is a function of wavelength. Minimum values in the middle of the visible are about 10 milliwatts for a 0.1-second exposure. A 1-milliwatt laser pointer has a beam that is about one square centimeter and is well collimated. So it produces about 1 milliwatt on your eye. You better blink fast if you are watching. The normal blink response is about 0.4 seconds. So watch out. If you are the presenter, keep the pointer aimed at the screen.

In spite of these calculations that show the concepts, it has been reported that no eye damage has been caused by "pocket lasers" with 3–5-milliwatt power.[55]

A different kind of damage has been reported. It is flash blindness of pilots caused by high-power green lasers during critical flight times.[56] They are more than 5 mW and located where the eye is more sensitive. These do not cause damage themselves but induce it by other means, such as distracting or disorienting the pilot. About 6000 of these dangerous attacks per year have been reported.[56]

4.23 LASIK

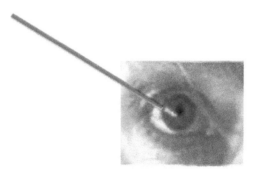

The acronym LASIK stands for "laser-assisted *in situ* keratomileusis," that is, "laser assisted, on the spot, reshaping of the horn."

Section 4.19 describes how the cornea does most of the focusing by the eye; the lens does the rest. If the cornea is not of the right shape, the eye can have myopia (nearsightedness), hyperopia (far sightedness), or astigmatism (distortion). Then the patient either needs glasses for correction or she can get the cornea reshaped. It is questionable how much good this can do for presbyopia (old-age eyes) since this condition is not based on the shape of the cornea.

---

[55]Johnson, D. "Can a pocket laser damage the eye?," *Scientific American* (December 1998).
[56]Laser Incidents – Federal Aviation Commission, www.faa.gov.

The process is simple but very technical and precise. First, the doctor must obtain a precise topographic map of the cornea to determine what to correct. This is done with a pachymeter, which is either optical or sonic. It uses either an optical pulse technique, such as lidar, or sound pulses, such as sonar.

The actual corneal reshaping process starts with generating a hole in the outer layer of the cornea, the epithelium, with either a scalpel or an excimer laser.[57] Then the shaping begins. Material is removed by short pulses from that excimer (excited dimer) laser at 193 nm. It removes the tissue without generating any heat or cutting, removing about ten micrometers at a time, by giving the molecules enough energy to jump out of their skin. (The photon energy exceeds the molecular binding energy.) The cornea is about 500 μm thick; these are small changes, about 2%.

There is a satisfaction rate of more than 90%. The reshaping process is quick and painless. The patients sit in a typical medical chair with a head restraint for about thirty minutes, and they then get up and walk out of the office. They can then sing, ♫ "I Can See Clearly Now ..." ♫

4.24 LIDAR

Lidar stands for "light detection and ranging." It is analogous to radar, which stands for "radio detection and ranging." It might also be called "laser detection and ranging," because all of them use a laser as a source. It is the same technology applied in two different parts of the electromagnetic spectrum. There are two different types of lidar ranging: Doppler and time of flight.

The Doppler version makes use of the shift in frequency caused by the (in-line) motion of the target, also called the red shift (Section 1.10). For many applications, that frequency shift is too small to be detected. The shift is proportional to the ratio of the speed of the object to the speed of light. For a car traveling at 100 mph, that ratio is approximately 10^{-6}, one part in a million. That frequency shift is too small for automobile applications, but it is very useful in a variety of astronomic and atmospheric investigations.

---

[57]I participated in this laser surgery performed by Robert Snyder, MD, at the University of Arizona Ophthalmology Department, and observed what he did.

Time-of-flight lidars send out pulses and measure the time it takes for a pulse to return. Light travels about one foot per nanosecond (186,000 mps × 5280 ft/mi = 9.8×10^8 feet per second. The reciprocal is 1.02 ns per foot). A pulse sent to a speeding car five miles away travels 52,800 feet in the round trip. That takes 52,800 ns, or 52.8 μs, enough for quite a few microsecond bursts (about 25 of them 1 microsecond apart).

Lidars can take images by flashing spot by spot over a specified area. The images can be in color with different laser sources. And they can measure the different distances based on the times of flight. So, these devices can generate multicolored, three-dimensional representations of their environment, even with colors we cannot see.

Lidars are used in autonomous vehicles, clear-air turbulence detection, and many satellite evaluations of our Earth. As with most every scientific application, lidars are getting better, smaller and cheaper since they are in development and in commercial demand.

4.25 Loupes

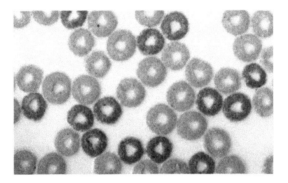

Loupes are simple optical devices used by jewelers and other artisans to get a good look at their craft and by surgeons in their work. The simplest loupe is a single lens for one eye.

Figure 4.25.1 Simple loupe.

The optics of the loupe are simple. The basic concept is shown in Figure 4.25.2. The loupe is arranged so that the object is at its front focal point. Then rays from it come to the eye parallel to each other. The object therefore appears at infinity and greatly enlarged.

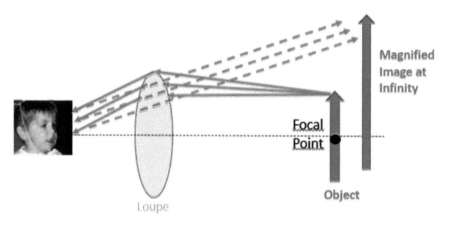

Figure 4.25.2 Loupe optics.

Although loupes may contain as many as five lenses, they very rarely have more than three, and most are singlets or doublets.

Many come as a binocular pair. They each have a single lens with magnification up to about three and a choice of object distance (working distance). Some combine with prescription spectacles; others are flip-in devices.

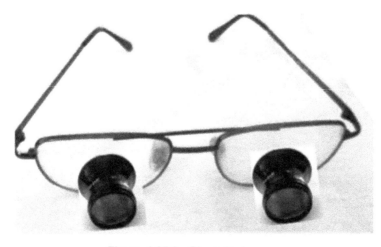

Figure 4.25.3 Binocular loupes.

4.26 Medical Thermographs

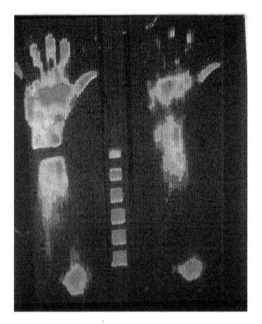

Infrared cameras that can detect small temperature differences can be used for a variety of medical purposes. This application is often called medical thermography, making a graph or picture of a thermal pattern for medical purposes.

The typical modern camera uses a silicon window, a chalcogenide lens (consisting of a mixture of sulfur, selenium, and tellurium), and an array of cooled mercury cadmium telluride detectors or microbolometer arrays and silicon or reflective optics. Modern cameras can image a person's entire chest with a spatial resolution of about 1 millimeter and a temperature difference sensitivity of about 0.01 K. Appendix A.4.26 describes the design of a generic thermograph.

One important example is screening for breast cancer. This is not recommended by the FDA as a screening procedure.[58] I think it is a legitimate screener, but I am not a doctor. X-ray and MRI techniques find lumps; thermography finds anomalous patterns of heat. These are usually associated with changes in blood flow, and they are often, if not always, associated with cancer and other lesions.

---

[58]U.S. Food & Drug Administration "FDA Warns Thermography Should Not Be Used in Place of Mammography to Detect, Diagnose, or Screen for Breast Cancer: FDA Safety Communication," https://www.fda.gov/medical-devices/safety-communications/fda-warns-thermography-should-not-be-used-place-mammography-detect-diagnose-or-screen-breast-cancer (February 25, 2019).

Excellent reviews of the history and effectiveness of the technique have been produced by W. Amalu and E. Ng.[59] They point out that thermography has been under investigation since 1965 by Gershon Cohen, although it was really a little earlier.[60] More importantly, they review 300 peer-reviewed studies of 300,000 women, all of which show that it has about an 80–90% true positive rate and about the same true negative rate. Ng reports 90% sensitivity and specificity (same indicators). Others deny that it is a valid screening test for breast cancer with considerable errors.[61] The FDA has approved it only as an adjunct diagnostic. After all these years, it is still controversial. Aetna considers it experimental (after 45 years).[62]

The process is simple: the patient sits topless in a cool room for about ten minutes to equilibrate and to optimize the environment for the thermogram, the thermal image. The operator then takes a picture with an infrared camera. There is no contact and no irradiation. The thermograph is typically at least a meter away. The diagnosis is based largely on the degree of asymmetry of the temperature of the breasts. A warmer temperature may indicate an abnormal degree of vascularity. Tumors need to be fed! And malignant ones need more than benign ones.

I have read some unsubstantiated claims that false positives turned out to be true positives several years later. If this is true, it is remarkable early detection.

Infrared imaging, medical thermography, is used in other medical investigations.[63] It has been shown that cancer of the cervix can be detected this way.[64] Investigations of burns to determine the extent of necrology and the limit of viable skin has been done. Malingerers who claim back pain can be exposed if there is no pattern of heat that indicates pathology. The partial closing of my wife's carotid artery by a tumor was visible on an infrared image that showed asymmetric temperatures in her face and neck. Infrared cameras readily show problems of circulation in arms and legs and can locate possible clogging points.[65]

---

[59]Amalu, W. C. "A review of breast thermography," *International Journal of Clinical Thermography*, https://www.iact-org.org/articles/articles-review-btherm.html; Ng, E. "A review of thermography as promising non-invasive detection modality for breast tumor," *International J. Thermal Sci.* **48** (5), 849 (2009).

[60]Wolfe, W. "Infrared imaging devices in infrared medical radiography," *Annals of the New York Academy of Sciences* **121**, 57 (1964); Barnes, R. "Thermography of the human body," *Science* **140**(3569), 870–877 (1963).

[61]Williams, K. et al. "Thermography in screening for breast cancer," *J. Epidemiol Community Health* **44**(2), 112 (1990).

[62]Aetna. Thermography Clinical Policy Bulletins, http://www.aetna.com/cpb/medical/data/1_99/0029.html.

[63]B. Lahiri et al. "Medical applications of infrared thermography: A review," *Infrared Physics and Technology* **55**, 221 (2012).

[64]W. Wolfe, unpublished investigations.

[65]Mashekova, A. et al. "Early detection of the breast cancer using infrared technology: A comprehensive review," *Thermal Science and Engineering Progress* **27**, 101142, Jan. (2022).

The introductory image is a before-and-after illustration of the effects of smoking on arm circulation. The left shows good circulation, red and orange indicators of normal temperatures in the hand, and the dark band of the wristwatch. The right shows dark, cold fingers and even reduced circulation around the wrist. It was the immediate result of a single drag on a cigarette.

Don't smoke!

4.27 Microscopes and Magnifiers

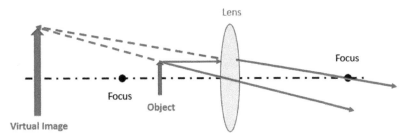

I may not be in line with common thinking, but I do not consider a magnifying **glass** a microscope. Some call it a simple microscope, and they call what I consider a microscope a compound one.

The **magnifying glass** is a simple instrument optically, but it can get complicated mechanically. The one I have shown above is just a convex lens with a handle. The one in Figure 4.27.2 is the one I use for tying trout flies. It is a larger convex lens but still just a single lens. Its "handle" is a complex, articulated, mechanical structure with a light that allows me to put the lens just where I want it, and it stays there.

They both work on the same principle. Arrange the object to be just inside the front focal point so that an enlarged, virtual image is presented to the eye from almost infinity. The magnifying glass does this with one simple lens. The microscope uses both an objective lens and an eyepiece lens as shown below.

Figure 4.27.1 Magnifying glass principle.

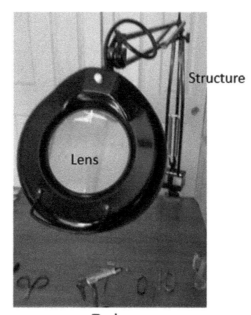

Tools

Figure 4.27.2   Fly-tying magnifier.

Tools

Figure 4.27.3   Bottom illuminating side.

Microscopes were probably invented by Hans Lippershey's kids fooling around with their father's lenses.[66] Hans Zacharisen made similar claims. Galileo inverted his telescopes (based on Lippershey) to enlarge small objects. Galileo used his telescope up close and even inverted it to get enlarged images of close objects. He also designed his *occhiolino*. Leeuwenhoek did much to advance the field of microscopy but did not contribute much to improve the design of the instrument.

The basic concept of the microscope is shown in Figure 4.27.4. The first lens forms an inverted image of the object on the left (the dark blue, vertical arrow). It is reimaged by the second lens. The two significant rays that can determine the first image are the solid ray parallel to the optic axis that goes to the back focus of the front lens, the little blue circle, and the ray that goes undeviated through the middle of that lens to the same point. The image is where they intersect. That image is reimaged by the second lens with two equivalent rays (dashed), through the middle and parallel to the axis and to the back focal point, but since that arrow is between the front focal point of the rear lens (gray) and the lens, the image is virtual. The final image is virtual and enlarged. A virtual image cannot expose a film, but it can be viewed through the other lenses and thereby recorded.

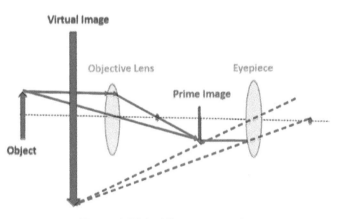

Figure 4.27.4 Microscope optics.

There are three major optical systems in a microscope: the illuminator, the objective lens, and the eyepiece. The eyepiece was essential to the early scopes, which were all viewed by the human eye. Modern versions usually project to some sort of electronic monitor, but they still do so through an eyepiece.

Proper illumination of the sample is important. August Köhler invented or developed an effective way to do it, called **Köhler illumination**.[67] He collimated the light onto the sample so that it would be illuminated evenly, as shown in

---

[66]Wolfe, W. *Rays, Waves and Photons*, International Organization for Physics (2020).

Figure 4.27.5. This process replaced **critical illumination**, which imaged the source onto the sample even though it required more optical elements. It is not good to see an image of a bulb filament on top of an amoeba!

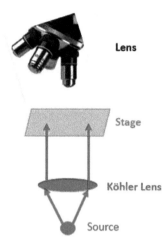

Figure 4.27.5   Köhler illumination.

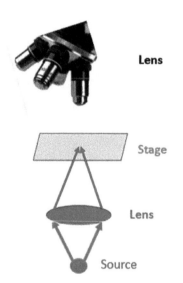

Figure 4.27.6   Critical illumination.

Dark field illumination is used mostly for transparent samples. Oblique illumination enhances the contrasts formed by their topography. It is accomplished by a disc that obscures the direct beams but transmits oblique

[67]Köhler, A. "Ein neues Beleuchtungsverfahren für mikrophotographische, *Zeitschrift für wissenschaftliche Mikroskopie und für Mikroskopische Technik*. **10**(4), 433–440 (1893).

illumination. The samples are illuminated; they are not left in the dark. It is just that the rays impinge at an angle shown in Figure 4.27.7.

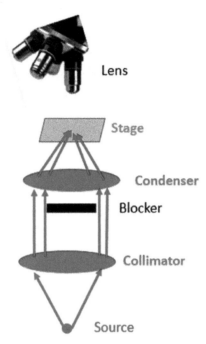

Figure 4.27.7 Dark field illumination.

Objectives come in many forms, but they may be divided into refractive and reflective ones. Most microscopes use refractive objectives partly from habit and partly because they are straight through without obscuration. Reflective ones are used mostly for ultraviolet and infrared applications, where glass is opaque.

Refractive objectives may have simple lens components, maybe two, or very complex ones with as many as four or five lenses. They vary from a magnification of about two to over 200. There are three objectives on a typical microscope: one with a magnification of $4\times$, the next at $10\times$, and the BIG DADDY at $100\times$ or a bit more with immersion.

An **immersion lens** microscope makes use of refraction to increase the magnification. An oil is placed between the specimen and the lens so that it makes contact with both. The general concept may be seen in Figure 4.27.8.

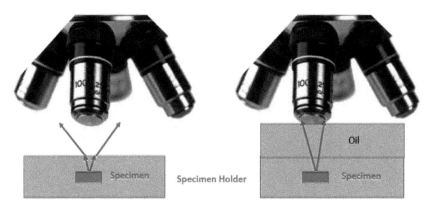

Figure 4.27.8   Immersion lens principle.

Without the oil immersion, the light that exits the glass cover refracts away from the normal as shown on the left (dense to less dense refraction). With an oil that matches the index of the cover glass, there is no refraction.

Eyepieces, as mentioned above, were originally meant to project the magnified image onto a person's eye. For that, they usually projected a collimated beam so the operator could focus on infinity, a relaxed focus, and they had to have enough eye relief (be far enough away from the lens itself) for comfort. That arrangement requires the correct placement of the exit pupil.

There are about a dozen eyepiece designs. The **Huygens** eyepiece was the first that used more than one lens, two plano-convex lenses with an air space between them, and the primary image as well. This allowed for some correction. It is not used today. The **Ramsden** eyepiece also uses two plano-convex lenses with the convex surfaces facing each other and the primary image in front of the eyepiece. The **Kellner** eyepiece may be considered a corrected Ramsden. It uses an achromatic doublet as the second lens. The **Plössl** eyepiece may be considered a corrected Kellner. It uses two identical achromatic doublets. I will stop there; I think the concepts are clear. There are modern versions of all of these versions with more lenses, more correction, more weight, and more expense. As eyepieces, they all provide eye relief and generate a collimated beam for the eye to focus. Versions that use cameras do the same, leaving it to the camera to do the imaging.

Probably the next major advance in microscopic instrumentation was the Zernike **phase contrast microscope**, invented by Frits Zernike in 1934.[68] It enabled investigators to see magnified images of live specimens without

---

[68]Zernike, F. "Phase contrast, a new method for the microscopic observation of transparent objects," *Physica* **9**(10), 974–980 (1942).

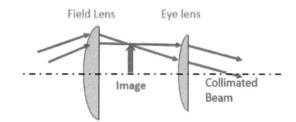

Figure 4.27.9   Huygens' eyepiece.

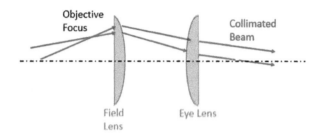

Figure 4.27.10   Ramsden eyepiece.

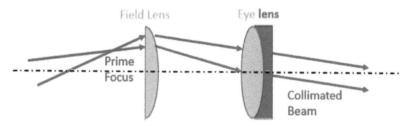

Figure 4.27.11   Kellner eyepiece.

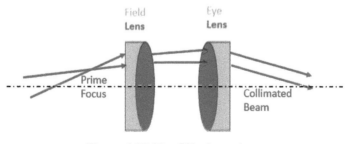

Figure 4.27.12   Plössl eyepiece.

staining or other techniques. It uses a standard optical arrangement but separates the light into two sections by an annular ring. One section is shifted in phase; the other is unaffected. When the two are combined, an interference pattern is generated that shows the relative phase shifts in the specimen. The phase shifts vary according to the thickness and density of the specimen. It has proven to be a major factor in microscopy.

I think Georges Nomarski was inspired by what Zernike did and produced his variation on it: **differential interference microscopy**. It is described in a patent by him in 1952.[69] It also uses a beam separated into two orthogonal polarizations, one portion of which passes through the sample, while the other is a reference. They are combined to obtain an interference pattern. It shows the phase difference as an amplitude difference, thereby revealing differing thicknesses and densities of the sample in a manner similar to the phase contrast microscope.

Stereo microscopes utilize the basic idea of generating stereo images, i.e., images presented from two directions (see Section 4.51). There are two ways this is done: one is to split the beam from a single aperture into two parts, the other is to use two completely different objectives. The first stereo microscope was used by a monk named Chérubin d'Orléans in the seventeenth century; he was not seeking three dimensions, just a better image by using both eyes. Ernst Abbe was the one who developed the theory of the microscope much later.

Confocal microscopes provide an image of only one part of the object at a time. This is accomplished by some form of small field stop at the focal plane of the instrument. A scanning technique moves this spot from point to point over the sample. This provides the opportunity to get a three-dimensional image of the sample. The microscope is focused at one distance, and the sample is scanned. The focus is changed slightly, and the sample scanned again. In this way, several images of the sample at different depths are generated. The device and technique were invented by Marvin Minsky in 1957.[70]

Ultraviolet or **fluorescence microscopy** is used to investigate the fluorescent properties of micro-organisms. Ultraviolet, or very violet, light is used to illuminate the sample. A color filter is used to transmit only the ultraviolet light onto the sample. A second filter is used near the eyepiece to block the ultraviolet light from the image so only the fluorescent light of longer wavelengths emitted from the sample is seen.

---

[69]Nomarski, G. "Interferential polarizing device for study of phase objects," US patent 2924142 (1960).
[70]Minsky, M. "Microscopy apparatus," US Patent 3,013,467 (1957).

4.28 Missile Guidance

Cruise missiles were guided by inertial guidance and map matching until recently, when GPS was used for the approach phase. Javelins are infrared guided missiles meant to destroy tanks. They are based on Sidewinder technology.

The early cruise missiles were aimed at their target from a long distance away. They flew at low altitude to avoid radar detection. The current Tomahawk flies at about 100 feet and has a range of about 1500 miles. Some are subsonic, some supersonic. They were guided by computer instructions and kept on course by iron gyros and later fiber optic gyros (see Section 4.17). When they got close, they compared an internal map with visible ground images. When they got a match, they fired. The modern Tomahawk has been upgraded to follow GPS instructions, possibly change targets during flight, and even loiter in the target area, awaiting new instructions.

Figure 4.28.1 Tomahawk missile.

The Javelin is an antitank, fire-and-forget missile that uses a 64 × 64-element mercury cadmium telluride detector array, reflective optics, and an infrared transparent dome, probably chalcogenide glass. The detector array is cooled to about 77 K by a bottle of argon gas that expands. It has two modes that are called "fast ball" and "slow ball." The first is a direct path from the launch area to the tank at low altitude. The second is a high arc that attacks the tank on the top where it is most vulnerable. I would have called them "low ball" and "high ball," or maybe "volley" and "lob."

Figure 4.28.2   Javelin missile.

The granddaddy of the air-to-air, infrared-guided missiles is the Sidewinder. It was developed in what is now the Naval Air Station at China Lake in the Mojave Desert and named after the rattlesnake that uses infrared sensing to find its prey. The optical portion of it is the guidance head shown as the black tip in Figure 4.28.3.

Figure 4.28.3   Sidewinder missile.

Figure 4.28.4   Sidewinder "namesnake."

It homes in on two sources from the opposing aircraft: its tailpipe and its jet plume. The tailpipe is a very good blackbody simulator at a temperature of 700–900 K, depending on the jet and afterburning. The plume is hot carbon dioxide and water vapor, with a little carbon monoxide and even carbon particles mixed in. The carbon dioxide radiates at 4.5 μm and the water vapor at 2.7 μm. They would be completely absorbed by the same constituents in the atmosphere except that they have broadened lines because they are so hot. Figure 4.28.5 shows how the radiation is reduced in the center of the spectral region by the atmosphere but is still abundant in the edges. This is the emission of a jet plume measured at a distance of 200 feet.[71] There would be greater absorption in the center at larger ranges. It would look something like that shown in Figure 4.28.6.

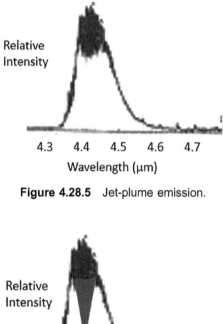

Figure 4.28.5 Jet-plume emission.

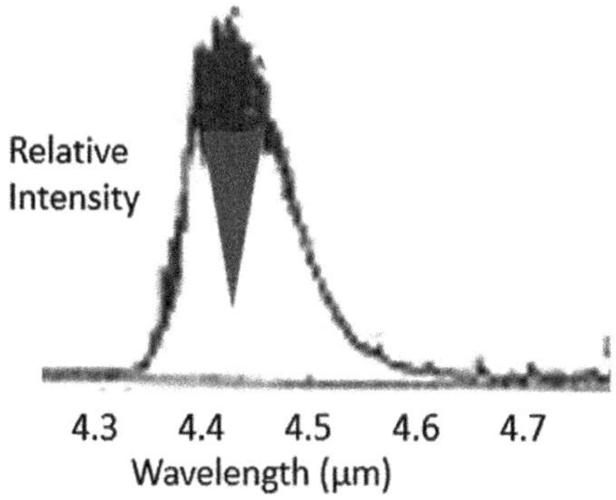

Figure 4.28.6 Farther away.

---

[71]Wolfe, W. and Zissis, G. *The Infrared Handbook*, US Government Printing Office (1978).

After a few modifications, the Sidewinder guidance head looked like the drawing in Figure 4.28.7.[72] An infrared transmissive dome of sapphire is on the left; a Newtonian telescope (parabolic primary and flat secondary) forms the image on the reticle. The image on the reticle is smaller than the blades. The reticle spins around and thereby provides information of both the elevation and azimuth angles. The blades also have the effect of smoothing out cloud interference and other extended sources. The current version is AIM 9X Block III. But that does not mean there were 9 times almost all the letters in the alphabet plus 3 versions, but there were many upgrades of detectors, detector arrays, reticle designs, windows, and coolants.

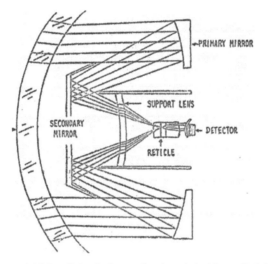

Figure 4.28.7   Sidewinder optics (reprinted from Ref. 72).

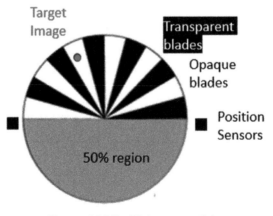

Figure 4.28.8   Rising sun reticle.

---

[72]Fender, J. "An Investigation of Computer-Assisted Stray Radiation Programs," Dissertation, The University of Arizona (1981).

One of the more recent innovations was the introduction of an optical relay system to move the detector array off the moving gimbal. Modern versions now have a detector array instead of a reticle, but they still need cooling and electrical connections to the missile's computer. It is awkward to make those connections if the gimbal is free to rotate. Thus, the system shown in Figure 4.28.9 has been introduced. The reticle is removed, and a lens collimates the beam that is directed to a set of four mirrors that reflect it to another lens and the array.[73] Now the only thing that passes from the moving gimbal to the fixed array and electronics is photons. This is the design of the AMRAAM® (Raytheon Missiles & Defense), the latest version of the venerable Sidewinder. This type of optical arrangement is called a coudé, after the French word for elbow. It can be designed with three mirrors to make it look more like a real elbow, but little is gained by that.

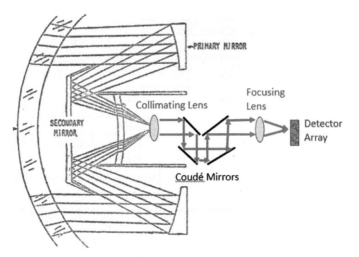

Figure 4.28.9 The AMRAAM® design with a coudé (adapted from Ref. 72).

---

[73]Wolfe, W. expert witnessing, Raytheon vs British Aerospace Engineering.

4.29 Multispectral Imagers

Multispectral imagers are devices that take images in several different colors i.e. in different spectral regions. They have been called multispectral imagers, hyperspectral imagers and imaging spectrometers. "Multis" are generally considered those with ten bands or fewer; "hypers" are those with more bands that are also contiguous. The term "imaging spectrometer" is a more general descriptor. Both kinds are used in satellites to evaluate the health of crops, locate forest fires or growth of cities, and in medicine for the location of tumors.

They consist of a device that produces the separate spectral regions and one that takes an image. There are several ways they do this.[74]

Many satellite devices use an array with different filters on each row and scan it over the part of the ground that is of interest, as is illustrated in Figure 4.29.1.

Figure 4.29.1   Multispectral crop scanning.

---

[74]Wolfe, W. *Introduction to Imaging Spectrometers*, SPIE Press (1997).

Others use a prism or grating to spread the spectrum over the detector array. There are other techniques, but these are the ones that are almost always used.

The satellite applications are described in more detail in Section 4.44; they are used to assess the heatlh and productivity of crops, waterways, and urban growth.

Multispectral imagers have also found their way into drones that fly over crops in the same way that satellites do but at a lower altitude, and with more specficity, more control by the farmer, and a smaller footprint.

Medical applications include the evaluation of skin cancers and similar maladies. One very personal example is the development of one to detect and define the extent of squamous cell carcinomas.[75]

The prototype camera we have developed looks like a regular camera with a 10-centimeter-diameter wheel in front of it for the illuminating LEDs (Figure 4.29.2). An eventual commercial version will probably have the same size and shape of most SLR cameras.

Figure 4.29.2   Multispectral cancer camera.

Almost all of these MSIs display the information in false-color images. That is, they use different visible colors to represent the images taken in different spectral regions, even those not in the visible. For instance, temperatures are often measured with the 12-μm infrared band and displayed as colors that range from blue to red to indicate higher and higher temperatures.

---

[75]My cancer and my project.

4.30 Ophthalmoscopes

It would seem that an instrument as simple as an ophthalmoscope would require little description, but there are many variations, and they require a bit of explication. The simple function of an ophthalmoscope is to provide an image of the interior of the eye to the doctor. This can be done with a simple source, a beam splitter, and a light projector and imager, but there are considerations of convenience, field of view, and so on. Figure 4.30.1 is a schematic of one that uses an amplitude-division beam splitter to get the light to the patient and the reflection back to the operator. Figure 4.30.2 shows a version with a spatial beam divider.

The tradeoff is not obvious. One can imagine a beam splitter with a 50% transmittance in the beam splitter version. Then the light intensity to the doctor is reduced to 25%. A simple mirror can be in the optical train as shown so that it loses no light as it illuminates the patient but has an obscuration up to 75% in the return to the doctor. But what about the struts that hold the mirror?

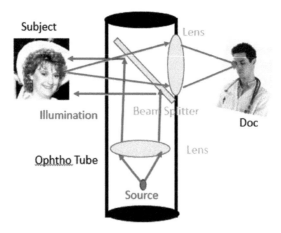

Figure 4.30.1 Amplitude division ophthalmoscope.

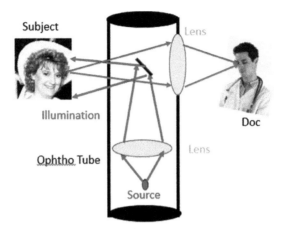

Figure 4.30.2   Spatial division ophthalmoscope.

The ophto, as I call it, was invented by Charles Babbage and by Hermann von Helmholtz independently a few years later. The earliest ophtos used external sources such as the sun for illumination. Helmholtz looked through a hole in a mirror (that required some very special seating of the patient). The advent of the incandescent bulb changed this, and the device became what the trade calls self-luminous. A battery and a bulb were put in the handle of the device that held the imaging lens.

Since the ophto is used by every physician, there has been much commercial attention and many variations: stereo, halogen bulbs, LEDs, fluorescence, miniaturization, wider fields of view, greater magnification, remote displays, and on and on. But the figures show how they work. Most ophthalmoscopes have the general configuration shown in Figure 4.30.3. There is a light source in the handle. The light is collimated by a lens and sent to a 45-degree mirror and optics in the head that direct the light to the patient and provide the image.

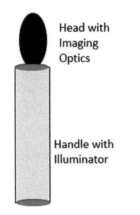

Figure 4.30.3   General ophthalmoscope configuration.

4.31 Otoscopes

Otoscopes are medical devices used for the inspection of the ear, especially the tympanum or ear drum, which is the membrane in the ear canal that separates the inside of our heads from the outside. The otoscope, or ear viewer (from the Greek words *ous* and *skopein*), is a simple device consisting of a handle and optics that project light into the ear and project a modestly magnified image back to the doctor. In its simplest form, it is just a lens that forms an image of the inner part of the ear using what light is available. It sends a collimated beam of the light reflected from the ear on the right to the eye of the doctor on the left in Figure 4.31.1. There is usually a cone called a speculum that protects the optics and positions the scope in the ear. It is shown greatly oversized in the figure.

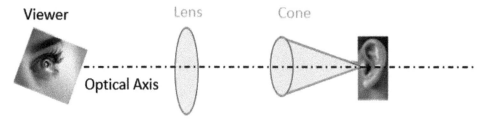

Figure 4.31.1   Simplest otoscope.

Since every doctor needs an otoscope, the market is large, and variations abound. Early versions used an incandescent bulb as a source to get a brighter image. Later versions used a brighter xenon lamp and the latest use LEDs. Some send the light to the ear by way of optical fibers to avoid any obscuration.

Magnifications range from about 2× to about 4× with slightly different lenses. Some expensive versions use television receivers and displays. This is for those elderly doctors with somewhat impaired vision and for recording and teaching purposes.

Most otoscopes have the configuration shown in Figure 4.31.2, wherein the light source is in the handle and a beam splitter is used in the head. Most manufacturers do a much better job of ergonomics than I have shown!

Video and three-dimensional versions have also been designed and marketed.[76] They incorporate much more complicated optics in the handle and the head.

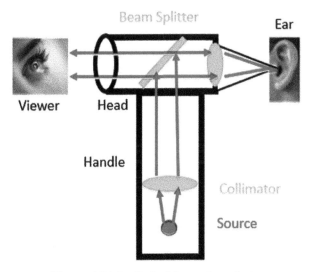

Figure 4.31.2 Optical form of most otoscopes.

---

[76]Bedard, N. et al. "Light field otoscope design for 3D in vivo imaging of the middle ear," *Biomed. Opt. Express.* **8**(1), 260–272 (2017).

4.32 Periscopes

The first periscope may have been invented and used by Johannes Gutenberg (of printing press fame) in a religious festival.[77] But the invention is usually attributed to Hippolyte Marié Davy in 1854, who first used it on a naval vessel.

Although most of us think of submarines when we think of periscopes, they are also used on tanks and other armored vehicles, by the artillery for the aiming of large munitions, and even by the infantry for spotting the enemy.

The first one was just two **45-degree mirrors**, as shown in Figure 4.32.1. It works, but the field of view is limited by the beam spread in the periscope tube. The long tube limits the spread of the beam.

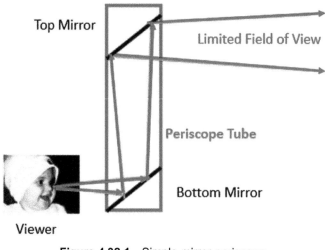

Figure 4.32.1   Simple mirror periscope.

---

[77]Wolfe, W. *Rays, Waves and Photons*, IOP Publishing (2020).

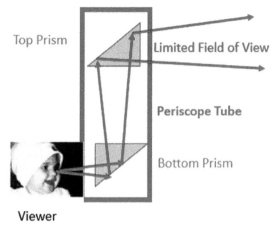

Figure 4.32.2 Prism periscope.

An alternative form uses **reflecting prisms** instead of mirrors. This provides better stability but a little more weight. The limitations on the field of view are the same, and the periscope of either design looks in only one direction.

Subsequent designs included the use of **field lenses** to relay the image down the long tube. Since every collimated beam is not completely collimated, there is always some beam spread. It is troublesome down such long tubes and solved by a system of field lenses. Each lens focuses the beam at the center of the next lens, as shown.[78] The entrance and exit lenses define the object and image fields of view.

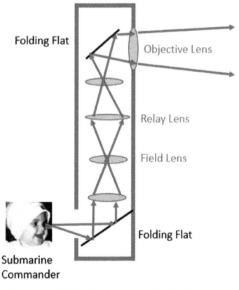

Figure 4.32.3 Periscope with field lenses.

---

[78]Smith. W. *Modern Optical Engineering*, SPIE Press (1990).

In 1936, Rudolph Grundlach patented a **rotary periscope**[79] so that the user could remain seated in one position and obtain a 360-degree view of the surroundings. The top was connected to the rest by a rotating mechanism. It seems obvious now, but so many things do after we learn about them.

One failure in the attempt to improve periscopes was a zoom system. Although the optical design was a success, the operator got confused and overestimated the distance to the dock.[80] Bam!

Periscopes *per se* are being phased out of the navy and replaced by **photonic masts**. These are not vertical columns of photons, as some oddball like me might envision, but thin, hollow posts that stick up from the sub and support mast heads that contain telescopes, night-vision cameras, and infrared sensors. Their images are relayed to the commander through cables (maybe optical cables, maybe even WiFi) down a tube that has a small area of penetration through the hull and small area above the surface.[81]

4.33 Photolithography

Photolithography has come to be the way that all electronics are made today. It was originally a way to reproduce works of art cheaply. It means "a picture on stone." The modern technique is not so different from the original. They formed an image with a greasy covering on a lime plate. Then they etched the surface away with a mild acid. The greasy image was protected. Water then washed away everything except the image, and inks could be applied.

---

[79]Zientarzewski, M. "Peryskop odwracalny wz," *Militaria i Fakty* **34**, 4 (2/2006); Grundlach, R. "Periscope for Armored Vehicles," US Patent 2130006 (1938).

[80]Fishcher, R. and Tadec-Galeb, B. *Optical System Design*, McGraw-Hill (2000).

[81]"Not your grandfathers submarine periscope," https://www.militaryaerospace.com/communications/article/16714309/not-your-grandfathers-submarine-periscope and wikipedia.com.

Modern methods start with a wafer of silicon that has been cleaned of all contaminants. A coating of so-called photoresist material is then applied evenly. It is a material that is sensitive to light so that regions that have been exposed to light can be washed away by a proper solvent. It can work the other way around, too, in which case the unexposed regions are washed away (this is the modern grease). A pattern of grid lines and other circuit elements, sometimes called a mask, is then projected onto the wafer.

Most operations use very good optics and step scanners because the resolution, the fineness of the lines, must be so good. Wide optical fields are not possible with this resolution. So very high resolution is used over relatively small fields of view, and the fields are exposed one step at a time, side by side.

Sources have evolved from mercury arc lamps with wavelengths of 400 nm to argon fluoride lasers at 193 nm. The shorter the wavelength, the narrower the line. Reflective optics are needed at these very short wavelengths; glass is opaque.

Two of the earlier optical systems used to form images on the silicon wafers were those of Dyson[82] and Ofner,[83] shown here in Figures 4.33.1 and 4.33.2. The Dyson drawing shows the mask that will be imaged onto the silicon wafer on the bottom. A beam splitter sends light from it through the optical system and down to the wafer. Dyson claims that the combination of lenses and mirror correct for all the first-order aberrations. The red arrows show the illumination. The blue ones show the imaging.

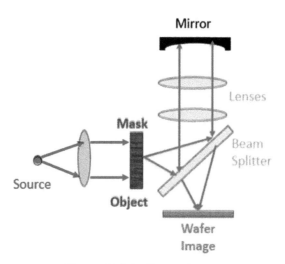

Figure 4.33.1 Dyson design.

---

[82]Dyson, J. "Unit magnification optical system without Seidel aberrations," *JOSA* **49**, 713 (1959).

[83]Ofner, A. "New concepts in projection mask aligners," *Opt. Eng.* **14**, 130 (1975) [doi: 10.1117/12.7978742].

The Ofner system reflects the image of the mask by way of an off-axis reflective optics system; the aberrations are corrected by the two off-setting angles. The same color conventions are used. The Ofner design has the advantage of an all-reflective optical system.

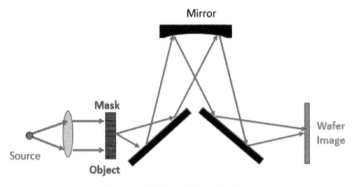

Figure 4.33.2 Ofner design.

Later systems used much more complicated optics with some twenty lens elements made of quartz. Glass is opaque in these very short wavelengths obtained with the sources described below, and all those lenses were needed to get that exquisite resolution over decent-sized fields of view.

Sources have evolved from the early mercury arc lamp to violet and ultraviolet lasers. Early systems used a mercury arc lamp as a source at a wavelength of 400 nm. Then a krypton fluoride laser at 250 nm was used, and the limit so far seems to be an argon fluoride laser at 190 nm. Each of these sources allowed for the deposition of smaller and smaller features on the wafer. They are now down to tens of nanometers.

Air becomes opaque at 180 nm. Future systems that use even shorter wavelength lasers and reflective optics will have to operate in a vacuum.

4.34 Photonic Greenhouses

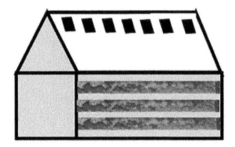

Photonic greenhouses have recently become viable. They are not greenhouses in the usual sense of glass-enclosed buildings in which to grow things. They are enclosed buildings in which to grow things, but they are not enclosed by glass. They consist of regular, opaque construction materials such as wood, concrete, and plaster. They are photonic plant nurseries, supernatural plant plants. They may even be greener greenhouses! And they can grow crops almost anywhere.

A photonic greenhouse is completely computer controlled and optically sensed. The light comes from carefully selected LEDs. They are selected based on the individual plant's growing requirements. The length of the "day" and of the "season" are determined by when the lights are on and when the crops ripen.

Depending upon the height of the crop, there can be three or four or more levels (maybe ten if they are radishes). The cultivated area is the area of a layer multiplied by the number of layers. A standard one-acre farm can become a five-acre plant factory. There are no storms inside the structure, and the computer-controlled watering system prevents drought. Any bugs—such as aphids, cutworms, locusts—can be dealt with in short order if they should somehow get in. Peter Rabbit will never get into this garden. Or Mopsy, Flopsy and Cottontail. The plants never freeze; the temperature is always in the right range. This setup does require energy, as opposed to the sun of glass greenhouses, but that can come from solar panels on the roof.

It has been found that ultraviolet light (<380 nm) is not good for plants.[84] Fortunately for them, glass houses do not transmit in this region, but neither do wood or concrete houses! Violet light (380 to 450 nm) enhances the aroma, taste, and antioxidants in some plants. Blue light (450 to 470 nm) enhances the production of chlorophyl and photosynthesis. Most plant leaves and stems reflect green light; that is why they look green. It has little effect, but it does have a small influence on their growth and chlorophyl production.[85]

Botanists and other scientists are gradually learning more and more about the exact colors, the wavebands, that enhance or deter different crops. We will eventually learn exactly what wavelength regions do what to which and when to apply them, and fully optimize these photonic growth factories.

I see this as the answer to the pessimists who claim that our population is outgrowing our ability to grow food.[86] Photonic greenhouses can be located anywhere there is sufficient power (including solar panels) and water available.

---

[84]There are several online sources for this, including Happyhydro.com.
[85]Ibid.
[86]The World Health Organization, news.un.org.

4.35 Plumbing Snakes

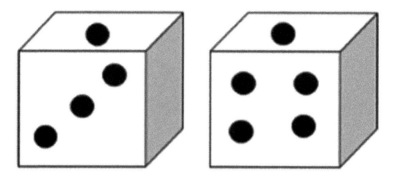

Some plumbing snakes are now equipped with optics that show what is in the pipe. These are snakes with snake eyes. They include residential, industrial, and municipal snakes or pipe probes.

In a sense, these are giant endoscopes. The optical principles are the same, but the details are a little different. Video plumbing snakes, as they are often called, consist of one or more LEDs that illuminate the area of the pipe and a digital camera with a million or so pixels and a field of view of about 20 degrees, as shown in Figure 4.35.1. The image obtained by the camera is relayed to the receiver by way of fiber optic cables because they are more efficient than copper and WiFi does not work in metallic pipes that have bends in them.

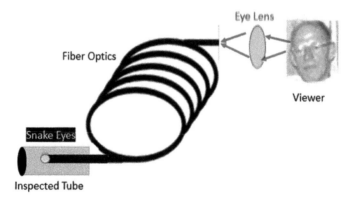

Figure 4.35.1   Plumbing snake representation.

The heads of these snakes also have transmitters so that they can be located. There are no rattles at the rear end of these snakes, just LED displays.

I have done a moderate search and find that they range in cost from under $100 to over $2500. But I do not know why. Those are economic, not optical, concepts.

4.36 Polarimeters

Polarimeters are used to measure the degree of rotation that is generated by certain so-called optically active substances. They are not used to measure the heat of a polar bear's seat, like bolometers.

Light from a source such as an LED is collimated and then polarized by one of the techniques discussed in Section 3.12. It is then passed through the sample and through another polarizer. This time it is considered an analyzer. The light is focused onto a detector. The analyzer is rotated until a maximum transmission is found. That determines the amount of angular polarization rotation that is generated by the optically active material. It is shown schematically in Figure 4.36.1. In the figure, unpolarized light is indicated by the combination of a dot and an arrow, indicating that light is polarized both into (dot) and in the plane of the paper (arrow).

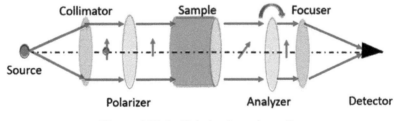

Figure 4.36.1 Polarimeter schematic.

Polarimetry is used in many industries, including pharmacology, dietetics, and chemical analytics. In dietetics, for instance, sugar is a known optically active material. It is either dextrose or levulose sugar—right- or left-handed sugar. Polarimetric tests can determine the amount of each in various foodstuffs.

4.37 Printers and Scanners

Printers, copiers and scanners all work on the same principle. They convert a two-dimensional, colored, digital representation to a two-dimensional, colored hard copy. A printer uses the digital information from computer memory. Copiers and scanners need to scan the original hard copy to get the digital information before they print it if they do.

A **scanner** uses linear arrays of detectors that scan over an illuminated hard copy, usually paper, with each of three different detectors recording the intensity of its color at each pixel. This is all recorded in digital form in memory to be used by the printer or saved in the computer. Figure 4.37.1 shows this diagrammatically. Three rows of detectors that sense in three different colors scan vertically over the text (that is actually the first part of this discussion). Each records the intensity of that color at that position and stores it in computer memory. A copier is the same as a scanner. Both a copier and a scanner save the information to memory. A **printer** operates with the information from computer memory. Figure 4.37.1 illustrates a scanner in operation.

Printers are now of two main types: ink jet and laser. (The dot-matrix variety are passé. They pounded an ink-soaked paper with the proper key. They were noisy, slow, and inefficient.) Laser printers use a photosensitive

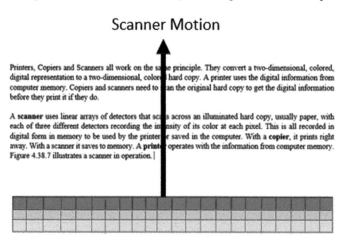

Figure 4.37.1   Scanner schematic.

drum and toner to fuse appropriate colors to the paper. Ink jet printers squirt colored inks on it.

Three-dimensional printers are of the ink jet variety. Imagine this: you scan a document and print it. Leave the paper in its same position and print something else on top of it. Repeat and repeat several times. You have built up a three-dimensional image with layers of ink in different positions on the paper at different levels. That's a three-dimensional image. It may be a thin one, but it is 3D.

But now, instead of ink, use plastic. After maybe 1000 layers, you might have made a toy car. That is the basis of three-dimensional printing, also called additive manufacturing. The inks have been replaced by plastics, aluminum, copper, resins, human flesh, and more.

There are some advantages in additive manufacturing on a commercial scale. One can go directly from a computer-aided three-dimensional design to a three-dimensional realization. No material is wasted, as with the usual design process, and no molds or other fixtures are required. Entire bikes, human ears, bikinis, Audi parts, and foods such as pizza and candy have been printed this way.

Don't squirm at food being ejected from a printer. It happens with McDonald's milkshakes all the time.

A scanner that is not attached to your computer is the one at the grocery store or almost any other store. It scans the product codes on your purchases, adds the cost to your total, and tells the store computer to reorder. It is called a bar code reader. It was invented by Bernard Silver and Norman Woodland in the sands of Florida. Woodland conceived of the idea of using narrow and wide lines to represent the dots and dashes of Morse code, and it went from there. The lines became circles so that they could be read from any direction, but they were still wide and narrow. The complicated versions like the one shown in Figure 4.37.2 read the corners for orientation and then black and white pixels like the Morse code. The patterns have been standardized, and the pattern is called a Universal Product Code (UPC).

I think that one day we will be able to find things in these stores using our smart phones to interrogate their inventories.

Figure 4.37.2   QR code.

4.38 Projectors

There are several types of projectors, although modern ones seem to have gravitated to the digital light projector (DLP) type discussed below.

The good old slide projector projected images of Kodak (or other) slides onto viewing screens. They consisted of a bright incandescent bulb, a lens to collimate the light from it onto a slide, and a lens to project an image of the slide on a screen. The collimating lens usually had a filter on it to block the copious infrared radiation from the bulb so that the slide did not get hot. The slides were introduced one at a time from a carousel or similar device. The optical system is portrayed in Figure 4.38.1.

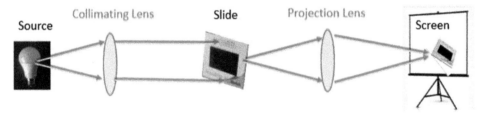

Figure 4.38.1   Slide projector schematic.

Computer projectors are mainly like the DLP devised by Larry Hornbeck of Texas Instruments in 1987.[87] It consists of millions of tiny micromirrors. Each pixel in the image corresponds to one of these micromirrors. The mirror is vibrated from the *on* position where it reflects light to the screen to an *off* position where it sends the light to a beam dump (light absorber). The ratio of the on time to the off time determines the intensity of light for that pixel. Color is produced either with a broadband source and a filter wheel or with three different colored sources. Early systems used the filter and a xenon lamp; later systems use three diode

---

[87]Hornbeck, L. "Spatial Light Modulator and Method," US patent 5,061,049 (1990).

lasers. The colors are displayed in sequence. This all happens within the integration time of the eye.

Classic movie projectors operated on the same principle as the slide projector A high-intensity source was collimated and shone on a cell of the movie reel. An image of that cell was projected onto the screen. The reels would spin; the cells would flash by faster than the flicker fusion rate, and a movie was produced.

Modern movie projectors are versions of the DLP. However, the computer feeds the projector one frame at a time faster than the eye can follow. Modern image and movie projectors do not use film. They project computer-produced digital images.

Most lecturers still request the *next slide*. Somehow *next video* does not have the same ring. It may just be habit.

4.39 Radiative Coolers

There are many applications and variations of radiative coolers. Some provide low temperatures on spacecraft to cool detectors; some control the entire craft; and some are on our terrestrial buildings.

One illustrative example is my **birdbath**, shown above. It freezes at night in the winter in my backyard even though the air temperature is about 40°F. That is because it is exchanging radiation with the sky above it, which is cold. In Tucson, there is low humidity, clear air and cold skies. This radiative phenomenon can be exploited to accomplish the applications cited above.

My **house** in Tucson is somewhat thermally controlled. It has solar panels on about one third of the roof. They absorb the sunlight and keep it from coming inside (and use it to keep my electricity bill low). Another third is red tiles. They are not helping, but the third section is painted white. White paint is really very good. In the daytime it reflects most of the sunlight, and at night it emits copiously in the infrared. Paint is highly reflecting in the visible but highly emitting in the infrared, where ambient temperature radiation peaks.

Figure 4.39.1 Home sweet home.

Radiative space coolers used to cool infrared detectors to low temperatures are of relatively simple design, as shown in Figure 4.39.2. A small patch views the coolth of outer space and is shielded from the sun and the spacecraft by a structure that resembles a simple baffle.[88] The scattered sunlight is not as critical in this application as it is in telescope baffles. There is enough cooling even if some sunlight is scattered to the cold patch. The outer cone keeps the sun off the inner cone that contains the cold spot. These devices can cool the cold spot to as low as about 80 K.

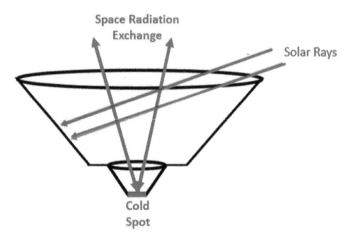

Figure 4.39.2 Radiative cooler concept.

---

[88]Wolfe W. and Zissis, G. *The Infrared Handbook*, US Government Printing Office (1978).

Spacecraft temperature control is performed in part by shields made of the proper materials. One way to shield materials that keep things cool is to reflect sunlight and emit infrared radiation, just like the paint on my roof. These were, unfortunately, materials with low alpha over epsilon α/ε ratios (unfortunate because Kirchhoff's law says that α/ε is 1). But this is a special α/ε. It is the absorption in the visible region and the emission in the infrared. One such material was developed by two engineers at Lockheed for use in a spacecraft, called the Optical Solar Reflector.[89] It was a plate of fused silica (silicon dioxide) with a reflective coating on the back, as illustrated in Figure 4.39.3. In the visible, the transparent silica transmitted the sunlight to the reflector that reflected it back to the sky. In the infrared, the opaque silica emitted the heat from the ambient temperature.

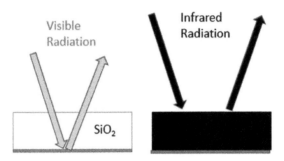

Figure 4.39.3 Optical solar reflector.

One commercial application of radiative cooling here on Earth is radiatively cooling air conditioners, cooling the rejection portion.[90] The basic concept is the same as one used by the Lockheed engineers.

A newer version of the paint was developed at Purdue University.[91] They claim 98% reflectivity throughout the visible and out to about 2 μm using a formulation of calcium carbonate. They also claim that a house with a 1000-square-foot roof will save some 10 kilowatts of power. That is like having an a/c unit at work.

Appendix A.4.39 supports this calculation using their values for the paint and some reasonable approximations.

---

[89]Marshall K. and Breuch, R. "Optical solar reflector - A highly stable, low alph/epsilon spacecraft thermal control surface," *J. Spacecraft and Rockets* **5**, 1051 (1968).
[90]Skycool, Skycoolsystems.com.
[91]Wiles, K. "The whitest paint is here – and it's the coolest. Literally," Purdue.edu.

4.40 Radiometers

These are not devices that measure radios; they are not radio meters. They measure radiation; they are radiation meters. There are two types: electrical substitution and comparison devices. The first compares the measured radiation with an electrically generated input that is part of the instrument. The second compares the input radiation with an internal source.

The ubiquitous Crookes radiometer, shown above, is sold in almost every planetarium. It is described as an educational aid by Wikipedia.[92] I once used it that way. But first let me describe it a bit more. It consists of a glass envelope that contains the structure. It has a set of vanes suspended on the top of a spindle that are black (absorptive) on one side and silvered (reflective) on the other. The vanes are free to rotate. They rotate in either of two directions, depending on the vacuum. The ones sold in the stores have a modest vacuum. As a result, the vanes rotate silver side first. The black sides absorb more of the solar radiation and get warmer. They emit more molecules, and those molecules are more active because they are warmer. They move around and bounce against the black side. They push the side forward. In a really good vacuum, such as a Nichols radiometer,[93] the rotation is in the other direction. There are no ejected molecules, and the rotation is caused by photon momentum exchange. When a photon hits the black side, it is absorbed and transfers one unit of momentum to the vane. If it hits the other, reflective side,

[92]Wikipedia. "Crookes radiometer," https://en.m.wikipedia.org/wiki/Crookes_radiometer.
[93]Nichols, E. and Hull, G. "The pressure due to radiation," *Astrophysical J.* **17**, 315–351 (1903).

it bounces back and provides two units of momenta. The reflective side is pushed forward more than the absorptive side.

My educational experience involved quizzing one of my students who expected to major in radiometry and already had some experience in it. I placed a Crookes radiometer on the windowsill and asked him to explain its motion. I inadvertently gave it a twist so that it rotated black side first. He started to give the photon momentum explanation when the vanes slowed, and stopped, and started to go the other way. He quickly said, _On the other hand...._ He knew both explanations.

The **electrical substitution radiometer** is shown in Figure 4.40.1. Radiation shines in from the left, past the shutter when it is open, and into the cavity where it is absorbed. The resultant increase in temperature is sensed by the detector. The shutter is used to then block the radiation. The cavity is usually conical in shape to aid in the absorption of radiation; it is coated with the best black available. After the input radiation level has been recorded the shutter is closed, and the heaters are used to generate the same electrical signal. It is the electrical equivalent of radiation.

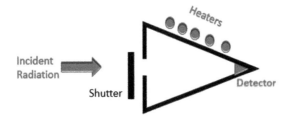

Figure 4.40.1 Electrical substitution radiometer schematic.

The **comparison radiometer** has much the same configuration, as shown in Figure 4.40.2 A chopper is inserted behind the aperture at an angle. There is no shutter. There is still a small aperture and a conical configuration to maximize the absorption of radiation. But now, the detector measures the alternate inputs from the outside, the unknown, and the internal calibration source, a small blackbody simulator.

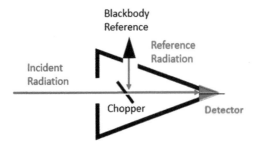

Figure 4.40.2 Comparison radiometer.

Electrical substitution radiometers have been used mostly for technical reasons. One such experiment was the measurement of the Stefan–Boltzmann constant.[94] Comparison radiometers come in many forms and have many uses. One is the forehead temperature thermometer. Many are used in industry to assess the health of machinery of all sorts. Malfunctions in machinery are often associated with an increase in temperature. Weather satellites use radiometers to measure the temperature of cloud tops. Remote sensing satellites use spectral radiometers to assess crops, oceans, and cities.

4.41 Rangefinders

Optical rangefinders for years have been based on triangulation. Only lately have they utilized pulsed laser systems. Triangulation devices are of two types: coincidence rangefinders and stereoscopic ones.

Coincidence rangefinders receive images from the object with mirrors about 1–2 meters apart. These two images come from two different angles and if simply superimposed would be indecipherable. But if one of them is rotated appropriately, they can be made to overlap and be congruent and clear. That rotation is a measure of the distance. It is accomplished by the compensator shown in light blue in Figure 4.41.1. The angular extent of its motion is enough to calculate the range.

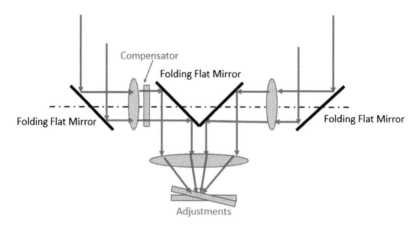

Figure 4.41.1   Triangulation rangefinder schematic.

[94]Blevin, W. and Brown, W. "A precise measurement of the Stefan–Boltzmann constant," *Metrologia* **7**, 1 (1971).

This is the same principle on which human stereoscopic vision is based. It is called parallax.

Stereoscopic rangefinders use the same procedure and principle, but they depend on the operator to superimpose the images.

So-called **stadia techniques** use the size of the image, its distance, and the size of the object. The ratio of the image height to the object height is the same as the ratio of the image distance to the object distance. Both are equal to the magnification of the optical system. This is shown in Figure 4.41.2. The image height h and image distance i are known. The stadia pole is marked; it is the object height H. It is also known. The object distance, the range, can be calculated from these three values. The word "stadia," which has come to mean graduated rod, comes from the plural of Greek *stadion*, a distance of about 190 meters, or 600 feet.

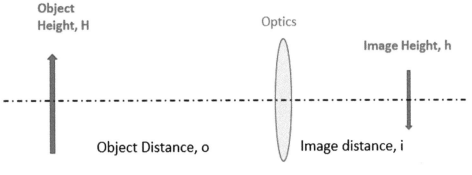

Figure 4.41.2 Stadia.

The modern technique uses a **pulsed laser**: shoot out a laser pulse and measure the length of time it takes to hit the target and come back. Then sense the return signal with an appropriate detector. This is also called **lidar** and is discussed in more detail in Section 4.24. These devices are gradually replacing the older ones since they are at least as accurate and much more compact. They use diode lasers. They range in cost from about one to several hundred dollars, depending on the accuracy, range, ruggedness and usage.

Another technique uses GPS to determine your position and that of your intended object. This is not as accurate as the other techniques. It is used by golfers, hunters, birders, and other outdoor people.

4.42 Reflectometers

Reflectometers measure the several different types of reflection from materials. They are used in the laboratory by radiometrists like me. They have no direct application to our everyday affairs, but they are essential to the design and development of the optical devices we use.

Some measure specular reflection (reflection angle equals incident angle); some measure diffuse reflection (all the light that goes all over the overlying hemisphere); and some measure goniometric reflection (all possible combinations of angles of incidence and reflection).

There are three main devices for measuring **specular reflection**: substitution, the John Strong VW method, and the much more complicated Bennett–Koehler instrument.

Specular reflection is the way light is reflected from a mirror. It obeys the law of reflection; the angle of reflection is equal to the angle of incidence.

The **substitution method** of measuring specular reflection is shown in Figure 4.42.2. A beam of light is focused onto a detector (on the right), and the radiation level is measured. The mirror to be tested is inserted, the detector is moved to the new location (below), and the level is measured again. The ratio is the specular reflectivity of the sample. Care must be taken to ensure that any background effects are eliminated or are the same in the two positions.

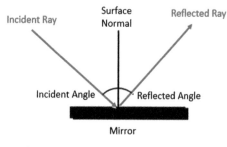

Figure 4.42.1 Specular reflection.

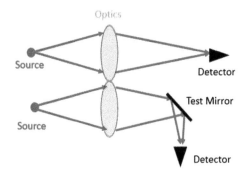

Figure 4.42.2 Substitution geometry.

The **Strong VW** method, named after John Strong, is almost as simple. As shown in Figure 4.42.3, light (in red) is shone from the upper left onto the instrument or reference mirror on the bottom (black) with no test mirror. The level is measured by the detector, shown as the black triangle. Then the test mirror (gray) is inserted, and the instrment mirror is moved to the position above. In the first case, the big red V case, the reflection measured is that of the instrument mirror. In the second case, the W case, the total reflection is once from the reference mirror and twice from the test mirror. The ratio will give twice the reflectivity of the test mirror. It is twice as sensitive as the simple substitution method, and the background remains constant.

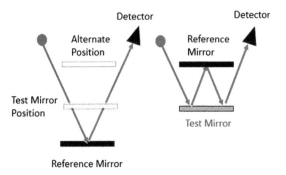

Figure 4.42.3 Strong VW schematic.

The **Bennett–Koehler** method[95] is much more complicated and involves over a dozen mirrors. It is said to be able to measure over a wider range of incidence angles, but it does not add to the comprehension of specular reflectivity measurements. Look it up if you really care.[96]

---

[95]Bennett H. and Koehler, W. "Precision measurement of absolute specular reflectance with minimized systematic error," *JOSA* **50**, 1 (1960).

[96]Wolfe, W. *Introduction to Radiometry*, SPIE Press (1998) [doi: 10.1117/3.287476].

Hemispherical reflectivity relates to the amount of light that is reflected into an overlying hemisphere. It almost always refers to the amount of light reflected into a hemisphere from a collimated or focused incident directional beam. It is therefore referred to as **directional**-hemispherical reflectivity.

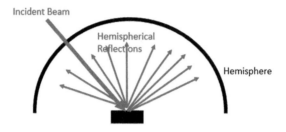

Figure 4.42.4   Directional-hemispherical reflectivity.

The **Coblentz hemisphere** is almost the instrumental personification of directional hemispheric reflection. It consists of a hemisphere in which there is a small hole for the entry beam, the sample (shown in blue in Figure 4.42.5) to be measured, and a detector (in black) right next to the sample. The light enters the hole and impinges on the sample. It is reflected in all directions to the hemisphere that is highly reflecting and specular and returns to the detector (since both are close to the center). The system must be calibrated by some known reflector to account for the reflectivity of the surface of the hemisphere. A series of holes or a slit is necessary to allow different directions of incidence. It is named after W. W. Coblentz, a famous scientist who once worked at the US National Bureau of Standards.

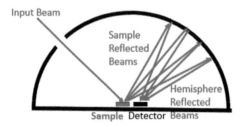

Figure 4.42.5   Coblentz hemisphere.

In what I call the **clamshells**, two paraboloidal reflectors are used together with a deflection mirror and a detector. The light enters from the left, at least in the diagram, and is reflected up to the top mirror and down to the sample (in **blue**). The light that enters is collimated. It is focused by the parabola on the sample. The sample reflects (scatters) the light in all upward directions to the same, top parabola. The sample is at its focus; it therefore reflects the light in a collimated beam to the lower parabola, which then focuses it onto the detector (in **black**). The position of the folding mirror determines the angle of incidence.

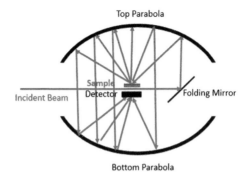

Figure 4.42.6 Clamshell arrangement.

The third of the standard methods of measuring directional hemispheric reflectance is the **integrating sphere** method. It may be the most frequently used. Light enters from the left and is incident upon the reference. It is diffusely reflected to the entire sphere, and the light is reflected by the highly reflecting, diffuse interior to the detector shown at the top. It is assumed that all the light that is reflected by the sample gets to the detector, i.e., that the interior coating is 100% reflective. Of course, this is not true, and calibration must be carried out to determine the extent to which the value is not 100%.

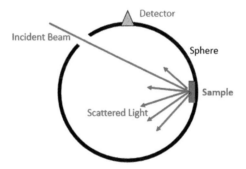

Figure 4.42.7 Integrating sphere.

In general, light can be incident from any pair of angles, and a certain amount of it is reflected into any other pair of angles. This four-angle, goniometric reflection is known as bidirectional reflectivity. The **bidirectional reflectance distribution function (BRDF)** was defined by Fred Nicodemus as the ratio of the reflected radiance to the incident irradiance.[97] It is a function of two directions (hence the name). The two directions are the azimuth and

---

[97]Nicodemus, F. "Directional reflectance and emissivity of an opaque surface," *Appl. Opt.* **4**, 767 (1965).

elevation angles of the incident ray and the azimuth and elevation angles of the exiting rays. It is two directions but four angles. It is the ratio of a radiance to an irradiance and therefore has the units of reciprocal steradians (but is still dimensionless), and it has the unusual property of a reflectivity that it can have a theoretical maximum value of infinity.

The first such goniometers were devices that mimicked the geometry of the definition. A schematic example is shown in Figure 4.42.8. The source can wander all over the hemisphere, as can the receiver, and they thereby generate all the required sets of angles. It is straightforward but tedious and limited to relatively small sources and detectors.

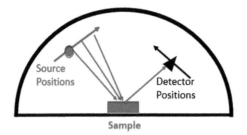

Figure 4.42.8   Definition-based goniometric reflectometer.

Larry Brooks designed and built a device based on a different principle that was suggested by Doug Goodman.[98] The source and receiver are both on a horizontal platform in a fixed position. The sample is turned in azimuth and elevation. It was necessary because our sources were large lasers.

The idea is shown conceptually in Figure 4.42.9. Imagine that the source and receiver are both located directly to the left. The top arrangement portrays

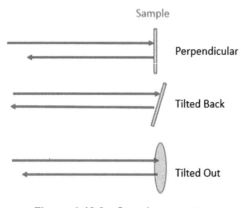

Figure 4.42.9   Sample geometry.

---

[98]Brooks, L. "Microprocessor Based Instrumentation for BSDF Measurements from Visible to Far IR," Dissertation, The University of Arizona (1982).

the sample perpendicular to the incoming beam. Then both the incidence angle and the reflection angle are zero (with respect to the surface normal). The middle arrangement shows the sample tipped back a few degrees. Then both the incident and reflected angles are about 10 degrees vertical with respect to the normal. The third arrangement shows the sample vertical but tipped with respect to the plane of the paper. Then the angles of incidence and reflection are about 10 degrees horizontal. It takes a bit of three-dimensional thinking, but all azimuth and elevation angles can be generated this way. That was Doug's contribution and Larry's implementation.

A schematic of how Larry did it is shown in Figure 4.42.10. We called it the AZSCAT, standing for Arizona Scatterometer, and we measured many space materials with it. The FIR laser on the lower platform emitted at 118 μm. The detector is not shown in its usual measurement position where it receives reflected light.

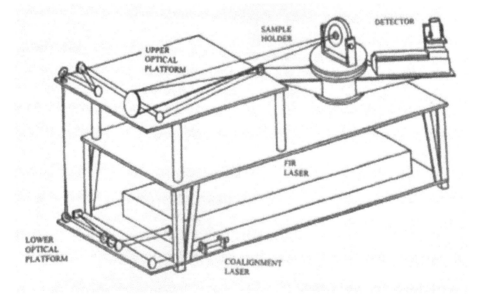

Figure 4.42.10 AZSCAT schematic (reprinted from Ref. 98).

The introductory photograph shows the top of AZSCAT with three lasers, attenuators, and directional mirrors, as well as the critical two-axis sample mount. Give Larry credit for the photo showing the red helium-neon laser shining on the sample.

4.43 Remote Sensors

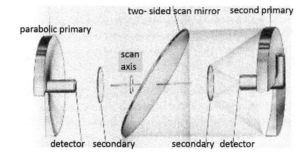

Almost everything we do is remote sensing. Taste and touch are not, but hearing, smelling and seeing are. A sixth sense can surely not be tactile. The phrase *remote sensing* has come to have a special meaning: the sensing of our natural environment from aircraft, satellites, or drones.[99] In this book, it is optical sensing by these means. Military reconnaissance is a different but similar category. It is chiefly interested in human activities and structures.

Satellite remote sensors are mostly multispectral imagers, but they also include sounders and lidars (see Section 4.29). There are many remote sensors high above our heads. Too many to list, but the basic concepts can be described. Most all of them are pushbroom scanners. See below for a discussion of the two types of broom scanners, push and whisk. They use several linear arrays that sense in several spectral bands. Table 4.43.1 shows the properties of five of them.

Table 4.43.1   Remote sensor properties.

Unit	Orbit	Altitude (km)	Footprint (m)	Swath width (km)	Visible bands	NIR	FIR
Landsat 8	polar	705	15–100	185	8	1	2
MODIS	polar	705	250–1000	2330	many	1	2
GeoEye	polar	770	0.46–1.84	100 × 100	5	1	0
IKONOS	polar	700	0.82/3.8	11.3	4	1	0
WorldView-4	polar	617	0.31/1.24/3.7/30	13.1	14	8	0

Landsat is the oldest of the remote-sensing multispectral imagers. The first was flown in 1972. The characteristics of the most recent version, Landsat 8, are as follows. It flies in a polar orbit at an altitude of 705 km. It has seven spectral bands in the visible ranging from 435 to 800 nm and a ground footprint of 30 km. It has one panchromatic band with a footprint of 15 km

---

[99]A Review: Remote Sensing Sensors | IntechOpen, intechopen.com.

and one in the very near infrared at 1375 nm. It has two thermal bands centered at about 11 and 12 μm with a footprint of 100 meters.

MODIS, the MODerate resolution Imaging Satellite, is really two satellites. Both fly in polar orbits: one passes the equator in the morning, the other in the afternoon. They completely cover the Earth every two days with 36 spectral bands from 0.4 to 14.4 μm and footprints of 250 m to 1 km, depending on the wavelength. It uses an afocal reflective telescope to direct the radiation to four different imagers, a two-sided flat scan mirror to generate a whiskbroom pattern, and an array of mercury cadmium telluride detectors cooled by a radiative cooler.

GeoEye was one of the first commercial remote sensors, made by General Dynamics in Gilbert, AZ. It is in polar orbit. It incorporates a panchromatic band and four visible spectrum bands with footprints of 41 cm. It operates in snapshot mode, scanning an area 100 by 100 km at a range of angles from the satellite.

IKONOS was another commercial remote sensor launched by Digital-View but now defunct. The name derives from the Greek word for image, *eikon*.

WorldView-3 was launched in 2014 by the Satellite Imaging Corporation. It is, like the others listed, in a polar orbit at about 650-km altitude. It has bands in the visible and near infrared. It incorporates the so-called cavis bands, clouds, aerosols, vapor, ice and snow in the 405–2225-nm spectral region. Its ground footprints vary from 0.31 m to 30 m at nadir. The larger one is for the cavis bands that are mostly meteorological and do not require high spatial resolution. The optics include a 1.1-meter-diameter collecting mirror, a panchromatic detector array of 35,000 pixels and 9300 for the non-visible ones. The scan is pushbroom. I note that the theoretical limit of resolution is 0.32 m at nadir for the middle of the visible. That corresponds nicely with their reported footprint. It is proportionately larger for longer wavelengths.

Drones also use regular and multispectral cameras. One application is farming. The drones fly over an acre at a time, recording such things as the chlorophyl content of leaves. The spectra of the two types of chlorophyl are shown in Figure 4.43.1, where the x axis is in micrometers (the visible spectrum), and the y axis is arbitrary, but the peaks are in proper proportion to each other. Measurements at a wavelength of one of the peaks, compared to one in between, will provide information on the chlorophyl content. Similar measurements can be made for moisture, mold, and other important aspects of crop growth, health, and disease.

A typical drone can fly on a single battery charge for about an hour. During that time, it can cover about 500 acres with a 10-centimeter footprint at an altitude of 400 feet. A 1-centimeter footprint is possible at the same altitude but with better optics. The drones come equipped with the sensors in a

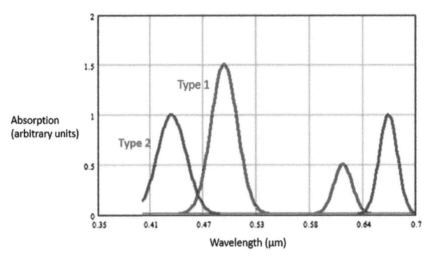

Figure 4.43.1   Approximate chlorophyl spectra.

price range of \$1500 to \$25,000.[100] I guess you get what you pay for. Better optics is one such case.

Aircraft mounted sensors are intermediate between the satellite and drone versions. They have advantages and disadvantages compared to drones and satellites.[101] They have, or are capable of, better ground resolution than satellites but about the same as drones. They can cover more territory in a single flight than drones but not as much as a satellite. They can be redirected during a flight, as can some drones, but not satellites (at least not easily and quickly).

All three types of remote sensor use one of three scan modes: snapshot, pushbroom, and whiskbroom. I think snapshot is obvious. It is what a camera does. It takes an instantaneous, two-dimensional picture. A pushbroom, illustrated in Figure 4.43.2, uses one or more linear arrays aligned perpendicular to the line of flight (blue arrow). It uses the motion of the vehicle shown by the blue arrow to generate the second dimension of the image. The whiskbroom arrangement uses one or more arrays oriented in the direction of flight that are swept side to side by the optics (black arrow) to get the second dimension. A whiskbroom scanner must rotate fast enough to make the consecutive swaths contiguous.

The introductory image is a representation of an air-to-ground scanner.[102] A single detector was used, and the two-sided mirror enabled the system to get

---

[100]Best Drones for Agriculture 2020: The Ultimate Buyer's Guide, bestdroneforthejob.com.

[101]Aerial Imagery Based on Commercial Flights as Remote . . . , ncbi.nim.nih.gov.

[102]Wolfe, W. *Opto-Mechanical Scanning Devices*, Contract report, University of Michigan College of Engineering (1958).

Figure 4.43.2 Pushbroom scan.

Figure 4.43.3 Whiskbroom scan.

two swaths perpendicular to the flight direction per mirror rotation. It did not use all 180 degrees of each pass, but it doubled the data acquisition rate. Modern versions use an array instead of a single detector.

4.44 Remote Thermometers

These are thermometers that take your temperature remotely, in the ear or from your forehead (not in your mouth or up your rear).

The ear device is double-cone shaped and is inserted part way into the ear canal. It measures the infrared radiation emission from the tympanum, the ear drum, and calculates temperature from it.

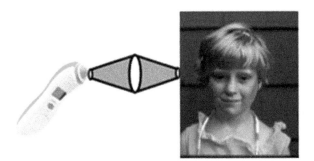

Figure 4.44.1   Representative remote thermometer.

The basic design is a simple optical system shown schematically in Figure 4.44.2. It may be just a reflective cylinder, cone channel condenser (Section 3.4), or similar reflector that collects the radiation and defines the field of view and an infrared sensor that may be a bolometer or thermopile (see Section 3.5).

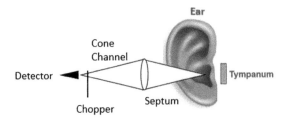

Figure 4.44.2   Representative ear thermometer.

The forehead thermometer makes contact or near contact with your forehead and operates in the same way. It can also act as an ear thermometer by removal of a cover.[103] These have been of great use during the COVID-19 virus pandemic.

All three are much easier and more comfortable to use than either oral or rectal thermometers. When used correctly, they are just as accurate as those invasive versions. Correct use, especially for the forehead thermometer, includes proper distance, perpendicularity, no sweat, and the elimination of thermal backgrounds such as sunlight.

One embodiment is a unit the size and shape of a hearing aid.[104] It is said to be useful for those who want to measure their temperature in their sleep or when they first wake up.

An average normal temperature measured orally is 98.6 °F. The rectal and tympanic versions are 0.5–1.0 °F higher. One taken on the forehead is 0.5–1.0 °F lower. It is generally considered that the rectal one is the most accurate, but it is also the most uncomfortable.[105] I am no doctor, but it seems to me that these can all be accurate, just different. It seems obvious that the temperature under your tongue is different from the one taken from your forehead or in the nether regions. So, for instance, a forehead normal temperature can be considered 97.6 °F.

4.45 Solar Panels

Solar panels are an efficient way to power your home. They make direct use of sunlight to generate electricity. They consist of panels of silicon or other

---

[103]Wohlman Non-Contact IR Thermometer, Amazon, amazon.com.

[104]Braun Thermoscan 5 Ear Thermometer – IRT6500, Braunhealthcare.com.

[105]C.S. Mott Children's Hospital. "Fever Temperatures: Accuracy and Comparison," University of Michigan, https://www.mottchildren.org/health-library/tw9223 (2020).

semiconductors and an electrical converter that changes the dc current generated by the panels to ac current that is used in our homes.

The electricity costs of my home, an all-electric 2000-square-foot home in Tucson, AZ, went from about $1500 per year to less than $250 per year with the advent of solar panels ten years ago. Almost all of those bills are administrative and tax costs.

Figure 4.45.1   My house.

It was shown elsewhere that the average amount of sunlight we receive at the surface of the Earth is just about 1000 watts per square meter. That means that we receive about 10^{20} photons per square meter per second. If every one of them turned into an electron, that is a big current: a current of 10^{20}

Figure 4.45.2   My housetop.

electrons per second! That would be about 50 amperes per square meter. But not every photon becomes an electron; only about 1 in 5, or 20%, do. That is still a little more than 10 amps per square meter, and that would be 15 amps for an average-size panel.

It may be simpler to think about it in terms of watts, but I like to describe the photon/electron picture because the numbers are awesome.

Panels are about 1.6 square meters in area; they are about 20% efficient, and there are 1600 watts of sunlight on them on the average sunny day if they face the sun directly. Each panel generates about 440 watts. The average household electricity use in the US is 914 kWhr per month.[106] That turns out to be about 1.3 kilowatts. On average, then, every house needs 3.5 such panels. But that ain't quite right. Half of the time, it is night, and some days the sun don't shine. It turns out that the average home needs 15 to 25 solar panels to fully meet its electricity needs (depending, of course, on its size, location, and usage).

Modern systems use two types of electrical devices to convert the direct current to alternating current that we use in our homes. By convention, a **converter** converts ac to dc, whereas an **inverter** converts dc to ac. They are both converters, but this language is used to differentiate which way they do it. These are inverters. I believe that someday we will have inverters but also direct-current bypasses to serve the LEDs we will have all around us. One type of inverter operates on the output of the entire array of panels as shown in Figure 4.45.3; the other one operates on each individual panel.

Figure 4.45.3 Inverter.

The basic operation of the panel itself is the conversion of photons to electrons. Photons with enough energy free the loosely bound electrons, which

---

[106]Electricity use in homes - US Energy Information, EIA, eia.gov.

then generate electrical current (see Section 3.5). The material commonly used is silicon because it is cheap and abundant, but it is not a perfect match to sunlight. Figure 4.45.4 shows the typical spectral response of silicon in blue and the normalized distribution of the sun in red.

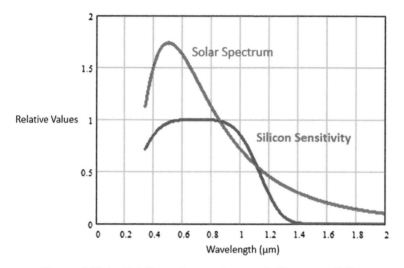

Figure 4.45.4 Relative solar spectrum and silicon sensitivity.

The refractive index of silicon is about 3.5, which means that its normal reflectivity is about 30%. Its absorption coefficient over the visible spectral region averages about 1 per cm. That means that a thickness of one centimeter absorbs only about 35% of the incoming light. For these two reasons and a few minor ones, solar panels are currently just about 20% efficient in turning sunlight into electricity. These inefficiencies of silicon sensors are gradually being overcome by the techniques described below.

One technique is to reflectorize the back of the panel to get what is effectively a double thickness. Another is the use of a thin, antireflection coating on the front. A third is the use of the heat not absorbed by the silicon, radiation with wavelengths longer than the 1.2-μm limit. Yet another is the mechanical procedure of following the sun, keeping the panel aimed perpendicular to the sun. Still others are described online.

The most drastic and perhaps most effective improvement is the use of a different material, one that matches the solar spectrum better. Perovskites are being investigated since they have adjustable bandgaps and can therefore be better matched to the solar spectrum.[107] They are readily manufactured and offer the promise of about 30% efficiency versus the 20% of silicon.[108]

---

[107]Boyd, C. B. and McGehee, M. D. "A match made in heaven: Stacking two solar cells boosts their efficiency," *The Science Breaker*, Oct. 29 (2020).
[108]*Optics and Photonics News*, OSA (November 2020).

4.46 Spectacles

Spectacles (**eyeglasses**, not chariot races), are a common optical element that almost all of us use daily. They were invented in ancient Italy in about 1200. They are used to correct image defects of the eye. These are mainly **myopia** (nearsightedness), **hyperopia** (far sightedness), **presbyopia** (old age eyes), and **astigmatism** (distortion). A myopic eye lens focuses the light in front of the retina; the hyperopic eye focuses behind it. The presbyopic eye does not accommodate the way it should; i.e., it can no longer adjust the figure of the lens to focus on nearby objects because the lens material has become too viscous and the muscles that adjust it too weak. The astigmatic eye is just not shaped right. The corrections for these are straightforward. For myopia and hyperopia, one needs a lens in front of the eye to adjust the combined focal length. For presbyopia, one needs bifocals. For astigmatism, one needs a specially shaped lens to correct for the misshapen lens.

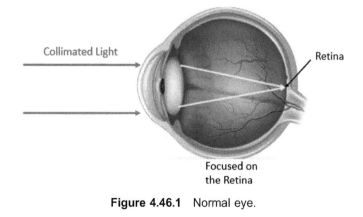

Figure 4.46.1 Normal eye.

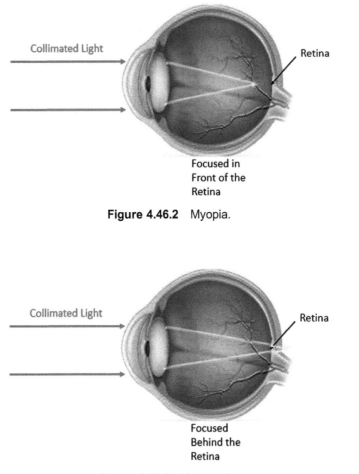

Collimated Light

Retina

Focused in
Front of the
Retina

Figure 4.46.2 Myopia.

Collimated Light

Retina

Focused
Behind the
Retina

Figure 4.46.3 Hyperopia.

The unit of **diopter** is used by eye professionals with good reason. It is the reciprocal of the focal length of the eye in meters. It is a descriptor of lens power. This is convenient since focal lengths combine as reciprocals, but lens powers in diopters add directly. A flat piece of glass has a focal point at infinity, does not focus, and has an optical power of zero diopters. It has no power to converge the light. Convex lenses have positive diopters, and vice versa. They are used to correct farsightedness by "bringing in" the focal point from beyond the retina to the retina. The relationship between focal length and diopters is shown in Figure 4.46.4. A more convenient relationship is that between the diopter and the magnification of the lens. This is shown in Figure 4.46.5.

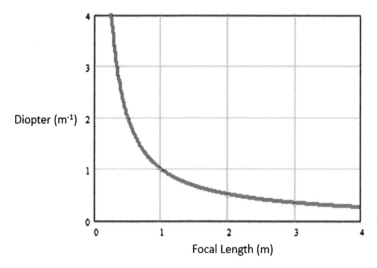

Figure 4.46.4 Diopter vs. focal length.

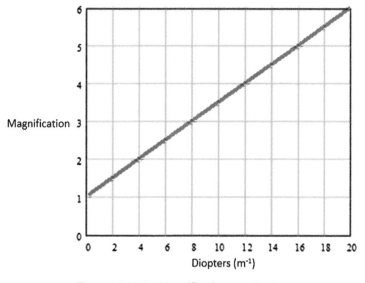

Figure 4.46.5 Magnification vs. diopter.

Bifocals were first introduced by Benjamin Franklin about 1776. They are simply spectacles with two focal lengths. The top, usually, with a low power, low diopter for distance viewing, if needed at all, and the bottom with more diopters for close viewing like reading.

Figure 4.46.6 Bifocals.

Trifocals were developed next, in 1827. The top section is used for distance viewing, the middle for intermediate viewing, and the bottom for reading.

Then came **multifocus, varifocal, progressive**, or **continuous focusing glasses** in 1907, in which there was a gradual change in focal length from top for distance to bottom for reading.

As a personal aside and maybe a useful tip, I have presbyopia and a little astigmatism. I use bifocals in general, but I have a pair of simple magnifiers for use at my computer. I found by trial and error that I needed readers with a magnification of three, that is, eight diopters, but just check the magnification at the store; leave the diopters to the optometrist. The magnification of the lower part of my bifocals was too much, and the top was too little. The readers are like baby bear's bed, "just right."

One fascinating development was **self-adjusting glasses**. They consist of a lens with flexible surfaces that contain a viscous substance. Applying pressure to such a lens can change its shape and therefore its focal length. This will not correct for astigmatism but can correct for both hyperopia and myopia. It can also correct for presbyopia by adjusting for reading after you sit down. It is especially valuable for those in regions where optometrists are few and far between. The relatively cheap glasses can be adjusted by the user and used immediately. It was invented by Joshua Silver, a British physicist.[109]

---

[109]Many online references, including thesciencebreaker.org and "Self-adjustable glasses: how one man's vision is helping others to see better," medicalnewstoday.com.

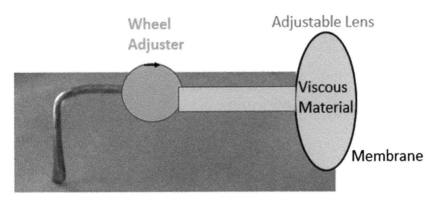

Figure 4.46.7 Self-adjusting glasses.

They have, as I wrote, a flexible outer membrane with a viscous material inside, shown in blue in Figure 4.46.7. A piston is shown in the same color and a wheel to push the piston against the membrane. Just turn the wheel back and forth until the image becomes sharp.

These have been combined with other lenses that correct for astigmatism. They are advertised on the internet as the ultimate solution. They are a combination of lenses that correct for astigmatism and the like and are combined with the self-adjusting lenses just described.[110]

Perhaps the most astonishing and most promising corrective eyeglass is not one at all. It is a device that provides vision to the blind. It is an electronic camera that sends its output to the optic nerve. It is a **bionic eye**. As of this writing, Second Sight has installed several of these in patients. They consist of a camera that takes an electronic picture that is transferred to electronics just outside the eye that sends the signal to a chip on the retina. This is the most astonishing thing to me. The chip now has 100 connections to the optic nerve. Although a 100-pixel camera is very modest, it is infinitely better than a zero-pixel camera or no eyesight at all. Recipients have reported such things as seeing an egg turn white as it cooks. They get glimpses of hallways and features of loved ones. The possible increase of pixels is a surgical challenge, the connection of more tiny wires to more optical nerve endings. It is beyond my ken, but I am optimistic that it will get better and may reach something like 20/200 vision. The internet has many videos and reports that describe this. I recommend *The Bionic Eye Prototype*, by Professor Rod Shepherd. He reports on developments leading to 1000 connections.

---

[110]Perlow, J. "Superfocus: The ultimate eyeglasses," ZDNet, https://www.zdnet.com/article/superfocus-the-ultimate-eyeglasses (May 31, 2012).

4.47 Spectrometers

Spectrometers, not specter meters, are meters that measure spectra, as the name implies. A spectrum is the intensity of light as a function of frequency. There are spectrometers, spectrographs, spectroscopes, spectroheliographs, and all sorts of spectral measuring instruments. A spectrograph is a spectrometer that uses photography or graph paper for its recording medium, whereas a spectrometer can use a dial or a digital readout, a meter. A spectroscope is a spectrometer that uses a telescope to view a spectrum. A spectroheliograph is a spectrometer that uses photography to measure spectra of the sun.

Most spectrometers are used in laboratories. They are rare in everyday use, although some smartphones now have attachments, mini-spectrometers, that can be used to assess the freshness of fruit or the amount of hemoglobin in your blood.[111]

The very first record of a spectrum may have been the rainbow that Noah saw after the flood subsided.[112]

Perhaps the very **first spectrometer** was that of Newton, who first showed that light consisted of colors. His spectrometer was a hole in a window shade, a prism, and a wall. He showed the colors, but he did not make any measurements:

> In a very dark Chamber, at a round Hole, about one third Part of an Inch broad, made in the Shut of a Window, I placed a Glass Prism, whereby the Beam of the Sun's Light, which came in at that Hole, might be refracted upwards toward the opposite Wall of the Chamber, and there form a colour'd Image of the Sun.[113]

---

[111]Several online sites, stellarnet.us, for instance.
[112]Genesis 9:13.
[113]Newton, I. *Opticks*, London (1704).

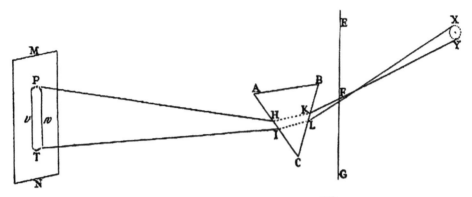

Figure 4.47.1 Newton's diagram.[112]

The standard, **classical spectrometer** has a source, a slit, a collimating lens, a prism, a focusing lens, and a detector, as shown in Figure 4.47.2. The white light (multicolored, polychromatic light) enters the slit and is sent by the lens as a collimated beam to the prism. The prism refracts the light different amounts according to its wavelengths. The resulting spectrum is directed by the focusing lens to the exit slit. Behind it is the detector. As the prism is rotated, the various colors pass through the exit slit, and the intensity of each color is recorded by the detector in sequence. I used one of these so-called single-pass, single-beam devices in the late 1950s. I remember adjusting the spectrometer for the slit opening to account for the change in intensity of the source with wavelength by a trial-and-error method of winding a string on a conical cylinder. It is tedious. It is the sort of procedure that only a grad student would do. I was that grad student.

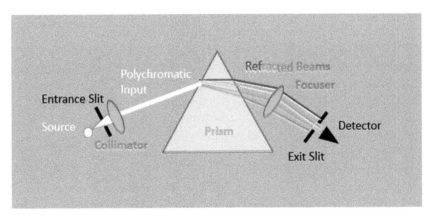

Figure 4.47.2 Classical spectrometer.

That is why they invented **double-beam** instruments. The prism spreads the spectrum and shines the beam shown in red in Figure 4.47.3 at a beam divider. It sends one part, about half, to the sample cell and the other to the reference cell and on to the detectors, where the ratio is obtained. No more string! This ratio adjusts for the variation in the source output and any environmental effects. The disperser can be a prism or grating; the beam divider can be a polygon as shown, a beam splitter, or a rotating sector with reflecting and transmitting sections. The reference cell is usually just air but can also be any inert gas.

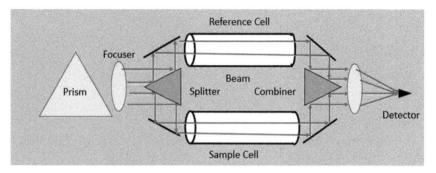

Figure 4.47.3   Double-beam spectrometer.

A more recent development is the **array spectrometer**. It does away with the exit slit by replacing it with a long, linear array of detectors. Then each individual detector senses the amount of radiation on it from the light of that appropriate color. Array spectrometers are the bases of many of the so-called multi- and hyperspectral satellite imaging systems.

Spectral lines are not lines. There is an excellent book on them called *The Shift and Shape of Spectral Lines*.[114] A true mathematical line does not have any shape; it is infinitely narrow. But in optics we call these finite-width peaks "spectra lines." The same fiction is used with point sources (which are not points).

The minimum line width, its diameter at half its maximum, is determined by the time of the transition of an electron or vibration of a molecule. The mathematics of this is in the Appendix, but the essence is this. The Heisenberg uncertainty principle states that the product of energy and time is a small constant about equal to the Planck constant. Since the energy of a photon is h times its frequency, then the product of the (uncertainties) in frequency and time is about 1. For instance, if the electronic transition takes 1 microsecond, the minimum frequency spread is about one million cycles per second.

The simplest line width is due to Doppler broadening of the charges moving back and forth with respect to the observer. This leads to the Gaussian distribution shown in Figure 4.47.4. The Lorentzian distribution is pressure broadening, the collision of the charges. It is also shown in Figure 4.47.4. All four types are shown in the figure. The theoretical one due to finite time and the uncertainty principle is in red. The Gaussian due to Doppler broadening is in **blue**. The Lorentzian is in green. Proximity broadening due to the interaction of other molecules is in **black**.

---

[114]Breene, R. *The Shift and Shape of Spectral Lines*, Pergamon (1961).

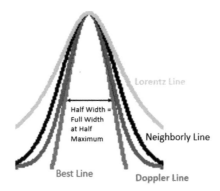

Figure 4.47.4 Spectral line shapes.

Spectral resolution is an important property of a spectrometer. As noted in Section 1.13, it is usually specified as spectral resolving power and is the central wavelength of the line divided by the width of the line at its half height $\lambda/\Delta\lambda$. It is an American definition: bigger is better.

Free spectral range is a measure of the distance between spectral lines. It is usually specified as FSR since $\delta\lambda$ and $\Delta\lambda$ have other meanings.

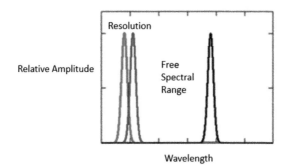

Figure 4.47.5 Resolving power and free spectral range.

Throughput and **multiplex advantages** are important in the consideration of more efficient spectrometers. The multiplex advantage refers to measuring the intensity of light at all wavelengths at the same time rather than wavelength by wavelength, as with the single-slit spectrometer. The throughput advantage refers to how much light you can get at a time. Obviously, the entrance slit of the classic spectrometer is a rather severe limitation. Multiplex and throughput advantages are sometimes referred to as the Felgett and Jacquinot advantages after the two men who pioneered them: Peter Fellgett[115] and Pierre Jacquinot.[116]

---

[115]Fellgett P. "Theory of Infra-Red Sensitivities and its Application to Investigations of Stellar Radiation in the Near Infra-Red," Dissertation, University of Reading (1949).

[116]Jacquinot, P. "The luminosity of spectrometers with prisms, gratings or Fabry-Perot etalons," *JOSA* **44**, 761–765 (1954).

An array spectrometer enjoys the multiplex advantage but not the throughput one.

The Fourier transform infrared (FTIR) spectrometer uses an interferometer as a spectrometer. The most common is a Michelson. It has both advantages at the expense of considerable calculation (which is cheap and easy today).

Consider how it works by examining what happens with just a few wavelengths in a Michelson interferometer. I have shown just red, green, and blue in Figure 4.47.6. The three of them go to the beam splitter and are separated into the two arms, one up and the other over. They are reflected back and joined by the beam splitter, which then acts as a beam combiner. They are combined below it, where they interfere. The mirror on the right moves. When it is in a position such that the red beam components constructively interfere, the other two colors do not. But the mirror moves so that in another position the blue beam components interfere constructively, and the others do not. This gives rise to a pattern called an interference pattern. It is a plot of the intensity of the combined beams as a function of the mirror position. It has all the information of what color beams have what intensity. It is shown for these three waves in Figure 4.47.7. It is just all mixed up. This pattern can be unscrambled by the use of a mathematical procedure called the Fourier transform, which is discussed in Appendix A.6.4.

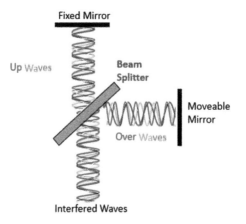

Figure 4.47.6   Waves in a Michelson interferometer.

Figure 4.47.7   The interferogram.

4.48 Spy Satellites

Spying on the enemy has a long history, but doing it from satellites started in earnest soon after Gary Powers was shot down in his U-2. The Soviets would not agree to President Eisenhower's "open skies," proposal, so the US flew U-2 aircraft at a 70,000-foot altitude over the Soviet Union, assuming they could not get to the planes. In 1960, they did; they shot down Francis Gary Powers with a ground-to-air missile. It really screwed up negotiations, so the US took to space.[117]

The CORONA program was started in 1959 (maybe in anticipation of the inevitable) by the NRO, the National Reconnaissance Office of the CIA. It utilized satellites with special photographic cameras to reconnoiter the Soviet Union and China.[118] They orbited at altitudes ranging from 75 to 100 miles, 120 to 160 km.

The cameras had a 610-mm focal length and used 70-mm film with a resolution of 170 line-pairs per mm. That translates to just under 5 microradians. At the two altitudes cited above, the ground footprints are about one half to three quarters of a meter. The lenses were made by Itek, the system by Kodak. They were initially a form of a Tessar triplet and then a Petzval version (see Section 3.10). They were big—almost 20 centimeters in diameter, and five times as long.

The exposed film was enclosed in a re-entry capsule that protected the film from the extreme heat and ablation of re-entry. Then a parachute was deployed, and the film was supposed to be snatched by an airplane of the USAF. That did not always happen. The last CORONA mission was labeled

---

[117]Wikipedia. "1960 U-2 incident," https://en.wikipedia.org/wiki/1960_U-2_incident.
[118]Wikipedia. "CORONA (satellite)," https://en.wikipedia.org/wiki/CORONA_(satellite).

KH-4B (for keyhole, a spying technique) in 1972, the source of the introductory image. It was replaced by the GAMBIT and HEXAGON (Big Bird) programs with satellites designated KH-7 and KH-8, but they were still-film systems that relied on bucket snatching. (KH-5 and KH-6 were short-lived film systems.)

The first digital system was KH-11, built by Lockheed in 1976.[119] Although the details and imagery are[highly classified], we can infer a great deal. There were only two facilities that were large enough to test the Hubble telescope in its entirety. Both belonged to NRO contractors. It can be reasonably inferred, therefore, that the NRO telescope was about as large as the Hubble telescope. The NRO also offered NASA two mirrors the size of the Hubble primary.[120] They also provided my college with six 72-inch (1.8-m) diameter surplus mirrors that became the Multiple Mirror Telescope.

I therefore assume that the latest Keyhole is very much like a Hubble turned upside down. If so, it is then a Ritchey–Chretien that is diffraction limited over a small field of view. It has a resolution of 0.25 microradians. If it flies as low as 100 miles (160 km), then its footprint is 4 centimeters, about 1.5 inches. Can you really read a license plate from space as Tom Clancy alleges?[121] Maybe so. Some recent photos of Korean ships show what may be legible characters, if I could read the symbols and they were originals.

Figure 4.48.1 Satellite imagery.

[119]Wikipedia. "KH-11," https://en.wikipedia.org/wiki/KH-11_KENNEN.
[120]Wikipedia. 2012 "National Reconnaissance Office space telescope donation to NASA," https://en.wikipedia.org/wiki/2012_National_Reconnaissance_Office_space_telescope_donation_to_NASA.
[121]Clancy, T. *Red Storm Rising*, Putnam (1986).

4.49 Stealth Optics

US stealth fighter planes and bombers were developed by Bill Perry in the Clinton administration.[122] They were radar stealthy. They were invisible or almost invisible to radars. The concept for doing this was twofold and simple: make the front ends of the planes of radar absorbing (non-reflecting) material and make them as streamlined as possible. This design reduces the reflected return of the radar signal to the receiver. It works.

Optical stealth, or **optical cloaking** as it is sometimes called, is not so simple. In the radar case, it is only necessary to reduce the reflectivity of the radar signal from its origin. In the optical case, the object reflects all the light from all the world from every direction to the receiver.

We can start with a simple example. A miserly Ebeneezer Scrooge wishes to hide his stash of jewels from the eyes of all who might take them. The arrangement shown is simple but cumbersome and works as long as the potential thief views it head on. As shown in Figure 4.49.1, he sees the worthless rocks instead of the expensive "rocks."

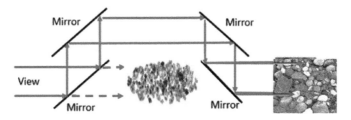

Figure 4.49.1 Simplistic stealth optics.

But this does not work if he views from a different angle, and it is very cumbersome. We can imagine replacing those big mirrors with many tiny ones that reflect in all the appropriate angles. That is one approach to optical cloaking with so-called metamaterials, i.e., materials fabricated on the same scale as molecules all with the desired shapes and angles.

---

[122]Personal discussions with Bill Perry, Gene Dineen, and Brian O'Brian, Sr., among others.

There have been several attempts at full-blown cloaking, but so far it has been accomplished in only a limited way. One approach by researchers at Purdue was an array of micro-needles spreading from a common point.[123] Its limitation is that it works only for a single wavelength. Another uses three or four lenses and is useful over the entire visible spectrum but for only a limited field of view and only between the lenses.[124]

A development with which I was involved was a stealth satellite. The idea was to launch a satellite and make it disappear to be used only in the case of some emergency, like war. This satellite had to be invisible throughout the visible, the infrared, and even radar regimes, and from all angles. Fortunately, I was on the Adversary Group. This was an advisory group chaired by Peter Franken to assess the possibilities. It was adversarial in the sense that it was criticizing every approach so that the best one would be pursued. I cannot go into all the details, but here are just a few of the problems in making any satellite invisible across the spectrum: You can paint it black so that it matches the night sky in the visible, but this enhances the infrared radiation. You can put a mirror in front of it, but that will only reflect light from one direction. You can put mirrors all around it, but that will reflect radar and visible beams in a variety of directions, and so on.

It became so difficult that I suggested that they leak the information that we really had such a stealth satellite. Naturally, this project would be called Project **Cry Wolfe**. They did not accept my suggestion, and I do not know the eventual outcome.

4.50 Stereoscopes

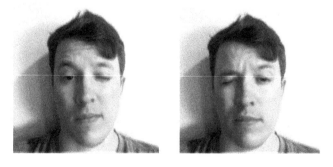

Stereopsis is the process of seeing in three dimensions (from the Greek words *stereo*, meaning solid, and *opsis*, meaning sight).

---

[123]Purdue University, "Engineers create 'optical cloaking' design for invisibility," sciencedaily.com.

[124]Choi, J. and Howell, C. "Paraxial ray optics cloaking," *Opt. Exp.* **22**, 9465–29478 (2014).

There are several different ways in which we discern the third dimension, which I consider distance or depth. One with which we are all familiar is triangulation. Our eyes are set apart by what is called the interocular distance, which is normally about 1.5 inches (4 cm). Accordingly, our eyes see slightly different views of distant objects (somewhat like seeing the two sides of an object).

This type of 3D viewing works only for a finite distance, about 100 meters. You can check this out by focusing on several different objects at different distances. Close one eye and then the other. If the object is close enough, the object will seem to move sideways in the two views. As the distance to the object increases, the movement gets smaller until there is none. For me, that is about 100 meters. That is the end of depth perception by triangulation, by parallax. This is illustrated in the introductory figure.

But the eye/brain system provides other clues. Distant objects are smaller. If we know what it is and how big it is, then if it appears smaller, it must be farther away. More distant objects are often partially obscured by closer ones. And as the eye focuses at different distances, the muscles that control that focus, that squeeze the optic lens, contract or relax, and your brain senses that.

Stereoscopic viewers, stereoscopes, use the triangulation method by presenting two different views of the object, taken from different angles with respect to our two eyes. The basic concept of the small handheld stereo viewers is shown in Figure 4.50.1. The same object is presented to the eyes from two different places. This simulates viewing the object from two different angles and enhances the separation.

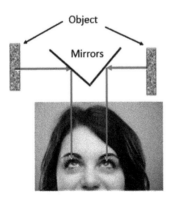

Figure 4.50.1 Handheld stereo concept.

Perhaps the most popular example of stereo viewing is the movie *Avatar*. It was done with polarized cameras and polarized glasses. Two cameras with orthogonal polarization took images of the scenes at appropriate angles. The polarized glasses we used to view these images presented our eyes with these images, as if our eyes were as far apart as the cameras. We had a big, synthetic triangulation baseline.

There are other methods for obtaining and presenting the alternate images. They include the use of different colors or rapid alternation of images, faster than the eye can see—faster than the flicker fusion frequency (see Section 4.19). However, polarization is the most popular.

The first stereoscope was apparently that of Charles Wheatstone in 1856,[125] although there are arguments.[126] It is pictured in Figure 4.50.2. Other famous names associated with stereoscopes are David Brewster and Oliver Wendell Holmes.

Figure 4.50.2 Wheatstone stereoscope (reprinted from Ref. 125).

David Brewster and Oliver Wendell Holmes each contributed by improving the instrument with lenses and prisms but still used the basic principle of parallax, triangulation.

The modern three-dimensional viewer is not a stereoscope at all; it is a hologram (see Section 4.18). It also provides parallax, i.e., the ability to move your head sidewise and see more of the side of the object.

Depth perception is closely related to stereoscopic viewing. Depth can be perceived by the parallax techniques described above, but it can also operate with only one eye. There are several clues that allow even a cyclops to perceive how far away things are.

One technique uses knowledge of how big a thing is. If we know how big it is and how big it appears, we have a good idea of how far away it is.

---

[125]Wheatstone, C. "Contributions to the physiology of vision—Part the First. On some remarkable, and hitherto unobserved, phenomena of binocular vision," *Trans. Royal Soc. London* **128**, 171–394 (1838).
[126]Wikipedia. Stereoscope, https://en.wikipedia.org/wiki/Stereoscope.

Another is if one object partially obscures another. We then know that the obscured one is behind, it is farther away than the one that obscures.

A considerably more subtle indication is that our eye focuses. It is possible to sense the tensing or relaxing of the muscles that focus the eye lens.

One technique that I do not recommend but that worked at the time was used by my friend, Howie Courtney. We were being examined for permission to fly in B-24s for some experiments. One was the stereo vision test: line up two indicators that are attached to strings by pulling on the strings until they match. I did it then it was his turn. He did it, too. Then they asked if he had any vision problems. "Yes, I am blind in one eye," he said. He managed to align the indicators by grabbing the string as I let it go.

4.51 Stroboscopes

Stroboscopes are devices that use intermittent, specially timed images to make various measurements. (from the Greek words $\sigma\tau\rho\delta\beta o\varsigma$ - *strobes*, meaning "whirling," and $\sigma\kappa o\pi\epsilon\tilde{\iota}\nu$ - *skopein*, meaning "to look at"). I guess the real meaning is to look at something that is whirling, such as the whirling of a hummingbird's wings.

One of the delightful uses is "stopping" a hummingbird (or "hummer") in flight to view its beautiful colors. A more practical use is adjusting the ignition timing of your car. The basic principle is to flash the light on the subject at just the proper frequency. If the hummer is flapping its wings at 100 rpm, then a flash rate of 100 rpm will show the wings always in the same place, as if they were stationary.

Police use flashing lights to disarm offenders. It has been shown that light pulses at a certain rate can confuse and even disable people.[127] It must be of high intensity and about 6 to 10 pulses per second to match the alpha rhythm of the brain.

---

[127]Rubtsov, V. "Incapacitating flashing light apparatus and method," US patent 7180426 (2007).

It was proposed but never used in warfare as the Canal Defense Light, a shuttered searchlight mounted on tanks[128] and "dazzle lights" meant to hide the Suez Canal. It may even have been the cause of recent sicknesses in US embassies in the form of an audio strobe at the right frequency. That would be sound waves at a frequency of 10 to 20 ppm, out of our range of hearing but close to the alpha frequency.

There are basically two ways to attain the stroboscopic effect, actively and passively. In the active mode, one opens the lens of a camera and pulses a light source at the chosen frequency. In the passive mode, the light is on constantly, and the camera shutter is opened and closed at the desired frequency.

Harold "Doc" Edgerton, an engineer at MIT, was the primary person who developed and applied the electronic flash to stroboscopic instruments.

One of the stroboscopic effects that you have seen and may have wondered about is the apparent motion of the wheels of a car in TV ads. The car moves forward, but the wheels seem to be rotating backward. A TV set is a form of stroboscope. It flashes images to you at 30 times a second. If the wheels of the car have not rotated a full 360 degrees in one thirtieth of a second, it looks like they are rotating backwards.

4.52 Submarine Communication

Communication with underwater nuclear submarines that have nuclear missiles is vital. In the event of an attack (God forbid), a message needs to be sent to them immediately to respond.

---

[128]Secret Strobelight Weapons of World War II, wired.com.

Communication with submarines when they are submerged is difficult. The sea does not transmit any type of radio waves any reasonable distance. It is also very opaque to infrared and ultraviolet radiation, but it has some modest transmission in the blue-green part of the spectrum.

Very-low-frequency radio waves, around 10–100 Hertz, transmit in seawater to a depth of about 20 meters. Long antennas have been dragged behind subs to receive these signals, but it is awkward, and the sub at this depth is easily detected.

The absorption of sea water in the visible and infrared regions of the spectrum is shown in Figure 4.52.1.[129] It shows clearly that it has a pronounced minimum at about 550 nm. The absorption coefficient there is 0.01 per meter, that is, 1% per meter.

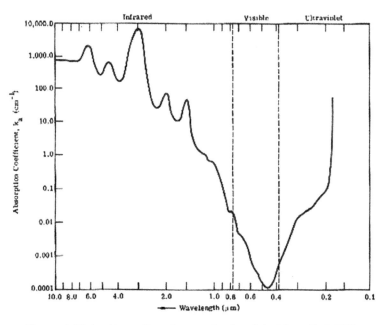

Figure 4.52.1 Absorption of seawater (reprinted from Ref. 129).

This results in a transmission shown in Figure 4.52.2. It is down to 5% at 300 meters range and gets worse after that. Underwater communications with visible light, even in that green part of the spectrum, are very limited in range. The transmission over a distance of one kilometer is 0.005%. The maximum useful output of a diode laser comunicator is about 10 milliwatts. It results in 500 nanowatts at the receiver. That is not useful.

---

[129]Wolfe W. and Zissis, G. *The Infrared Handbook*, US Government Printing Office (1978).

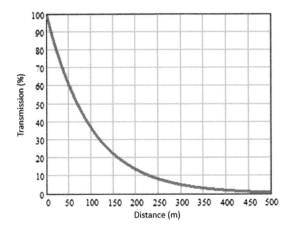

Figure 4.52.2   Transmission of seawater.

Communication between satellites and submarines seems even more problematic due to the ocean waves and other factors. It was proposed in 1972 in a study at the Naval Postgraduate School in Monterey, CA.[130] It was the subject of a patent dispute many years later.[131] It was unresolved in the first case but taken under study to obviate the backscatter from the ocean's surface. The patent was rejected in the second case for the wrong but similar reasons by the judge. You can read his ruling online[132] if you have the patience and time and a masochistic inclination.

A satellite in polar orbit passes over the sub very briefly and about once a day. One in geosynchronous orbit can be over it at all times, but its signal will cover a wide area and be susceptible to interception. The surface waves refract and scatter the beam in many directions and the seawater absorbs at a great rate.

It is a puzzlement.

---

[130]I chaired the study.
[131]Pfund, C. US patents 4,279,036, 4,664,518, and 4764982 (I advised the government).
[132]276-87C and 592-88C • Charels E, Pfund vs the United States . . . , case-law.vlex.com.

4.53 Teleprompters

Teleprompters are a combination of optical projectors and beam splitters or so-called one-way mirrors. (check Sections 3.11 and 4.38 if you like). The projector is usually down low and aims up at a simple pane of glass, as shown in Figure 4.53.1. The glass is set at an angle such that its reflection is readily seen by the speaker who sees the reflection up close with about 4% reflection; the audience looks right through it with about 96% transmission. Although the prompters are off a bit to the sides, this diagram shows the concept with them straight on. They are usually arranged a little higher than the speaker who looks up at them. They are often higher than the spectators, so they see mostly sky or ceiling through them, and they can see the speaker under them.

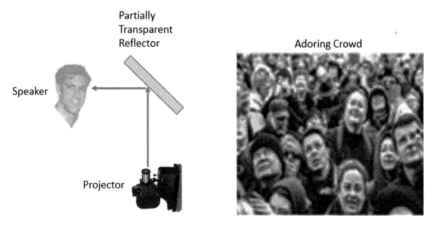

Figure 4.53.1 Teleprompter.

Some teleprompters have enhanced reflection, and some are even opaque. The transparent or semi-transparent versions allow the audience to see the speaker through them. Of course, the opaque ones partially block the view.

Some prompters, as they are also briefly called, require that the projector has reversed text because there is a reflection; others just add another reflection.

4.54 Telescopes

It all started with Hans Lippershey or his kids when they were fooling around with his lenses.[133] Soon, Galileo designed his own version that was ten times better. These were refractive telescopes. A more comprehensive and more technical description is given by Lloyd Jones in the *Handbook of Optics*.[134]

The **Galilean telescope** design of 1609 is shown schematically in Figure 4.54.1. It had a positive objective lens that formed an inverted image of an object at infinity at its focal point. I could not show the object way out there. It is in close, in green, but its rays are parallel to the axis. The first image is at the focal point of the second lens, a negative one. The second lens, which Galileo chose to be a concave lens, formed a virtual, enlarged erect image shown by the dashed lines and the erect, light green arrow that the eye sees at infinity.

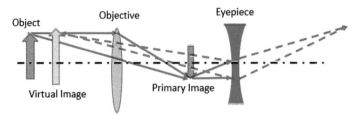

Figure 4.54.1 Galilean telescope.

---

[133]Wolfe, W. *Rays, Waves and Photons*, IOP Press (2020).
[134]Jones, L. *Handbook of Optics*, Bass, M., ed., McGraw-Hill Chapter 18 (1995).

The **Keplerian telescope** improved it with a wider field of view by using a convex lens as the eyepiece. It works on much the same principle, but it has an inverted virtual image, as shown in Figure 4.54.2.

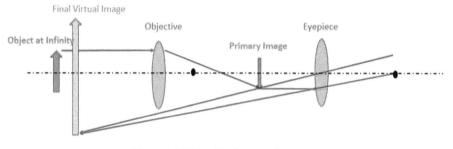

Figure 4.54.2 Keplerian telescope.

Newton designed what is often said to be the **first reflecting telescope** because he believed that you could not correct for chromatic aberration with glasses (but he did propose a lens arrangement with water and glass[135]). He may have been following the work of James Gregory, described below. Figure 4.54.3 is an illustration from Newton's book. The primary is a sphere of speculum metal, a copper-tin alloy that polishes well. There is a prism used as a reflector to get the focal region out of the way of the incoming beam. He noted that the image formed by the spherical mirror is inverted but that the prism sides could be curved appropriately to erect it. That would be an unusual but useful lens.

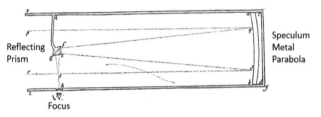

Figure 4.54.3 Newton's telescope (reprinted from Ref. 135).

James Gregory, a Scotsman, designed what is probably the first reflecting telescope, but he never built one. His design was published in 1663, several years before Newton published his work. It was built by Robert Hooke in 1673, after Newton.

The **Gregorian telescope** uses an ellipse as a secondary to preserve the on-axis quality and to get the image back behind the primary in an accessible

---

[135]Newton, I. *Opticks*, London (1704).

area.[136] It makes use of the property of an ellipse that all rays from one focus go to the other focus. This is illustrated in Figure 4.54.4.

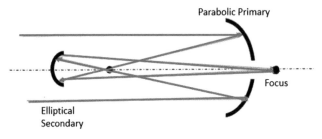

Figure 4.54.4 Gregorian design.

Laurent Cassegrain invented the **Cassegrain telescope** in 1672, somewhat after the design of Gregory was published. It uses the same parabolic primary but a hyperbolic secondary.[137] It uses the property that all rays from one focus of a hyperbola converge to its other focus. Somewhat like a convex ellipse.

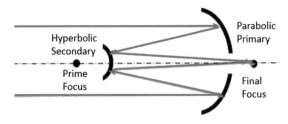

Figure 4.54.5 Cassegrain telescope.

An extension of the Cassegrain is the Ritchey–Chretien, by George Ritchey and Henri Chretien, that optimizes the two surfaces as general aspheres and not just conic sections. It was originally two hyperbolas but is now considered to be general aspherics. This is the design of most large, modern astronomical telescopes.

Bernard Schmidt was searching to find a design for larger fields of view.[138] The above designs are great for objects that are on axis, but they blur objects that are off axis. The basic concept of the **Schmidt telescope** is to make everything on axis. He did this by using a spherical mirror and placing the aperture stop at its center of curvature. Figure 4.54.6 shows the three different beams: the red beams, bordered by rays, all have centers that coincide with

---

[136]Chambers, R. *A Biographical Dictionary of Eminent Scotsmen*, Blackie and Son (1870).

[137]Cassegrain, L. *Journal des Scavans* (April 25, 1672).

[138]Schmidt, B. "Ein lichtstarkes komafreies Spiegelsystem," *Central-Zeitung füik und Mechanik* **52**, 25 (1931).

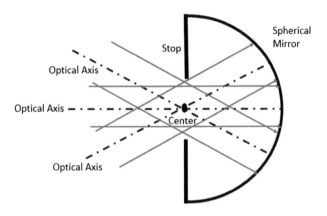

Figure 4.54.6 The Schmidt concept.

diameters of the sphere, the black dashes. They are on axis. The image surface is curved and suffers from spherical aberration.

He corrected the spherical aberration by placing an aspheric corrector at the aperture. Its shape is too subtle to show. This is a good wide-angle system, but it is limited at large angles by the inclined refraction of the corrector plate, which, in effect, changes its shape.

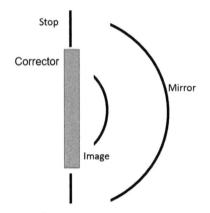

Figure 4.54.7 Corrected (complete) Schmidt.

There is also an all-reflective version of the Schmidt, shown in Figure 4.54.8. It uses a 45-degree flat at the center of curvature that serves as the stop and has the reflective equivalent of the refractive Schmidt corrector on it. The modest version has a circular symmetric figure; the better version uses ellipses. I have no reference for it, but I have known about it for at least 60 years.

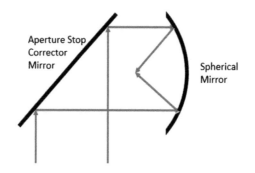

Figure 4.54.8 All-reflective Schmidt.

Perhaps with similar reasoning, Albert Bouwers[139] in the Netherlands and Dmitri Maksutov[140] in Russia independently came up with a similar solution during WWII. They, too, put the stop at the center of curvature, but they used concentric, spherical corrector plates to correct spherical aberration. They provided the options of placing the correctors behind the stop, in front of the stop, or both.

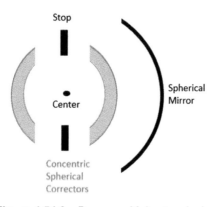

Figure 4.54.9 Bouwers–Maksutov design.

Herschel had a simple solution for getting the image plane out of the path of the incoming beam. He tilted the mirror.[141] This meant that there were off-axis aberrations, but he could live with them for large enough focal ratios.

---

[139]Bouwers, A. *Achievements in Optics*, Elsevier (1950).

[140]Maksutov, D. "New catadioptric meniscus systems," *JOSA* **34**(5), 270–284 (1944).

[141]Maurer, A. and Forbes, E. "William Herschel's astronomical telescopes," *J. British Astronomical Association* **81**, 284 (1971).

Figure 4.54.10   Herschelian system.

An alternate approach was to use an eccentric pupil., i.e., use only part of the mirror. Then the image plane is available without a secondary, but it requires a mirror about twice the size of the aperture. It is called an eccentric pupil but describes only a few of mine.

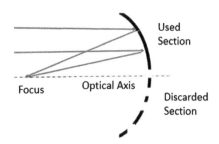

Figure 4.54.11   Eccentric pupil.

The Mersenne telescope[142] is afocal: it converts a collimated beam into a collimated beam of a different diameter. It has been used as the basis of three-mirror anastigmats described below. There are two versions of the Mersenne, both with parabolic mirrors, as shown in Figures 4.54.12 and 4.54.13. They might be considered the Gregorian and Cassegrainian, or concave and convex, forms. But the secondaries are parabolas, not ellipses or hyperbolas. The product of the beam diameters, and its angular spread remains constant. The smaller exit beam spreads a little more.

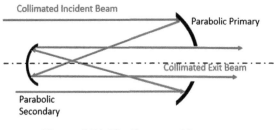

Figure 4.54.12   Concave Mersenne.

---

[142]Mersenne, M. *Harmoniques*, Jacques Villery (1634).

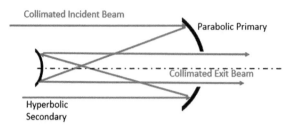

Figure 4.54.13   Convex Mersenne.

Three-mirror anastigmats (optics corrected for spherical aberration, coma, and astigmatism) use three mirrors rather than the one or two described above. This allows them to eliminate those aberrations. The earliest design was probably by Maurice Paul.[143] It is shown in Figure 4.54.14. The first two mirrors comprise a Mersenne afocal system. The third is spherical and simulates a Schmidt system with the stop at the center of curvature, i.e, at the convex mirror. Only spherical aberration needs to be corrected, and that is done with the other surfaces.

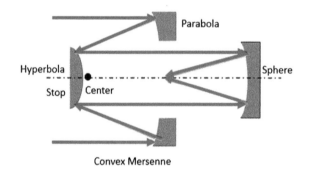

Figure 4.54.14   Paul anastigmat.

It seems clear from the diagram that some means must be introduced to make the focal plane accessible. One could make the final mirror a combination like a Cassegrain, but the procedure was to design off-axis systems, as described below.

During the Cold War, there was considerable investigation of systems that would not scatter sunlight to the focal plane of ICBM detectors. They were generally called off-axis telescopes. The eccentric pupil and Herschelian systems were two of these, but better performance was needed. This search resulted in the design of off-axis three-mirror anastigmats with accessible

---

[143]Paul, M. Systèmes correcteurs pour réflecteurs astronomiques, *Revue d'optique théorique instrumentale* **14**(5), 169 (1935).

focal areas. One design by Lacy Cook is shown in Figure 4.54.15.[144] It is a variation of the Paul–Baker design. It is much like one half of a Paul with adjusted mirrors shown schematically in Figure 4.54.16. Lacy had the advantage of a modern program that could optimize the spacings and the mirror figures.

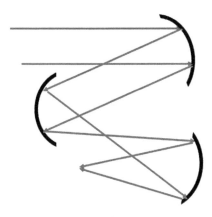

Figure 4.54.15 Cook anastigmat.

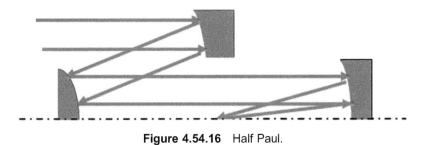

Figure 4.54.16 Half Paul.

A significant and different type of improvement in telescope design was the adaptive optics mirror described in Section 3.11.

Perhaps the most important advance in telescope design in modern times was the multiple-mirror telescope (MMT),[145] invented by Aden Meinel and Frank Low. They decided/hoped they could combine the outputs of six mirrors to a single focus and thereby gain the advantage of area and span. They obtained six surplus mirrors from the National Reconnaissance Office in 1974 and ground and polished them in the optical shop of the James C. Wyant College of Optical Sciences of the University of Arizona.

---

[144]Cook, L. "Three-mirror anastigmat used off-axis in aperture and field," *Proc. SPIE* **1083** (1979) [doi: 10.1117/12.957416].

[145]Beckers, J. B. L. Ulich, and T. J. Williams. "Performance of the Multiple Mirror Telescope (MMT) I. MMT: The first of the advanced technology telescopes." *Proc. SPIE* **322** (1982) [doi: 10.1117/12.833497].

The MMT saw first light atop Mount Hopkins in southern Arizona in only five years from its start. A cross-section showing two of the mirrors and their combination and the installation on Mount Hopkins are shown in Figures 4.54.17 and 4.54.18.

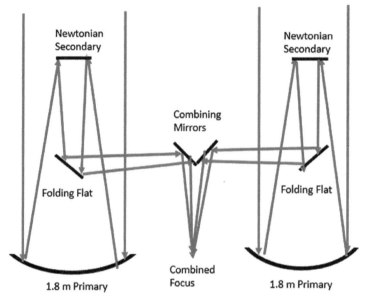

Figure 4.54.17   MMT cross-section diagram.

Figure 4.54.18   The MMT on Mt. Hopkins.

There are two advantages with MMTs. The first is the increased collecting area, which is just the sum of the areas of the individual mirrors. More collecting area means more sensitivity, since more photons are available for analysis. The other is the span, the increased resolution which is determined by the distance from the outer edge of one mirror to the outer edge of its opposite.

The Large Binocular Telescope is in operation atop Mount Graham, also in southern Arizona. It uses two of the 8-meter-diameter mirrors made at the University of Arizona and two adaptive optics secondaries in a Ritchey–Chretien form.

Figure 4.54.19 Large Binocular Telescope.

The Giant Magellan Telescope is a project of many countries and institutions. It will consist of seven 8.4-meter-diameter mirrors in the Ritchey–Chretien configuration with their outputs combined. A model is shown in Figure 4.54.20. Note the scale of the model shown by the person indicated by the arrow.

Figure 4.54.20 Giant Magellan Telescope model.

The Hubble Telescope was also a Richey–Chretien configuration with a 2.4-meter-diameter mirror. It orbited at 347 miles altitude. Although it was a Mr. Magoo at first because of a one per cent error in its figure, later corrective lenses in each of the measurement units provided astounding results. Figure 4.54.21 shows the lens cap and the baffle tube. Inside is the pair of mirrors and the instrument bay, along with the solar panels that power it.

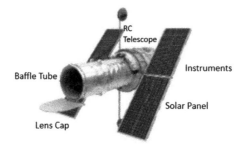

Figure 4.54.21   Hubble Telescope.

There are some astounding, yes, astronomical, astronomical telescopes in the offing: the James Webb, the Extremely Large, the Thirty Meter, and the Large Synoptic Survey Telescopes. These will be situated in space and high in the Andes. They all make use of the multiple-mirror and adaptive optics techniques and embody the concepts just discussed. But watch for what they will be able to do!

Terrestrial telescopes provide an erect image to the observer. They are used for hunting, sports, bird and elk watching, and other outdoor activities, even for the Abominable Snowman. Almost all are refractors.

The Galilean design described above was like this. It is still in use, as is that of Kepler. Most terrestrial telescopes are based on the Keplerian design (with some kind of image inverter).

A typical optical system design is outlined in Figure 4.54.21: the objective focuses the light onto an aiming reticle; a second lens passes the light through a device that inverts and reverts the image to its original position; and the eyepiece sends the combined image to the eye, target image, and reticle. The image rectifier may be a pair of lenses or Porro prisms, as in binoculars, or some other device. The objective may be a fixed or a zoom optical system.

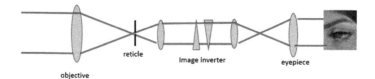

Figure 4.54.22   Typical terrestrial telescope.

4.55 Television Sets

I have lived through the entire age of television, from the 12-inch black-and-white sets of the 1940s to the Jumbotrons of today. Today is better.

Those old sets used cathode ray tubes (CRTs). For nostalgia's sake, I will review the concepts of their operation before a discussion of the versions of today. There are still some CRT sets around.

The **pickup tube** or **video recorder** unit in the broadcast station of those old devices started with an electron gun, which I have shown on the right of Figure 4.55.1. It has a negative applied voltage. The grid to the left of it in blue is positive and attracts the electrons (in the z direction, to the left). The electromagnetic deflection plates direct the electron current in the x and y directions (up and down, in and out). The optical system focuses an image on the grid that generates electrons by energy conversion in a photoelectric material, one that exchanges electrons for photons. The number of electrons in each spot dictates the final current by repelling the electrons from the gun different amounts and thereby decreasing the current.

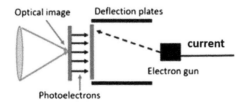

Figure 4.55.1   CRT schematic.

The **display tube** operates in much the same way. The same diagram applies, but now the input current is modulated by the recorded signal. The electron beam is still attracted by the gun and oriented by the defection plates. The electrons now impinge on a phosphor that emits light of different intensities according to the intensity of the input current, the number of electrons per second.

The complete TV system consists of both the video camera in the studio in the football stadium or elsewhere and the display device that we have in our homes, usually simply called the TV set.

Most modern TV cameras use a prism to separate the incoming image into three colors, sometimes with an additional filter. Each of the colored images is recorded by a photodetector array and recorded electronically.

The sets in our homes receive and interpret the electronic signals and display it in one of two different ways: either liquid crystal display (LCD) or organic light-emitting diode (OLED). As I wrote, CRTs are now passé, and so are plasma sets.

The LCD requires backlighting, since it is a form of light modulator. The illuminants can be LEDs, electroluminescent panels, or fluorescent ones. They are mostly LEDs or fluorescents.

The LCD light modulator consists of a linear polarizer followed by a liquid crystal followed by a second linear polarizer oriented at 90 degrees to the first. The liquid crystal is a liquid in which there are crystals. In this case, they are spiral (twisted) crystals that can rotate the direction of polarization. When the crystal is not activated, i.e., when a voltage is not applied, the light that is polarized vertically, as shown in Figure 4.55.2, passes through the liquid crystal unaffected and is cancelled by the polarizer oriented in the opposite direction, horizontally in the figure. Remember that light is indicated as polarized in the plane of the paper by an arrow and perpendicular to it by a circle, and that only light polarized perpendicular to the polarizer grid direction is transmitted. When a voltage is applied, the crystals get oriented and rotate the direction of polarization so that light is transmitted by the second polarizer, as illustrated in Figure 4.55.3.

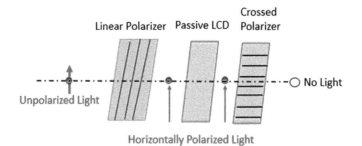

Figure 4.55.2   Passive LCD.

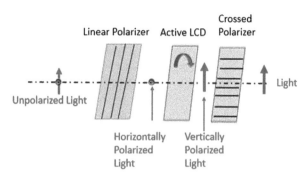

Figure 4.55.3 Active LCD.

The voltage is applied to each pixel, and the degree of polarization transmission is gradual.

An OLED, or organic LED, generates its own light. It is an electron-to-photon transition as in inorganic LEDs. But OLEDs are much more efficient and cheaper than their inorganic counterparts. The OLEDs are not as bright, but they have better contrast and blackness. Both are bright enough and are otherwise equivalent, but OLEDs are still more expensive.

The **standard sets** we had for so many years had 480 lines 720 pixels wide. A color set therefore has just over a million pixels ($480 \times 720 \times 3$, or 1,123,200).

High-definition sets now have 1920×1080 pixels, a total of about 6.2 million.

The latest **4k sets** have about 4000×3000 pixels; it is still varying. That totals about 36 million pixels when you consider the three colors as well. This equals the resolution of the human eye for some larger sets that are way across the room.

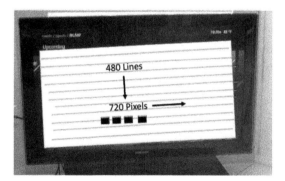

Figure 4.55.4 Lines and pixels.

A fascinating feature for football fans is the scrimmage and first-down line markers on the set. Both of these are marked in yellow, respectively, and they **seem** to be marked on the field. Notice that the players obscure them. But they are generated in the computer. This is a complex process. The geometric arrangement of the field needs to be calculated and the distances ascertained and arranged in the correct orientation. Then the two lines and their distances need to be established. The lines are then laid down pixel by pixel but only if the original pixel is the color of the turf. In that manner, if a ref or player obscures the intended line, his color is maintained. You can see these features in Figure 4.55.5.

Figure 4.55.5   Yard markers (figure credit: Pixabay).

What can we expect in the way of improvements? As I gaze into my crystal ball, all I can see are reduced prices and holographic, three-dimensional TV.

4.56 Theodolites, Transits, Sextants, and Octants

A **theodolite** is an instrument that measures angles. It is used by surveyors to plot out different areas. In my first days at Bucknell University the civil engineers were all over campus with them. I never used one. They are simply a terrestrial telescope mounted on a divided circle. You aim the telescope at one point and note the angle on the circle. Then aim to another point. Note the other angle and subtract. In good theodolites this can be done both horizontally and vertically.

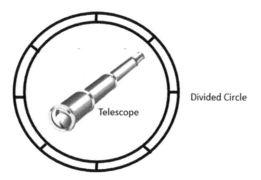

Divided Circle

Telescope

Figure 4.56.1 Simple transit.

A **transit theodolite** has a telescope that can rotate through the zenith angle, more than just vertical.

A **sextant** may be considered a poor man's theodolite. It has a telescope that is mounted rigidly to the frame and is aimed at the horizon. A mirror moves vertically on a divided circle of only 60 degrees. It reflects light from a known star into the telescope by way of another mirror. In this way and with knowledge of the time of day, a latitude may be calculated.

An **octant** covers 90 degrees and rotates a mirror to bring the image of one object in coincidence with that of another. The rotation of the mirror is a measure of the angle between them. The objects are often the horizon or the North Star.

4.57 Underground Object Detection

Several types of underground objects can be detected by optics. These include land mines, gas lines, and ancient, buried cities.

Infrared detection is based on several thermal properties of the sun and the Earth. The sun rises in the east and begins to warm the Earth at 1000 watts per square meter. That heat gradually warms the soil and the objects in it. They warm up according to their heat capacity, which is the amount of energy, the number of watt seconds, it takes to heat a unit mass one degree. Artifacts, such as pipes, mines, and cities, have heat capacities that are different from those of soil. They therefore heat more or less rapidly than their environment. The same process occurs in reverse after sunset. They cool at different rates. These heat differences are conducted by the soil to the surface and can then be measured as local differences in the amount of infrared radiation emitted from the surface.

We at the Wyant College of Optical Sciences have carried out two tests on the detection of **land mines** by this technique. The first was the use of an infrared camera aimed from the opening shown by the yellow line in

Figure 4.57.1 on a simulated buried land mine indicated by the orange line. The second was a similar test at the Marine base in Twentynine Palms, CA against real but disarmed mines. Both were successful.

Figure 4.57.1 First test.

Integral pipes, those without flaws, without leaks, can be detected. The fluid in them is almost always at a temperature that is different from the surroundings. They are cooler in the summer and warmer in the winter. This thermal difference is conducted by the soil to the surface and results in a slightly different surface temperature locally.

Leaky underground pipelines were detected in Libya, where they were constructing the Great Manmade River Project to bring water from southern Libya to the coastal regions where most residents lived. The actual pipes were 4 meters in diameter and 6000 kilometers long. These preliminary tests were conducted to determine if infrared instruments could be used to detect leaks in a timely fashion.[146] Leaky pipes wet the soil and increase the thermal conductivity, thereby further increasing the observable temperature difference. They buried 20-mm-diameter pipes 110 mm deep in the sand. They used two different infrared cameras, an IRISYS and a Flir A310f. They could detect both the pipeline and the leaks and determine the difference.

One application of detecting a subsurface object was a conceptual planning one. It was to determine if you could find the location of a **Peacekeeper missile** in its tunnel. The Peacekeeper was an ICBM capable of carrying 11 nuclear-armed re-entry vehicles. It would be used in retaliation as part of the obsolete MAD program. One concept was to

---

[146]Shakmak, B. and A. Al-Haibaibeh, "Detection of water leakage in buried pipes using infrared technology," *2015 IEEE Conf. on AEECT*, Nov. (2015).

house them in tunnels that were about a mile or so long in the desert southwest. They could reside anywhere in the tunnel, hidden from attack. I was tasked with calculating whether their position could be detected by a reasonable infrared device. The missile would generate heat by virtue of its standby requirements. (You cannot just turn it off.) The answer was yes, you could if you could fly over it at an altitude of a couple thousand feet and no one disturbed the soil above it. There was never a deployment. And thank God, MAD is history. The concept of Mutual Assured Destruction was madness.

Ancient cities have been mapped by lidar systems in helicopters. The famous city of Angkor Wat in Cambodia has been mapped by lidar (as well as an even older city near it).[147] The lidar sends laser pulses through the overhanging canopy to reflect from the ancient artifacts. The timing of the return determines the depth of penetration of the canopy. The same is true of a sprawling Mayan civilization in South America.[148] An archeologist by the name of Sarah Parcek has found underground structures in Scotland by examining the pattern of foliage on the surface.[149] In areas where there are artifacts, the plants do not grow so well. She used a technique similar to detecting land mines, i.e., detecting the difference between the heat capacity of adobe structures and that of loose soil.

4.58 Submarine Wake Detection

Infrared submarine detection was and is an important function in naval warfare. Submarines now have nuclear-tipped missiles and can stay submerged for extended periods of time since they are nuclear powered.

---

[147]Lawrie, B. "Beyond Angkor: How lasers revealed a lost city," bbc.com (2014).
[148]BBC News. "Sprawling Mayan network discovered under Guatemala jungle," Feb. 2 (2018).
[149]A. Tucker. "Space archaeologist Sarah Parcak uses satellites to uncover ancient Egyptian ruins," *Smithsonian Magazine* Dec. (2016).

Sonar detection by other subs or by surface ships is limited in range and must be in contact with the water. Detection of the wake of a submarine is a partial solution that can be done with an aircraft and over large areas.

Studies have been done on the detection of submarine wakes that have led to mixed results.[150] Submarine wakes are disturbances on the surface of the ocean that result from the passage of the sub underwater. These wakes can be of either a higher or a lower temperature than the surround because they can be either the disturbed, colder water churned up to the surface or the heated trail from the exhaust of the nuclear engines. They can also be as a result of just the right mixing of the two. But usually there is a wake that can be detected by appropriate aircraft infrared sensors with a spatial resolution of about one foot and a sensitivity of a small fraction of a degree. The wakes are typically about one degree different from the surround and 100 feet across and 600 along the track, and they last about an hour.[151] As noted in Section 4.53, water has an emissivity that is very close to unity in the infrared.

There are many airborne infrared scanners that can do this. One example is the AN/AAD-5, the design of which I supervised in the late 1960s. The AN/AAD-2, an earlier version, was used in some successful submarine-wake-detection trials in Chesapeake Bay.[152] Both warm and cool wakes were observed. A radiometer made by the Barnes Engineering Company with a temperature sensitivity of 0.003 °F was also used in these tests.

4.59 Warehouse Optics

In our modern world of BIG BOX and online stores, there must be massive warehouses, and there are. Amazon alone has over 100 fulfillment centers,

---

[150]Personal summer study involvement.

[151]Moser, P. *Technical Note Infrared Wake Detection*, US Naval Development Center NADC AW N5917 (1959); *Infrared Wake Detection*, DTIC, online.

[152]Moser, P. "Submarine Wake Detection Program," US Naval Air Development Center, Anti-Submarine Laboratory (1962).

each approaching one million square feet. These are not manned just by people but also by robots that move things around day and night, 24/7. They are guided largely by optical means.

One technique relies on indoor **drones**. They fly to designated locations and read barcodes on cartons to check inventory or to retrieve a product for delivery. Another is automated **forklifts** that use cameras to follow lines on the floor to designated locations to store or retrieve products. Some of them can lift major loads meters in the air to overhead storage bins. Some measure the areal and depth dimensions of the storage racks with a pulsed time-of-flight camera.

It is illustrative to trace the travel of a single product. After it has been made, it is delivered to the warehouse for storage until it is ordered. Its bar code is read by an optical scanner. It is then moved to its designated place in the warehouse by a robot that follows a line on the floor with a camera or follows GPS instructions. It is lifted to a storage area. When it gets purchased, a robot goes to its spot according to its instructions, reads its bar code with a scanner and follows the yellow brick road to the loading dock. Its bar code is probably read once more, and its shipping destination printed on the box. And away it goes.

4.60 Weather Satellites

Weather satellites provide imagery of storms and clouds from above. They are instrumental in weather prediction since they locate storms and how they are moving.

The first weather satellite was the **Television Infrared Observational Satellite (TIROS)**. It was an experiment to see if weather satellites would be useful. They were, and they are. The first TIROS, in spite of its name, had two television cameras but no infrared one, and it orbited in a polar orbit. There were ten TIROS satellites, the later ones incorporating the infrared camera, and there were a dozen TIROS-N satellites. Several **Advanced Technology Satellites** were flown before the first geostationary one, the **Synchronous Meteorological Satellite (SMS)**.

The first **Geostationary Operational Environmental Satellite (GOES)** satellite was launched in 1975, and there have been about 15 since; each was a little better than the previous. That first one had two mercury telluride detectors, one as a backup (see Section 3.5). They were cooled by a radiative cooler. They can operate at temperatures higher than 77 K, the design value, but they are not quite as sensitive. The telescope slowly scanned an area the size of the continental US with a flat scanning mirror in front of it. The design, as I remember it, was the flat in front of a parabolic mirror, with a single detector cooled by a radiation cooler, as I have diagrammed in Figure 4.60.1. The resolution did not have to be very good. It was mapping clouds. If the resolution spot is 1 km and it scanned the continental US every 15 minutes, each pixel will take about 90 microseconds, which is okay since the detector response is about 1 microsecond.

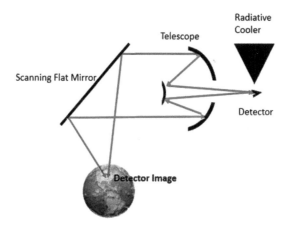

Figure 4.60.1 GOES optics schematic.

It took about 15 minutes for a full scan, but that was fast enough for meteorological events. It scanned in the 8-12-μm spectral region to measure the radiation from cloud tops. Higher clouds are more menacing and colder at the top. It was at an altitude of 35,786 kilometers (22,236 miles), the so-called geosynchronous altitude. It moved in its orbit at the same rate as the Earth turns. It was therefore always over the same place on Earth. It was geostationary. Its full scan covered several different regions, each about the size of the continental US.

The present GOES, GOES-R, is much more sophisticated. It has two visible channels, four near-infrared channels, and ten far-infrared ones. It does far more than weather observation. Some of the channels monitor soil surface temperature, some monitor fire activities, and others monitor vertical temperature. They operate in the same way as the remote sensors in orbit do (see Section 4.44).

The lightning detector is based on some earlier work by Mike Nagler.[153] He showed that all the flashes of lightning on the face of the Earth could be detected by the application of two techniques. One was to use the very-near-infrared part of the spectrum, where the sun is not quite as bright and does not reflect quite as much off clouds. The other was to use a frame-to-frame processing technique. In this, all the pixels of one frame are compared to all of those in the next. Only those with a change, one that was caused by lightning, show up. His prototype for airborne investigations is shown in Figure 4.60.3 and is about the size of a breadbox. His calculations for a similar device in space resulted in a design of about twice the size.

Figure 4.60.2   Prototype lightning sensor.

It is not clear to me why the GOES instrument is so much larger than Mike's. There is plenty of radiation in lightning strikes. It may be for spatial resolution.

---

[153]Nagler, M. "Design of a Spaceborne Lightning Sensor," Dissertation, University of Arizona (1981).

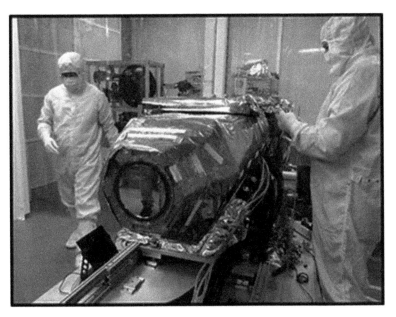

Figure 4.60.3   GOES lightning sensor.

Figure 4.60.4   GOES lightning sensor.

The so-called **Sounder** is a sophisticated device that measures the vertical temperature profile of the atmosphere. It does so by inversion of the radiative transfer equation of a partially absorbing medium. Imagine a vertical column of atmosphere that is partially absorbing. Divide it conceptually into small, numbered segments, as shown in Figure 4.60.5. It is actually continuous, but this assumption permits a convenient description.

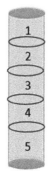

Figure 4.60.5   Numbered column of CO_2.

The radiation that comes out of the top—since the satellite instrument is looking down—will be a combination of all these cells. It will be the emissivity times the blackbody radiation at its temperature of cell number 5 times the transmission of all the cells above it. The same for cell number 4, and so on. This calculation is made for different temperature distribution until there is a fit to the measured data. It is also done for several different wavelengths in the 12-μm spectral region, where carbon dioxide has different absorption coefficients at slightly different wavelengths. The emissivity and transmissivity of each cell is known since the absorption coefficients, density, and composition of carbon dioxide are known. It works through the courtesy of extremely efficient computer calculations! The inversion technique may also be called trial and error or by guess and by golly. Current performance is an uncertainty of 1 km over a height range of 1–6 km.[154]

The **POES** satellite is the polar version of a weather satellite that orbits perpendicular to the Earth's rotational motion over the poles about 14 times a day at an altitude of about 850 km. This means that the paths precess, thereby covering a slightly different part of the Earth on each pass, eventually covering it all. Just like the GOES, there are several POES. Whereas each GOES covers an area about the size of the continental US, Europe, or the South Pacific all the time, POES satellites cover the entire Earth sequentially.

---

[154]Huang, H., Smith, W. L., and Woolf, H. M. "Vertical resolution and accuracy of vertical infrared sounders," *J. Appl. Meteorol.* **31**(3), 265–274 (1992).

4.61 Windows

You may not think of windows, ordinary house windows, as optical instruments, and neither did I at first. But they are. One definition of an instrument by Merriam Webster is *a means by which something is achieved, performed,* or furthered." Certainly, windows achieve our ability to look out but keep the weather from coming in. They can be very complicated in both form and function. (And this does not include Windows by Microsoft.)

First, a short discussion of the **transfer of heat**. There are three ways heat is transferred: conduction, convection, and radiation. Conduction is the most efficient and is most efficient in metals that have free electrons that do the job by moving. Insulators do it by the rubbing of the molecules on each other. It is sort of like elbowing in a crowd. Convection is the currents of gas. The heated molecules bounce off each other, tending to expand the gas and therefore the heat, and to move around. It is the next most efficient. Radiation is the flow of photons, each of which carries a small bit of energy. It is the least efficient but maybe the most significant. It is the way we receive sunlight!

Ordinary windows are just single panes of glass in frames of wood, plastic, or metal. The glass is transparent from the near ultraviolet to the near infrared, from about 350 to 2500 nanometers. It transmits about 96% of the incident visible light from the sun and atmosphere. Each pane does not **conduct** much energy; it has a thermal conductivity value of 0.8 compared to metals in the hundreds. And it surely does not **convect** any gases through it. It **transmits** visible light, which has energy. It absorbs the infrared radiation of sunlight and re-emits it at its own temperature that is often some tens of degrees higher than ambient. Touch your window and see, especially in Tucson.

Insulating windows consist of two or three panes of glass separated by an inch or so of an inert gas. It can be air, but more often it is argon. The better types have plastic or wood frames to reduce conduction by the frame. The argon is there to reduce the modest convection between the panes. The windows can also be coated with a material that transmits visible photons but reflects the infrared ones that heat but do not illuminate (and the ultraviolet ones that fade our furniture).

A typical window is shown schematically in Figure 4.61.1. It shows the frame in brown. It should consist of an insulating material such as wood or plastic and not metal, which is sometimes the case. A metal frame conducts the heat and reduces the overall efficiency of the window. There are two glass panes with an inert gas such as argon between them. Such a gas is a poorer conductor and convector of heat than air (especially humid air). In good windows, the outer pane has a thin film of metal that prevents the transmission of ultraviolet radiation. That is the light that fades the fabrics of the chairs and rugs and things inside. Some have that film on the inside surface of the outer pane to protect it.

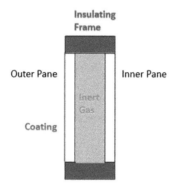

Figure 4.61.1   Double-pane insulating window schematic.

They come in double-pane and triple-pane varieties. Although triple sounds better than double, it is the thermal resistance value that counts. Some triple-pane windows are no better than some double panes, but they are usually more expensive. I saw one ad that proclaimed you could get triple-pane windows for the price of double panes. But were they any better? Figure 4.61.2 shows a comparable triple pane window. It is identical to the one above except that it has a third pane of glass.

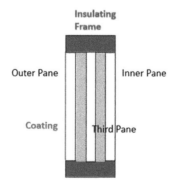

Figure 4.61.2   Triple-pane insulating window schematic.

The only difference is that a glass pane has replaced the same thickness of inert gas. Glass is a better conductor of heat than the gas, but it could be better if coatings are on the third pane.

The key to this is the R and U values. The R values and U values are used to rate the thermal insulating properties of materials and systems respectively. They are not reciprocal, but they are complementary. The R value is a measure of the heat transfer of a material, such as glass, wood, or plaster (like that of the pane **or** the frame). The U value is usually that of a system (like the pane **and** the frame). They are reciprocal only if applied to the same thing, but not generally. The values are power transfer per unit area per degree temperature difference. It can be stated in either English or metric units, that is, BTU per square foot per degree Fahrenheit or watts per square meter per degree centigrade. For better insulation, better heat barriers, lower Us, and higher Rs are better.

R you ready? The choice of double-pane or triple-pane windows is all up to U!

Photochromic windows reduce their transmission when sunlight shines on them. Certain materials, photochromic ones, change their light-absorption properties when light is incident upon them, especially more energetic ultraviolet light. One example of this is silver halides. They are the materials in photographic film that darken upon exposure to light. One might crudely imagine a piece of photographic film covering the window. As the sun shines on it, it darkens as the transparent silver halide turns to opaque silver.

In a bit more technical description, energetic incident photons release an electron from silver chloride, which is transparent. That electron combines with the silver ion to make a silver molecule. It is opaque. Ag^+Cl^- goes to $Ag + Cl$. Silver chloride separates into silver and chlorine.

I would not choose these for the windows for my home. I want control. I would choose electrochromic ones.

Electrochromic windows change their transmission when electricity is applied to them. This is accomplished in several different ways. The main ones are the application of thin layers of conducting material to make electrodes separated by an appropriate distance. When a voltage is applied to the ions, the electronic charges on the inside go to the outside and increase the reflectivity. A second method uses similar electrodes to orient liquid crystals in the glass. In one orientation, they attenuate light; in the other, they transmit it.

Stained glass windows are windows made of colored glass sections. The appellation derives from the former technique of applying a silver stain to the outside of the window that when fired turned a yellowish color. Perhaps a better name for the modern creations is colored glass windows. The small sections of different colors are formed into the desired shapes to portray flowers or butterflies or angels. The colored glass is produced by inserting dyes into the glass while it is still in its semi-liquid stage. They are then assembled and held in place with thin lead beads or strips. Two are shown in the introductory photograph (in my living room).

There are many more types of windows: single hung, double hung, bay, transom, oriel, casement, and so on. They are different mechanical and geometric configurations and do not entail any additional optical concepts.

Chapter 5
Optical Experiments

This chapter is about several significant optical experiments that helped explain our universe and the nature of light. Galileo observed the phases of Venus to establish the Copernican concept of a heliocentric universe, or at least the solar system. John Mather carried out the COBE-DIRBE experiment using the FIRAS instrument to confirm the "big bang" theory of the origin of the universe. Newton used a prism to show us that white light consists of a spectrum of colors. Young, Arago, and Fizeau collectively proved that light was a wave motion and not corpuscular. Lummer, Kurlbaum and others carried out careful experiments on blackbodies that allowed Planck to quantize emissions and start quantum mechanics. Einstein interpreted the photoelectric experiments of Heinrich Hertz to show that light was also particles, later named photons. Michelson and Morley carried out a test to establish the luminiferous ether and proved the opposite. It was one of the events that lead to relativity. The eclipse experiment headed by Arthur Eddington was one of several experiments that established the general theory of relativity.

The timeline of the epochal events is unusual. Two of them happened in the 1600s. Then nothing of import occurred until the 1800s and 1900s, when there was a flurry of activity. Perhaps it took two centuries of renaissance to get there. Finally, in the modern age when we could launch satellites, the experiment to confirm the big bang theory of the origin of the universe was performed.

Here is a chronological table of the experiments:

1610 – Galileo used his new telescope to show that our solar system is centered on the sun.

1660 – Newton used a prism to show that white sunlight consists of colors.

1660 to 1800 – Nothing notable in optics happened for 140 years.

1800 – Herschel discovered the infrared part of the spectrum.

1807 – Young performed his double-slit experiment, giving the first proof that optical radiation is waves.

1817 – Arago showed that there is a diffraction spot; the second proof of waves.

1851 – Fizeau showed that light travels slower in media such as water than in air, the end to corpuscles.

1887 – Michelson and Morley disproved the existence of the luminiferous ether, paving the way for relativity.

1887 – Heinrich Hertz discovered the photoelectric effect that allowed Einstein to later infer photons.

1900 – Several experimentalists made blackbody spectral measurements that caused Planck to start quantum theory.

1919 – The solar eclipse headed by Arthur Eddington added to the proof of general relativity.

1989 – The COBE-DIRBE experiment headed by John Mather confirmed the "big bang" theory.

These seminal events are arranged in chronological order in the diagram below.

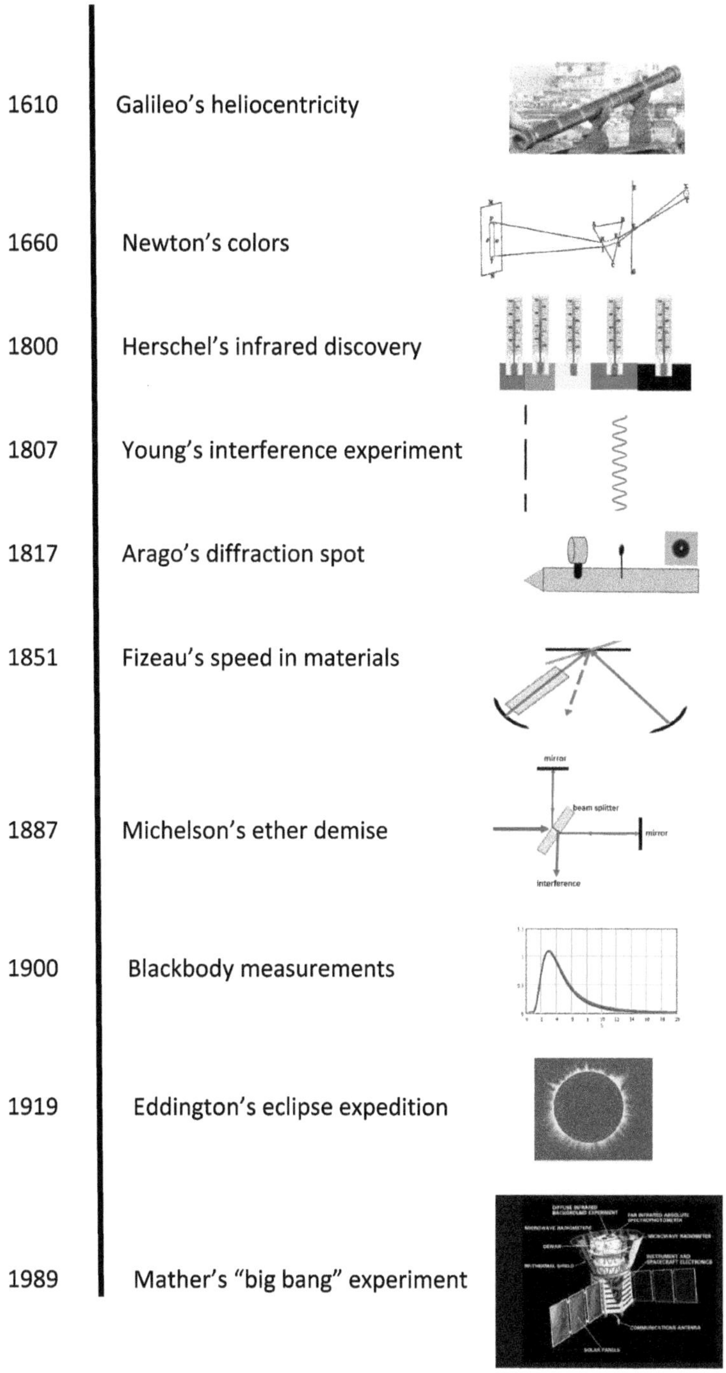
1610 Galileo's heliocentricity
1660 Newton's colors
1800 Herschel's infrared discovery
1807 Young's interference experiment
1817 Arago's diffraction spot
1851 Fizeau's speed in materials
mirror
beam splitter
mirror
interference
1887 Michelson's ether demise
1900 Blackbody measurements
1919 Eddington's eclipse expedition
1989 Mather's "big bang" experiment

5.1 Galileo's Heliocentricity

Galileo Galilei is known for his proof of the heliocentric world, but he did much more than that. We all know about his work with falling bodies from the tower in Pisa. He was the first to identify the concept that all motion was relative. That concept later led Einstein to his famous relativity revelations. He was also the first to enunciate that a body in motion tends to stay in motion, one of Newton's laws. In this section, however, we deal with his observational proof of the heliocentric world. It was based in part on his observations of the phases of the moon and of Venus.

The Ptolemaic system was a geocentric one. The Earth was the center of the universe. This started in the first few centuries and was accepted until Nicolaus Copernicus suggested in 1543 that it was really a sun-centered, or heliocentric, system rather than a geocentric system.[1] The Ptolemaic system was egotistical. **We** had to be the center of the universe. The Church believed the sun revolved around the Earth partly as a result of some Bible verses that seem to say that the sun moves: "And the sun stood still, and the moon stayed..."[2]

Galileo learned of the invention of the telescope by Lippershey (or his kids) and improved upon it.[3] He then used it to explore the heavens. He observed the rings of Saturn, satellites of Jupiter and phases of Venus. Venus looked different at different times, with different geometric arrangements of the sun itself and our Earth.

We are all familiar with the phases of the moon caused by the reflection of sunlight from different angles as the moon rotates around the Earth. Figure 5.1.1 reminds us of the geometry (not to scale). The sun shines on us from the left. The moon revolves around the (green) Earth, and the reflections at half-moon (top and bottom) are crescents. Full moon is when all of the sun reflects back to us from the moon shown on the right. New moon is when the moon obscures the sun, and no reflected light gets to us.

---

[1]Copernicus, N. *De revolutionibus orbium coelestium*, Johannes Petreius, Nuremberg (1543).
[2]Joshua 10:13.
[3]Wolfe. W. *Rays, Waves and Photons*, IOP Publishing (2020).

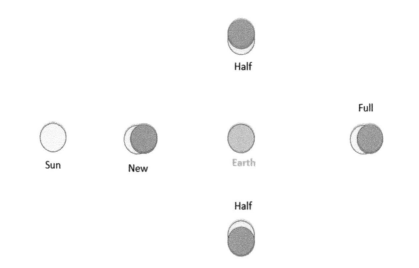

Figure 5.1.1 Phases of the moon.

The planet Venus also has phases, but the geometry is not the same (Figure 5.1.2).

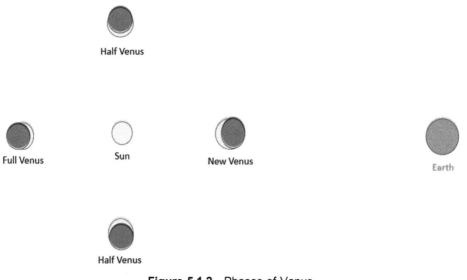

Figure 5.1.2 Phases of Venus.

The red planet orbits the yellow sun inside the orbit of the green Earth. The full Venus is shown to the left; the two crescents on the top and bottom and the reflection of the sun on the right-hand position are completely obscured.

The geometry with the Earth and Venus is different because Venus does not orbit the Earth and is closer to the sun than the Earth. There is enough difference in the geometry and the timing to conclude that the sun is not orbiting the Earth. If everything revolved around the Earth, according to the geocentric theory, then full Venus would have to be on the other side of the sun in a larger orbit and the new Venus on the right. The geometry and the apparent sizes would indeed be different.

But we know and Galileo knew that this was not so. The relative arrangement and the timing would both have to be different.

5.2 Newton's Colors

Sir Isaac Newton was one of the giants of physical science. His laws of motion are historic, although Galileo may have preempted him on at least one (a body stays in motion). His excursions into optics have been somewhat questionable with his ideas about corpuscles, but his determination that the light from the sun consists of many colors is well known. This idea is described by Newton himself in his book *Opticks*, which is available online:[4]

> *In a very dark Chamber, at a round Hole, about one third Part of an Inch broad, made in the Shut of a Window, I placed a Glass Prism, whereby the Beam of the Sun's Light, which came in at that Hole, might be refracted upwards toward the opposite Wall of the Chamber, and there form a colour'd Image of the Sun.*

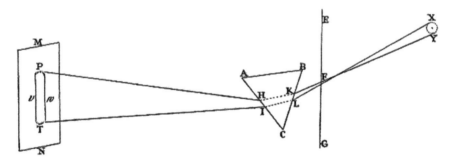

Figure 5.2.1 Newton's diagram (reprinted from Ref. 4).

He arranged a prism in a room with a sun-facing window and put a hole in the window shade (which he called a shut). It was not a slit, as I once thought, but an oval, as shown in Figure 5.2.1. It was, as he wrote, one-third of an inch wide, and from his figure it looks to be about one inch long. He therefore produced an image of the sun—a multicolored, elongated image of the sun. It

---

[4]Newton, I. *Opticks, A Treatise on the Reflection, Refractions, Reflections and Colours of Light* (1704); available online via The Project Gutenberg eBook of Opticks, https://www.gutenberg.org/files/33504/33504-h/33504-h.htm.

was a form of a pinhole camera with a prism. In a way, this may have been the very first color camera. He describes the solar image as oblong, and this makes sense with an oblong "pinhole." He also concludes (correctly) that the width of his image subtends the solar angle of about one-half degree.

This was the first demonstration that white light is composed of many colors, and it was an experimental arrangement that was adapted by others for similar measurements (notably the discovery of the infrared part of the spectrum). Figure 5.2.2 is a more dramatic illustration in technicolor.

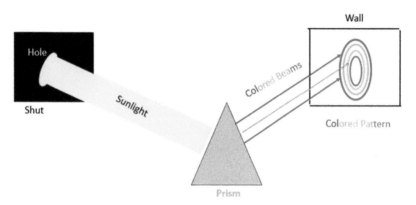

Figure 5.2.2   Representation of Newton's experiment.

5.3 Herschel's Infrared Discovery

Sir William Herschel discovered the infrared part of the electromagnetic spectrum in 1800, although that was not his quest.[5] He was just trying to optimize solar spectrum viewing. He was also a devotee of Newton and followed the same path (see Section 5.2). He put a slit in a sun-facing window and caused the light to shine upon a prism, just like Newton. He searched for the best color with which to view the sun without injury. In the course of these trials, he sensed that some colors warmed him more than others. "Could there be a relationship between light and heat?" he wondered.

According to the *American Scientist*, his arrangement looked like the image shown in Figure 5.3.1.[6] This illustration is apparently from Herschel's notes and is cited all over the internet.

[5]Herschel, W. "Investigation of the powers of the prismatic colours to heat and illuminate objects," *Philo. Trans. Royal Soc.* **90**, 255–283 (1800).
[6]White, J. "Herschel and the Puzzle of Infrared," *Am. Scientist* **100**, 23 (2012).

Figure 5.3.1 Herschel's experimental arrangement. (reprinted from Ref. 6)

Based on this figure, the slit in his shut (window shade) had to be horizontal. He was an ardent follower of Newton who used a hole in his shut too. The spectrum spreads horizontally, and one thermometer is moved from color to color. The other two are used as references for the ambient temperature.

He added a thermometer to his prismatic Newtonian setup. He placed the thermometer in one color after another, noting the temperature of each, starting at the violet end. He found that the temperature increased more the farther he got into the red, and even beyond. In fact, he had not reached a maximum in the visible portion of the spectrum. It has been reported that he was more interested in the properties of this "invisible light" than its existence.[6]

The popular presentation is more like that in Figure 5.3.2, which is also somewhat easier to understand.

The discerning reader (you) may note a conundrum here: the solar spectrum peaks at about 550 nm. Thus, the thermometers should record a maximum at that wavelength. But the maximum was out somewhere in the infrared. This is a consequence of the nonlinear dispersion of a prism.

Nonlinear dispersion means that the light in a prismatic spectrum is not uniformly distributed in space. As represented in Figure 5.3.3, blue light takes up more space than yellow light, which in turn takes up more space than red light. The longer wavelengths of light are more concentrated in space. This

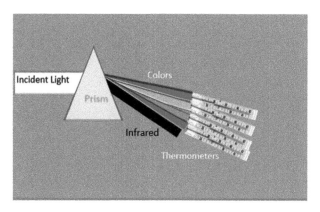

Figure 5.3.2 Herschel experiment schematic.

means that the thermometer of a fixed size collects more photons over a given area (spectral width) in the red than in the blue because there are more photons concentrated there. The concentration of light (i.e., of photons or energy) is sparser in the blue than at longer wavelengths, as represented in Figure 5.3.3, which also shows his use of thermometers.

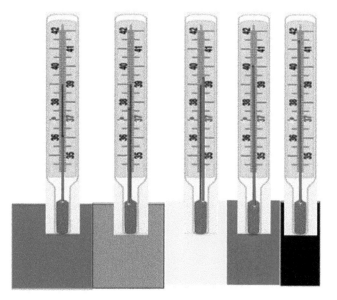

Figure 5.3.3 Herschel's measurement.

The somewhat deeper significance of this experimental result is the realization that light is not only that which we can see but also radiant electromagnetic energy at invisible wavelengths. Thank goodness that Herschel did not use a grating as a spectral disperser; it has linear dispersion! We might have never known about the infrared, and I might never have had a job.

5.4 Young's Double-Slit Experiment

This experiment was one of the turning points in the understanding of light (which is still not yet fully understood). The prevailing concept at the time was that light consisted of corpuscles. This was due to the famous Isaac Newton, who believed that they were small particles and had many others convinced. But there were some—including Huygens, Fresnel, and Young—who believed that light consisted of waves. This was one of the three experiments that showed light to be waves.

Thomas Young was a medical student at Göttingen University, but he was a very inquisitive genius. He studied optics, Egyptology, and other sciences. The presentation of the results of his double-slit experiment was on November 24, 1803 before the Royal Society of London. The essence of the experiment is to divide a wave into two parts and then combine them to show that the parts interfere. The irony is that he did not use slits at all. Young's double-slit experiment was done without slits!

Young performed his experiment in much the same way that Newton showed that white light consisted of colors. He made a small hole in a window shade and directed the beam of light that came through it to a small card.[7] He held the card edgewise to the beam, thereby splitting it into two parts. The interference pattern appeared on the opposite wall. The experiment was later repeated with real slits.

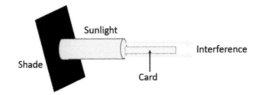

Figure 5.4.1 Young's arrangement.

It is easier to understand what he did by reference to the real double-slit arrangement. In his day, there were no lasers. He needed a single source so that all the waves would be emitted in phase. He did this with a single slit. We can imagine that source on the left and an optical system to generate plane waves. They are incident on the double slit. Waves emanate from the two slits shown as circular crests that overlap, but are really spheres. Where they overlap, there is constructive interference, as diagrammed in Figure 5.4.2. The red line shows one locus of constructive interference. There are others where the crests of the waves coincide.

---

[7]Young, T. "Experimental demonstration of the general law of the interference of light," *Philo. Trans. Royal Soc. London* **94** (1804); Shamos, M., ed., *Great Experiments in Physics*, Holt Reinhart (1959); Young, T. *A Course of Lectures on Natural Philosophy and the Mechanical Arts*, Joseph Johnson (1807).

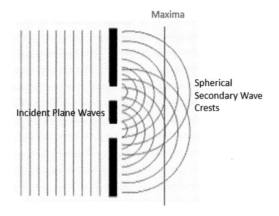

Figure 5.4.2   Young's double-slit schematic.

The double-slit experiment has been performed more recently with a quantum-mechanical twist. A single photon was sent to a single slit. It behaved as expected and went right through in a straight line to the screen. But when the experiment was repeated with two slits, the single photon went to what would be a maximum of the interference pattern. It somehow sensed that there were two slits. There are three different explanations for this by three different Nobel laureates: (1) there is a pilot wave associated with the photon (deBroglie–Bohm); (2) the photon has enough width that it senses both slits (Heisenberg); and (3) the photon first explores all possible paths, and as it does so, it senses the slit (Feynman). I think this underlines the fact that we really do not know what light is.

5.5 Poisson/Arago Diffraction Spot

This was another of the critical experiments that disproved Newton's corpuscular theory of light and supported the wave theory. It has a fair amount of irony and good scientific method in its tale.

It occurred when the war between the corpuscles and the waves was in full tilt. Only verbal shots were fired. There was a competition hosted by the French Academy of Sciences in 1818 on just this subject. Augustin Fresnel submitted his theory that light consisted of waves. He asserted that the waves diffracted, in a way similar to the way sound does. Denis Poisson, a member of the evaluation committee who believed in corpuscles, seized on this as a sure sign that light was **not** waves.

Poisson did the calculations based on Fresnel's theory and showed that there would be a bright spot of light a specific distance behind an opaque obstruction if light were waves. This simply could not be. After all, how could particles be caused to swerve in behind such an obstacle? But Dominique Arago, head of the committee, decided it was only fair to test the theory. He did, and he did it very carefully because everything had to be just right. He found the spot.[8] He used a circular obstruction that had to have a very smooth edge and had to be larger than the wavelength of the monochromatic light but smaller than the observation distance in order to produce Fresnel diffraction. And he had to put the screen in the right place.

I call this the Arago spot, after the one who was successful. Poisson could not care less what I call it now, but he would probably not like to have his name associated with such a failure.

The arrangement is shown schematically in Figure 5.5.1. The light, in green, is shown blocked by the occulting disc according to the corpuscular theory, but the diagram also shows the little dot in the middle of the diffracted area.

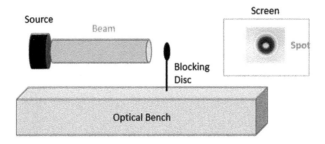

Figure 5.5.1 Arago spot diagram.

The obscuring disk must have very smooth edges, and its diameter must be the square root of the product of the wavelength of the light and the distance between the obstacle and the screen. It was not an easy experiment.

Figure 5.5.2 is my attempt to show how the individual waves add together on the screen behind the obstacle. Recall that with an aperture or an obstacle, an entire new set of waves proceed beyond it (see Section 1.6). In this case, a host of new waves, subsidiary waves, all emanate from around the obstacle

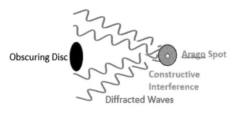

Figure 5.5.2 Formation of the spot.

[8]Arago, D. "Rapport fait par M. Arago à l'Académie des Sciences, au nom de la Commission qui avait été chargée d'examiner les Mémoires envoyés au concours pour le prix de la diffraction," *FO* **1**, 79–87 (1819).

and converge with coincident maxima at a particular spot behind it. I have only shown four of them.

This is a wonderful example of how science should be done: test a theory with an experiment, especially if there are competing theories. By the way, Fresnel won the prize. And we know light is waves.

Or photons.   Or waves.   Or either.   Or both.

5.6 Fizeau's Speed of Light in Materials

The third experiment that caused the demise of the corpuscular theory of light was the measurement of the speed of light in a medium, not in a vacuum. Newton's theory requires that light travel faster in a solid material such as water or glass than it does in air or in a vacuum. Fizeau set out to determine whether this was true. It wasn't.

A monochromatic source is collimated and caused to pass through two tubes. Water, or an equivalent liquid, flows in opposite directions in the tubes. A second lens causes the two beams to merge and interfere based on any phase difference between them. The interference pattern can be interpreted to determine which beam had a phase shift in which direction.

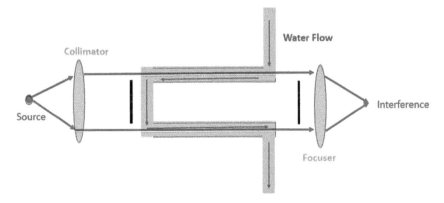

Figure 5.6.1   Fizeau schematic.

Fizeau found that light travels slower in a denser substance such as water than in air or in a vacuum. This was probably the final nail in the corpuscle coffin.

5.7 Michelson Morley Experiment

In the late 1800s, little was known about the nature of light. The wave theory had supplanted corpuscles, but it was not yet understood that light was electromagnetic radiation. Hertz had not done his experiment, and Maxwell was diligently working on his famous results. It was not yet understood how light (electromagnetic) waves propagated.[9]

Ocean waves travel in water. Sound waves move in air. The propagation is mechanical, from atom to atom. "There must be some sort of material in which light waves propagate;" was the prevailing thought at the time. The material proposed was the **luminiferous ether** or the "light-bearing upper air." It had to be more rigid than steel to support the high-frequency light waves but also massless and with no viscosity so that planets could pass through it unhindered. It had to exist in vacuum and throughout all space. That means as we spun around the sun on a daily basis, the stationary ether would seem to be a wind.

It was also a figment of the imagination.

Albert Abraham Michelson and Edwin Williams Morley set out to prove its existence.[10] They figured they would show that there was an ether wind by making measurements in two perpendicular directions—north and south versus east and west. Light that travels with the flow of the ether takes longer to go a given distance than light traveling perpendicular to the same flow. This can be measured by an interference effect in the Michelson interferometer (see Section 4.21). The Earth spins through the ether as sketched in Figure 5.7.1. It creates a wind by that relative motion, as shown schematically in Figure 5.7.2. This is analogous to the wind dogs love as you drive through still air.

Figure 5.7.1 Spinning Earth.

---

[9]Wolfe, W. *Rays Waves and Photons*, IOP Publishing (2020).
[10]Michelson, A. and Morley, E. "On the relative motion of the Earth and the luminiferous ether," *Am. J. Sci.* **34**, 203 (1887).

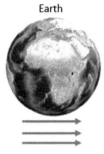

Earth

Apparent Ether Wind

Figure 5.7.2 The ether wind effect.

Michelson at that time was a professor at the Case School of Applied Science, and Morley was at Western Reserve University (WRU). They collaborated with an instrument in the basement of a stone dormitory at WRU to be in a stable environment (a peaceful, stable dormitory? At least it was not the basement of a fraternity house!). The instrument was mounted on a thick stone block about 8 feet by 4 feet on top of a big pool of mercury in a closed room in the basement.

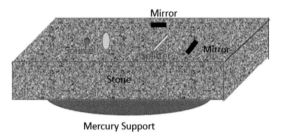

Mercury Support

Figure 5.7.3 Representation of the Michaelson–Morley setup.

Their optical bench was much like what we use today. We mount our instruments on large stone blocks, about 3 feet by 6 feet and 1 foot thick. We now use pneumatic pistons as legs for isolation.

Their version of the Michelson interferometer for this experiment used tilted mirrors to increase the pathlengths to 11 meters to increase the sensitivity. Only the horizontal portion is shown in Figure 5.7.4; the vertical arm is the same. After the beam is split, it reaches a tilted mirror that reflects it to the next, and the next, and the next, until it reaches a perpendicular mirror and is returned upon itself (then to the beam combiner and interference with the other beam). They created a path length of 11 meters.

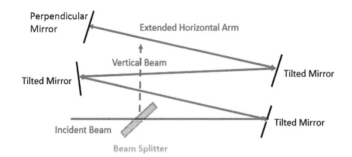

Figure 5.7.4   Michelson's extension representation.

The result has been described by some as the most famous **failed** experiment in history. But it paved the way for relativity. They repeated it in different seasons at different times of the day, almost any way they could think of, over the course of a year. But there was no difference: no ether wind, and no ether. It may be the most **successful** negative result in history. That is the way science should be done: develop a theory based on observations, design an experiment to test it, and then accept the result.

As a footnote, I might add an empathetic note on the choice of the stone dormitory basement. We have trouble making measurements in the fourth basement of our college, when there is traffic on Speedway Avenue, a thoroughfare about a mile away. The Sperry plant on Long Island where I once worked tipped back and forth with the tides. Optical measurements can sometimes be very, very touchy.

5.8 The Photoelectric Effect

When light of short enough wavelengths is shone on a material of the right consistency, it emits electrons. As we say today: photons in, electrons out. That is the photoemissive or photoelectric effect. Incident photons with enough energy emit electrons. They are photoelectrons.

The phenomenon was first noted by Heinrich Hertz in 1887 when he illuminated metal plates with ultraviolet light.[11] He noted a change in the length of a spark that was a receptor of radio waves. The emitted electrons increased the conductivity of the air and permitted a longer spark. Somewhat later, J. J. Thompson observed that it was electrons that were being emitted from the metal that caused such changes.[12]

For some time, until 1905, the classical explanation of the phenomenon was accepted. That explanation stated that more intense light generated more

---

[11]Hertz, H. "Über einen Einfluss des ultravioletten Licht auf die electrische Einladen," *Annalen der Physik und Chemie* **267**(7), 421–44 (1887).
[12]Thomson, J. "A theory of the connexion between cathode and Röntgen rays," *London, Edinburgh, and Dublin Philo. Mag. J. Sci.* **45**(273), 172–183 (1898).

energetic emitted electrons; i.e., they would move faster. And the emission was independent of wavelength.

It was another of the marvelous insights of Einstein in 1905 that ended this notion.[13] He noted that the effect only occurred if the wavelength of the incident light was short enough, and the more intense the light was, the more electrons were emitted. In addition, the higher the frequency of light was, the more energetic the emitted electrons. This behavior was consistent with a particle model. Einstein concluded that light was quantized, and later, in 1926, Gilbert Lewis dubbed these particles of light "photons."[14]

This was just five years after Planck had determined that the emission of light was quantized. He did not accept that light itself was quantized, just the vibrations that caused it (see Sections 1.1 and 5.9).

As we now know, each photon has an energy proportional to its frequency. It is then easy to understand that the more photons that are incident on the surface, the more electrons will be emitted, and the higher the energy of the photons is (the shorter the wavelength), the higher the energy of the emitted electrons. It is a straightforward particle model. It makes so much sense now that we know it!

5.9 Blackbody Spectra

The story of blackbody radiation and the origin of quantum mechanics has been told and retold many times because it is interesting and significant.[15] This is the same story from a different perspective, from an experimental perspective rather than the theoretical.

It has been noted by many authors that at a meeting of the German Physical Society in October 1900, Lummer, Kurlbaum, Rubens, and Paschen reported some new experimental evidence of the spectrum of blackbody radiation[2] that did not agree with the current theoretical predictions. Planck did an arbitrary curve fit that made the correction but did not have the theory. After an agonizing period of analysis, he introduced the quantization of emitters. That is the oft-told story. But what about those measurements? By the way, he designated *h* as a helping value (*hilfenwert*).

Rubens and Kurlbaum used a platinum box with a blackened interior and small aperture as a blackbody simulator.[16] As described in Section 3.3, blackbody simulating devices are usually long cylinders with conical ends housed in thermally controlled housings and heated to high temperatures. They

[13]Einstein, A. "Über einen die Erzeugung und Verwandlung des Lichtes betreffenden heuristischen Gesichtspunkt," *Ann. Physik* **17**, 132 (1905).
[14]Lewis, G. N. "The conservation of photons," *Nature* **118**, 874–875 (1926).
[15]Wikipedia. "*Max Planck*," https://en.wikipedia.org/wiki/Max_Planck.
[16]Rubens, H. and Kurlbaum, F. "Über die Emission langer Wellen durch den schwarzen Körper," *Verhandlungen der Deutschen Physikalischen Gesellschaft* **2**, 181 (1900); ibid, "On the heat radiation of long wavelength emitted by black bodies at different temperatures," *Astrophys. J.* **14**, 335 (1900).

have small apertures and large interiors. Rubens and Kurlbaum probably ran it at as high a temperature as possible to get the highest signal. At that time, it was probably about 1000 K. The blackbody peak would then be about 3 μm, and the tail would taper off to 10 or 20 μm, diminishing in amplitude.

They would have used a spectrometer to measure the spectrum. These measurements were made in the late 1800s. The infrared part of the spectrum was discovered less than 100 years earlier (see Section 5.3). Thermocouples and bolometers were discovered in about the mid-1800s. Photoconductors in the 1870s (see Section 3.5). Prisms were available, but the infrared spectrometers they must have used to make the measurements were still in their infancy (or at least their teens). The prisms were probably sodium chloride, i.e., very soluble table salt, but in the form of a solid triangular prism. Each investigator had to make his own spectrometer.

I used a salt-prism spectrometer as part of my graduate work. It was probably not much different from theirs, but mine was a commercial product. The salt prism had to be kept warm to avoid misting over. The slit width had to be calibrated to account for the blackbody spectrum variations with wavelength. There are lots of details to attend to. Fortunately, their experiment was a relative one. Absolute values are much harder to measure. They needed to measure the relative spectral distribution of the blackbody curve. And, as it has been reported, it was a little different from the currently accepted theory, the Wien expression. My Mathcad calculations of the two blackbody curves, the correct one by Planck in red and the approximate one in blue, are in Figure 5.9.1. The difference between them is shown in the next figure and the relative difference in the final one. The relative difference increases monotonically with wavelength. Measurements in the longer wavelengths become more important. The absolute values in Figure 5.9.1 show that the power density is about 0.1 watt per square centimeter per micrometer at 10 μm and even several times less at 20 μm. These measurements were not easy to make, but they were highly significant. They led to quantum theory just as much as Planck's theory did.

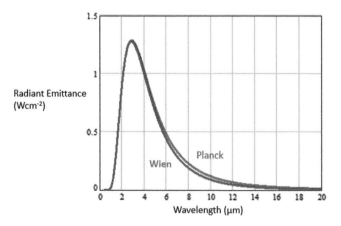

Figure 5.9.1 Planck and Wien blackbody curves.

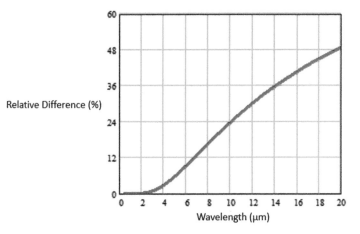

Figure 5.9.2 Relative difference.

5.10 Relativity Tests

There have been several experiments that have proven the validity of both the special and the general theory of relativity. By now, we should call them the **laws of relativity**. All of them relied on optics one way or another.

The **special theory** states that time goes slower in moving frames of reference. It is something we do not experience in our daily lives. Even if we go 2000 mph in an obsolete SST, time only slows down 0.0000016 percent of its rest value. The validity of the special theory had to be verified by atomic clocks taken on flights around the world. The differences between the clocks in the planes and on the ground over periods of many hours could be measured, although they were very small—just tens to hundreds of nanoseconds.[17]

The **general theory** states that time slows down in regions of higher gravity. Einstein proposed three tests: the precession of the perihelion of Mercury, gravitational lensing, and the gravitational red shift.

One test was already available. All planets have elliptical orbits. The part of the ellipse closest to the sun is called the perihelion. It was observed that over the years, the position of the **perihelion of Mercury** did not agree with the calculations of Newtonian mechanics, even after including the gravitational attractions of the other planets. The perihelion was not where it was calculated to be at the right time. Einstein calculated the effect using the time dilation of general relativity.[18] It should be no surprise that his calculations agreed with the measurements.

---

[17]Wikipedia. "Hafele–Keating experiment," https://en.wikipedia.org/wiki/Hafele%E2%80%93Keating_experiment.

[18]Einstein, A. "Der Grundlagen der allgemeinen Relativitatstheorie," *Annalen der Physik* **49**(7), 769 (1916).

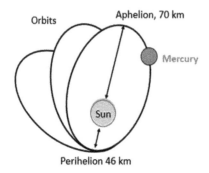

Figure 5.10.1 Perihelion of Mercury.

Perhaps the most famous test of the general theory is the eclipse expedition headed by Sir Arthur Eddington in 1919 at Principe Island and Sobral, Brazil.[19] They found that starlight was refracted by the gravity of the sun by the exact amount that Einstein had predicted, one more time.

They measured how far light from a star is deviated due to the gravity of the sun. It might at first seem that it is the gravitational attraction of the sun on the photons that causes this behavior, but photons have no mass. Light (photons) travels slower in higher gravitational fields because time goes slower. The effect is much like that of a mirage, in which there is a speed-of-light gradient due to a density gradient (see Section 2.6). In this case, there is a speed-of-light gradient because there is a time gradient caused by a gravity gradient.

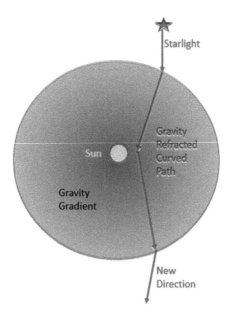

Figure 5.10.2 Gravitational refraction.

---

[19]Wikipedia. "Eddington experiment," https://en.wikipedia.org/wiki/Eddington_experiment.

The gravitational red shift is related to the Doppler shift only in the sense that they are both shifts in the frequency of light. The Doppler red shift is caused by the fact that the source and receiver are moving apart. The gravitational red shift is caused by the photon expending energy as it moves against a gravitational attraction. The energy of a photon is the constant h times the frequency. The constant does not change; energy was lost. Therefore, the frequency is reduced, and the wavelength gets longer—to the red. The effect has been measured with stars and on Earth.[20]

One experiment that is not widely reported was carried out by one of my colleagues at Washington State University. For reasons that are a bit beyond me, he was a collector of atomic clocks. He took three of them with him when he camped high up on Mount Washington, while he left the others at home. He compared them at the end of the weekend. Sure enough, the ones at home were measurably slower than those he took camping The difference in time, based on the difference in altitude that resulted in the difference in gravity, could only be measured by atomic clocks since it was so small. Nanoseconds again!

5.11 The COBE-DIRBE Experiment

This experiment was done to test the "big bang" theory of the origin of the universe. The basic idea is that our universe began with a "big bang" that happened 3.8 billion years ago. It was a giant ball of intense plasma at an incredibly high temperature. A temperature so high that temperature has no meaning. Through the years, it cooled until now the outer reaches of our universe are at a temperature of about 3 degrees Kelvin above zero (2.72548 ± 0.00057 K, to be exact). Those outer reaches where there are no stars or planets (or anything) should radiate as a blackbody at that temperature. This phenomenon is called the cosmic background radiation.

The Far-Infrared Absolute Spectrophotometer (FIRAS) was the instrument John Mather used to perform this experiment. It measured the spectrum of that primordial radiation and found it to match that of a blackbody at 2.725 K. Mather used a Michelson interferometer spectrometer that operated in the far infrared at a temperature of 1.5 K to keep its self-radiation from interfering with the measurements.

The Michelson interferometer spectrometer is described in Section 4.47. This one used wire grids as polarizing beam splitters. The measured

---

[20]Hetherington, N. "Sirius B and the gravitational redshift - an historical review," *Quarterly J. Royal Astro. Soc.* **211**, 246 (1980).

spectrum looked exactly like the theoretical one. All the data points coincided with the calculated curve! The results resembled those shown in Figure 5.11.1 but with more coincident data points and in terms of frequency.

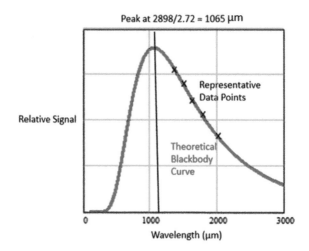

Figure 5.11.1 COBE-DIRBE results, wavelength style.

Appendices

This section of the book provides mathematical support for many of the descriptions provided in the main body of the text. It is organized in the same way as the earlier chapters and sections. For instance, Chapter 1 is supported by Appendix A.1, and Appendix A.1.5 supports Section 1.5. Not all sections in the main text have a corresponding section in the appendix (it was not always necessary). Section A.6 covers various fundamental scientific principles that are applicable throughout the book and does not correspond to specific sections in the chapters.

Appendix 1
Optical Phenomena

This section of the appendix is the mathematical support for the descriptions of the basic concepts of the nature of light, and how it interacts with itself and the rest of the world. It cannot be extensive in its limited space. I have assumed that the reader of this appendix has the mathematical background to fill in the details. Refer to the classic by Emil Wolf for more information.[1]

Section 1.1 includes vector calculus that might be a bit much for a high school or a humanities group. You might try asserting the wave equation in one dimension, which is given toward the end of the section, and show that the sine function is a solution by plugging it in.

The section on Polarization includes some matrix manipulations. Consult Appendix A.6.5 on that.

The other sections use only moderate mathematics, or as my friend John Strong would say, "mathematics of moderate rigor."

A.1.1 Light

Light is an electromagnetic wave in a band of frequencies or wavelengths that is visible to the eye. The equation for such an electromagnetic wave can be derived from the wonderful equations of J. Clerk Maxwell.[2] They are discussed extensively online.[3] Probably the best reference is Born and Wolf.[4] Maxwell's equations are given as

$$\nabla \cdot \mathbf{E} = \frac{\rho}{\varepsilon_0},$$

$$\nabla \cdot \mathbf{B} = 0,$$

---

[1]Born, M. and Wolf, E. *Principles of Optics*, Pergamon (1959).
[2]Maxwell, J. "A Dynamical Theory of the Electromagnetic Field," *Philo. Trans. Royal Soc. London* **155**, 459–512 (1865).
[3]Wikipedia. "Maxwell's Equations," https://en.wikipedia.org/wiki/Maxwell%27s_equations.
[4]Born, M and Wolf, E. *Principles of Optics*, Pergamon (1959).

$$\nabla \times \mathbf{E} = -\frac{\partial \mathbf{B}}{\partial t},$$

$$\nabla \times \mathbf{B} = \mu_0 \left(\mathbf{J} + \varepsilon_0 \frac{\partial \mathbf{E}}{\partial t} \right).$$

In free space, where there are no charges ρ or currents $\mathbf{J}$, the equations reduce to

$$\nabla \times \mathbf{E} = -\frac{\partial \mathbf{B}}{\partial t},$$

$$\nabla \times \mathbf{B} = \mu_0 \cdot \varepsilon_0 \frac{\partial \mathbf{E}}{\partial t},$$

One can take the curl of the third equation and use a vector identity (see A.6.5) to obtain the wave equation, which is in vector differential form:

$$\nabla^2 \mathbf{E} - \mu_0 \varepsilon_0 \frac{\partial^2 \mathbf{E}}{\partial t^2} = 0.$$

It was known at that time when Maxwell was developing his equations that the speed of light c was equal to $\sqrt{(\varepsilon_0 \mu_0)}$ so the equation becomes

$$\nabla^2 \mathbf{E} - c^2 \frac{\partial \mathbf{E}}{\partial t^2} = 0.$$

The vector equation can be separated into its coordinate parts. The expression for the x component in rectangular coordinates is

$$\frac{d^2 E_x}{dx^2} - c^2 \frac{d^2 E_x}{dt^2} = 0.$$

There are identical solutions for the other rectangular coordinates.

The solution for this equation is

$$E_x = E_x \sin(\omega t - kx),$$

where E_x is the amplitude of the wave, ω is its angular frequency ($2\pi f$), and k is its wave number ($2\pi/\lambda$). Similar results are obtained for spherical and cylindrical coordinates, but the functions are not simple sines.

In spherical coordinates, the expression for the radial coordinate is

$$\mathbf{E}_r = \frac{E_r}{r} \sin(\omega t - kr),$$

where **E** is the vector field, E is its amplitude, and r is the radial distance from the source. Note that the radial coordinate gives rise to an intensity that is proportional to the inverse square of the distance.

Light is also photons. These are massless particles with energy given by hf, $h\nu$, or hc/λ and momentum by h/λ. These are very perplexing particles. They have a frequency of oscillation. They are polarized so that the vibration is linear in a particular direction or circular. They have a volume. They may be accompanied by pilot waves.

A.1.2 Polarization[5]

Since light is a transverse wave, the electric and magnetic field components can vibrate in all sorts of directions perpendicular to the direction of travel. Up and down is vertical polarization; left and right is horizontal. But polarization can also be at any angle, or circular, or random.

The two methods of characterizing these vibrations are the Stokes and Jones vectors, and the method for describing and calculating their actions in optical systems are the Mueller and Jones matrices.[6] See Section A.6.5 for a discussion of matrices.

The Jones vector is just the two perpendicular components of the electric field:

$$\mathbf{J} = \begin{bmatrix} \mathbf{E}_x \\ \mathbf{E}_y \end{bmatrix}.$$

I think it is obvious that horizontally polarized light is written as

$$\mathbf{J} = \begin{bmatrix} 1 \\ 0 \end{bmatrix},$$

and vertically polarized light has the digits switched. Other forms are discussed later.

The Stokes vector, which describes the state of polarization, is not so simple. It is written as

$$\mathbf{S} = \begin{bmatrix} S_0 \\ S_1 \\ S_2 \\ S_3 \end{bmatrix},$$

---

[5]Shurcliff; W. *Polarized Light,* Harvard University Press (1962); Chipman, R. et al. *Polarized Light and Optical Systems,* CRC Press (2019); Wolfe, W. *Optical Engineer's Desk Reference,* OSA and SPIE (2003).

[6]Jones, R. "A new calculus for the treatment of optical systems, IV," *JOSA* **32**(8), 486–493. (1942).

where S_0 represents unpolarized light, S_1 represents horizontal linearly polarized light, S_2 indicates linearly polarized light at 45 degrees, and S_3 is right circularly polarized light. It is obvious that unpolarized light has the following Stokes vector:

$$\mathbf{S} = \begin{bmatrix} 1 \\ 0 \\ 0 \\ 0 \end{bmatrix}.$$

Simlarly, a matrix with 1 in the second row and the rest as zeros represents horizontally linearly polarized light, and so on, by defintion.

Table A.1.2.1 gives the Stokes and Jones vectors for some common states of polarization.

Table A.1.2.1   Stokes and Jones Vectors.

State	Image	s_0	s_1	s_2	s_3	Jones
Linear horizontal	—	1	1	0	0	$\begin{bmatrix} 1 \\ 0 \end{bmatrix}$
Linear vertical	|	1	-1	0	0	$\begin{bmatrix} 0 \\ 1 \end{bmatrix}$
+ 45 degrees	/	1	0	1	0	$\begin{bmatrix} \surd2 \\ \surd2 \end{bmatrix}$
-45 degrees	\	1	0	-1	0	$\begin{bmatrix} \surd2 \\ -\surd2 \end{bmatrix}$
Right circular	⟲	1	0	0	1	$\frac{1}{\surd2}\begin{bmatrix} 1 \\ -i \end{bmatrix}$
Left circular	⟳	1	0	0	-1	$\frac{1}{\surd2}\begin{bmatrix} 1 \\ +i \end{bmatrix}$

The Jones matrices that show how the state of polarization is affected by various optical elements are four-element matrices of complex elements. Then the process for calculating the polarization change in a system is multiplication of the Jones vector by the Jones matrices as many times as there are elements in the system.

The final Jones vector **J** is the original one multiplied by all the matrices $\mathcal{J}$:

$$\mathbf{J}_{\text{final}} = \mathcal{J}\mathcal{J}\mathcal{J}\mathbf{J}_{\text{initial}}.$$

The general Jones matrix is a 2×2 array, as shown here:

$$\mathbf{J} = \begin{bmatrix} a & b \\ c & d \end{bmatrix}.$$

The simple, linear polarizer is $a = 1$ and b, c, and $d = 0$. Then,

$$\mathcal{J}\mathbf{J}_{\text{initial}} = \mathbf{J}_{\text{final}},$$

$$\begin{bmatrix} 1 & 0 \\ 0 & 0 \end{bmatrix} \begin{bmatrix} 1 \\ 1 \end{bmatrix} = \begin{bmatrix} 1 \\ 0 \end{bmatrix}.$$

The matrix that describes right circular polarization is

$$\frac{1}{2} \begin{bmatrix} 1 & j \\ -j & 1 \end{bmatrix},$$

and that for left circular polarization reverses the signs of the j.

The Mueller matrix for horizontal linear polarization is

$$\frac{1}{2} \begin{bmatrix} 1 & 1 & 0 & 0 \\ 1 & 1 & 0 & 0 \\ 0 & 0 & 0 & 0 \\ 0 & 0 & 0 & 0 \end{bmatrix}.$$

The matrix for vertical linear polarizations replaces the 1 in the first row, second column, and in the first column, second row, with –1.

Additional matrices for these calculi can be found in the references and on the Internet.[7]

A.1.3 Refraction

Snell's law, the law of refraction, can be derived from the electromagnetic boundary conditions. The tangential electric fields must be equal on both sides of the boundary. Thus,

$$E_i \exp j(\omega_1 t - k_1 \cdot r) = E_t \exp j(\omega_2 t - k_2 \cdot r).$$

---

[7]Wikipedia. "Mueller calculus," https://en.wikipedia.org/wiki/Mueller_calculus; "Jones calculus," https://en.wikipedia.org/wiki/Jones_calculus.

For the equality to hold, the exponents must be equal:

$$\omega_1 t - \mathbf{k_1} \cdot \mathbf{r} = \omega_2 t - \mathbf{k_2} \cdot \mathbf{r}.$$

The frequency does not change so that $\omega_1 t = \omega_2 t$, and they can be subtracted from both sides. Thus,

$$\mathbf{k_1} \cdot \mathbf{r} = \mathbf{k_2} \cdot \mathbf{r}.$$

Since k is $2\pi/\lambda$, and the dot product means to multiply by the sine, and the wavelength in a material is the vacuum wavelength divided by the refractive index, $\lambda = \lambda_0/n$:

$$2\pi/(\lambda/n_1) \sin \theta_1 = 2\pi/(\lambda/n_2) \sin \theta_2.$$

Canceling the 2's and π's, and putting the n's up where they belong results in Snell's law:

$$n_1 \sin \theta_1 = n_2 \sin \theta_2.$$

$$\text{Gimme a QED!}$$

When light enters a denser material from a less dense one, it bends toward the normal, as shown in the main text. When light exits a denser material into a less dense one, it bends away from the normal until it is totally reflected, when it cannot bend away any further. The angle at which this occurs is called the **critical angle** and is found when the sine of the exiting angle is equal to one because it cannot get any larger:

$$n_2 \sin \theta_2 = n_1 \sin \theta_1 = n_1 \times 1,$$

$$\sin \theta_2 = n_1/n_2,$$

$$\theta_2 = \arcsin(n_1/n_2) = \theta_{\text{critical}}.$$

The critical angle as a function of refractive index is shown in Figure A.1.3.1 for refractive indices from 1.5 to 4.0.

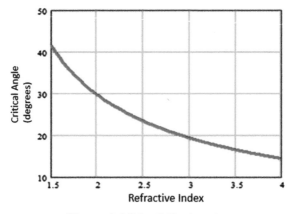

Figure A.1.3.1   Critical angle.

It is more difficult to derive the expressions for transmission and reflection. I just quote them here. See Born and Wolf if you must see the derivation.

The expressions for transmission amplitude into the material into which it is refracted for the two polarizations are

$$t_p = \frac{\sin 2\theta_i \sin 2\theta_t}{\sin^2(\theta + \theta)\cos^2(\theta_i - \theta_t)},$$

$$t_s = \frac{\sin 2\theta_i \sin 2\theta_t}{\sin^2(\theta_i + \theta_t)}.$$

It can also be obtained as the one's complement of the single surface reflectivity for a material with no absorption. It is shown in Figure A.1.3.2 for light polarized in the **plane of incidence** and for that polarized perpendicular to it.

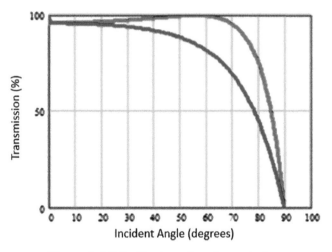

Figure A.1.3.2 Single-surface transmission.

Refraction depends upon the refractive index which, in turn depends upon the interaction of light with matter. This is the subject of dispersion theory. The refractive index n is given by

$$n^2 - 1 = \frac{Nq^2}{m\varepsilon} \cdot \frac{1}{\omega^2\omega_0^2 - j\omega g},$$

where m and q are the mass and charge of the electron, ε is the dielectric constant, N is the number of charges, ω is the variable frequency, ω_0 is the resonant frequency of the molecule, and g is the damping constant. This is of the form of a function that gradually decreases with frequency until it matches the resonant one where it peaks.

Several approximations have been made to this theoretical treatment to be able to fit data for lens design. These include the Cauchy, Sellmeier,[8] and Herzberger[9] versions. They all take into account that there are several resonant frequencies.

The **Cauchy formula** is

$$n = A + \frac{B}{\lambda^2} + \frac{C}{\lambda^4} + \cdots,$$

where A, B, and C are fitted constants, and λ is the wavelength in micrometers. For borosilicate glass, the constants are A = 1.5046 and B = 0.0042. This is a very limited fitting equation, based on wrong assumptions and valid only in the visible spectrum.

The **Sellmeier formula** is

$$n^2 = 1 + \frac{A\lambda^2}{\lambda^2 - \lambda_1^2} + \frac{B\lambda^2}{\lambda^2 - \lambda_2^2} + \frac{C\lambda^2}{\lambda^2 - \lambda_3^2} + \cdots,$$

where again A, B, and C are fitting constants, and so are the λ_i.

The Sellmeier constants for borosilicate glass are A = 1.03961212, B = 0.231792344, C = 1.10146195, $\lambda_1^2 = 0.00600069867$, $\lambda_2^2 = 0.0200179144$, $\lambda_3^2 = 103.560653$.

The **Herzberger formula** is

$$n = \frac{A}{\lambda^2 - 0.028} + \frac{B}{(\lambda^2 - 0.028)^2} + \frac{C}{(\lambda^2 - 0.028)^4} + \frac{D}{(\lambda^2 - 0.028)^6}.$$

Figure A.1.3.3 is a plot of the Cauchy and Sellmeier curves using the reported constants. It shows the Sellmeier curve gradually decreasing towards the next absorption curve as it should, but it is not so with the Cauchy curve.

Joe Nissley has reported on the use of the Herzberger and Sellmeier equations to represent the dispersion curves of several infrared materials.[10] The agreements are generally good with a few discrepancies that are probably small measurement errors.

---

[8]Sellmeier, W. "Über die durch die Aetherschwingungen erregten Mitschwingungen der Körpertheilchen und deren Rückwirkung auf die ersteren, besonders zur Erklärung der Dispersion und ihrer Anomalien," *Annalen der Physik und Chemie* **223**(11), 386–403 (1872).
[9]Herzberger, M. and Salzberg, C. "Refractive Indices of Infrared Optical Materials and Color Correction of Infrared Lenses," *JOSA* **52**, 420–427 (1962).
[10]Nissley, J. "Dispersion Curve Fitting in the Infrared," Thesis, University of Arizona (1979).

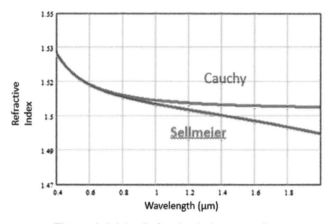

Figure A.1.3.3   Refractive-index curve fits.

A.1.4 Reflection

Specular reflection from non-conducting surfaces obeys the laws of Fresnel. The amplitude of reflection of light polarized in the plane of the paper is

$$r_p = \frac{n_2 \cos\theta_1 - n_1 \cos\theta_2}{n_2 \cos\theta_1 + n_1 \cos\theta_2} = \frac{\tan(\theta_1 - \theta_2)}{\tan(\theta_1 + \theta_2)}.$$

The amplitude of the reflection perpendicular to the plane of the paper is

$$r_s = \frac{n_1 \cos\theta_1 - n_2 \cos\theta_2}{n_1 \cos\theta_1 + n_2 \cos\theta_2} = \frac{\sin(\theta_1 - \theta_2)}{\sin(\theta_1 + \theta_2)}.$$

Both equations are obtained by solving the boundary conditions of the electromagnetic waves. The tangential components of the electrical and magnetic fields must be continuous. This detailed calculation is nicely done in the references.[11] The results were shown graphically in Section 1.4.

We can use the first form of either expression to get on familiar turf. When light is incident perpendicularly, there is no difference in polarization, and the angles are zero, so the cosines are 1:

$$r_s = r_p = \frac{n - 1}{n + 1}.$$

The usual expression is for the intensity of reflection and is the square of this equation.

As described in the main text, there are multiple reflections in a transparent medium. This gives rise to a different total reflection and

---

[11]Jenkins, F. and H. E. White, *Fundamentals of Optics*, McGraw-Hill (1957); Born, M. and Wolf, E. *Principles of Optics*, Pergamon (1959).

transmission of a transparent plane parallel plate than the simple expression. The ray reflects off the first surface with a reflectivity denoted as r_1. It transmits into the plate as $1 - r_1$. Assuming no absorption, it then reflects off the second surface as r_2 and transmits out as $1 - r_2$. It has successive reflections as indicated in Figure A.1.4.1.

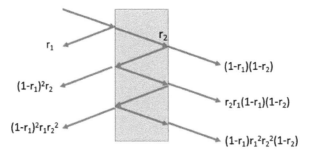

Figure A.1.4.1 Multiple reflections.

$$R = r_1 + (1 - r_1)^2 r_2 + (1 - r_1)^2 r_2^2 + \ldots,$$

$$T = (1 - r_1)(1 - r_2) + r_1 r_2 (1 - r_1)(1 - r_2) + \ldots.$$

If the same medium is on both sides of the plate, then $r_1 = r_2 = r$, and the expressions simplify:

$$R = r + (1 - r)^2 r + (1 - r)^2 r^3 + (1 - r)^5 r \ldots = r + (1 - r)^2 (r + r^3 + r^5 + \ldots),$$

$$R = r + \frac{r(1 - r)^2}{(1 - r^2)} = r + \frac{r(1 - r)}{(1 + r)}.$$

The transmission expressions are similar and lead to

$$T = \frac{(1 - r)}{(1 + r)}.$$

A uniform **diffuse reflecting surface** reflects equally in all directions in the hemisphere above it. It is often called a Lambertian reflector in honor of Johann Heinrich Lambert, and it is often characterized as the cosine law. The distribution of the intensity is cosinusoidal. A better way to characterize it as equal radiance in all directions.

The basic equation of transfer is

$$P = L A_i \cos \theta_1 A_2 \cos \theta_2 / r^2.$$

The differential area of a sphere of radius R is $R \sin\theta R d\varphi$, and the differential solid angle is this value divided by R^2:

$$d\Omega = \sin \theta d\theta d\varphi.$$

The integration is over the projected area $dA\cos\varphi$:

$$P = Ld\Omega dA\,\cos\theta d\theta d\varphi = \int\int L\,\sin\theta\cos\theta dAd\theta d\varphi$$

$$= LdA\int\int \sin\theta\cos\theta d\theta d\varphi,$$

where the φ integration is from 0 to 2π, and the θ integration is from 0 to $\pi/2$. Then,

$$M = P/A = L\Omega$$

$$= \int_0^{2\pi}\int_0^{\pi/2} \sin\theta\cos\theta d\theta df = 2\pi L\int_0^{\pi/2}\sin\theta\cos\theta d\theta = \pi L\int\sin^2\theta d\theta = \pi L.$$

The irradiance on the overlying hemisphere from a Lambertian (isotropic radiance) source is therefore

$$M = P/A = \pi L.$$

It is not $2\pi L$, as many think because there are 2π steradians in a hemisphere. One way to understand it is that the average area of the emitter is half its physical area because it emits in all directions. Another is that the average value of the cosine over half a period is one half.

The general case of reflection is incidence at any pair of angles and reflection at any pair as well. These angles are each both azimuth and elevation. The bidirectional reflectance distribution function (BRDF) is defined as the reflected radiance in the specified direction divided by the incident irradiance from its direction. It therefore depends on four angles and has the unusual unit of reciprocal steradian, although it is dimensionless. A steradian is a ratio.

To make this concept a little more concrete, consider two extreme and unrealistic examples:

- Assume perfectly collimated light is incident perpendicularly on a perfectly diffuse material so that the light is reflected uniformly in all directions of the overlying hemisphere with a reflectivity of ρ and an irradiance of M. The BRDF is ρ/π. Since the reflected radiance of a Lambertian surface is M/π, that is the result.

- Assume the reverse is true. Then the BRDF is infinite because the radiation is reflected into a zero solid angle (theoretically).

All BRDFs will fall somewhere between these unrealistic extremes.

A.1.5 Interference

This is the phenomenon of the interaction of two light waves, two electromagnetic waves. We can represent them mathematically in terms of sines. That is more convenient in this case than the exponential notation. Then the sum is

$$E_{\text{sum}} = E_1 + E_2 = A_1 \sin(\omega_1 t - k_1 r + \varphi_1) + A_2 \sin(\omega_2 t - k_2 r + \varphi_2),$$

where ω is the circular frequency $2\pi f$, t is time, k is $2\pi/\lambda$, and φ is the phase term. We can then consider increasingly complex situations. The simplest is when $\omega_1 = \omega_2$, $k_1 = k_2$ and $\varphi_1 = \varphi_2$. Then the interference is constructive, and the result is

$$E_{\text{sum}} = E_1 = (A_1 + A_2) \sin(\omega_1 t - k_1 r + \varphi_1).$$

If, however, the second wave is shifted in space by one half a cycle, there is destructive interference since $\sin(x + \pi) = -\sin(x)$, so the sum is zero. If the frequencies are the same (both ω and k) but the phase φ varies, then the sum will be the sum of the two waves with one displaced from the other. The more complex situation is when the frequencies ω and k (since $k = 2\pi/\lambda = 2\pi f/c = \omega/c$) are not equal. Then, if the phases are equal, the result is

$$E_{\text{sum}} = \sin(\omega_1 t - k_1 r + \varphi) + \sin(\omega_2 t - k_2 r + \varphi).$$

The pertinent trig identity is

$$\sin(x) + \sin(y) = 2 \, \sin[(x + y)/2] + 2 \cos[(x - y)/2].$$

The result involves the sum and difference frequencies in a complicated way:

$$E_{\text{sum}} = 2 \, \sin \tfrac{1}{2}(\omega_1 + \omega_2)t - \tfrac{1}{2}(k_1 + k_2)r + 2 \cos \tfrac{1}{2}(\omega_1 - \omega_2)t - \tfrac{1}{2}(k_1 - k_2)r.$$

This was shown in Section 1.5 but will be expanded upon here. Figure A.1.5.1 shows two waves with frequencies two percent different. The next figure shows their sum. It is another sine wave that gradually decreases. The third figure shows two waves with 10% frequency difference. By comparison, the waves start to get out of synch much sooner. They destructively interfere and then add together constructively, as shown in the final figure.

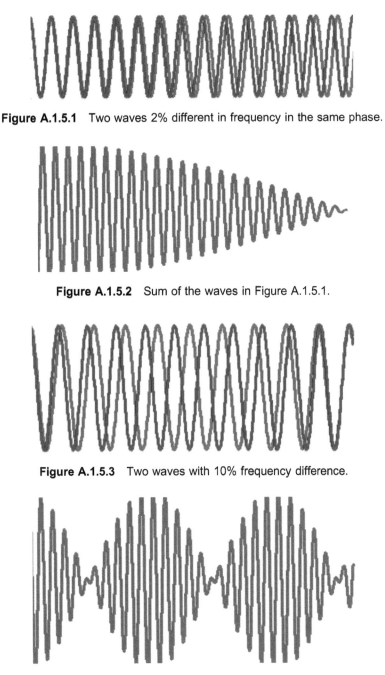

Figure A.1.5.1 Two waves 2% different in frequency in the same phase.

Figure A.1.5.2 Sum of the waves in Figure A.1.5.1.

Figure A.1.5.3 Two waves with 10% frequency difference.

Figure A.1.5.4 Sum of the waves in Figure A.1.5.3.

The result is similar when more than two waves interfere. Figure A.1.5.5 shows three waves that have slightly different frequencies. Figure A.1.5.6 show their sum. It is also a gradual increase and decrease as the waves synchronize and then desynchronize.

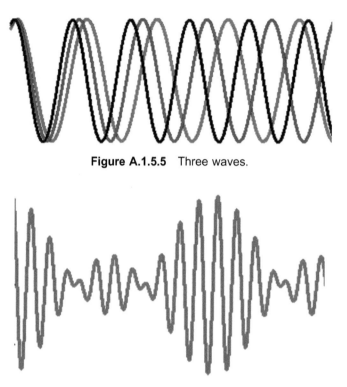

Figure A.1.5.5   Three waves.

Figure A.1.5.6   Sum of the waves in Figure A.1.5.5.

This is the basis of interference filters, discussed in Section 3.8. Add the right number of waves of the right frequencies to get maxima and minima you desire.

A1.6 Diffraction

Diffraction is the process by which light waves restart after they are interrupted by an aperture or obstacle. The waves start all over again as if emanating from new sources all along the wavefront that has been blocked. We do not know why, and the business of optics is not to deal with why, but what and how.

The diffracted light consists of the many new waves that are all summed up to get the total. This is accomplished with the Fresnel–Kirchhoff diffraction integral for unit amplitude, which is[12]

$$E(p) = \frac{jA}{2\lambda} \iint \frac{e^{jk(r+s)}}{rs} (\cos\alpha - \cos\beta)dS.$$

This equation gives the expression for the electric field at point P somewhere to the right in Figure A.1.6.1. In it, j is the imaginary number, λ is the wavelength, k is $2\pi/\lambda$, α and β are the angles that r and s make with the

---

[12]Born, M. and Wolf, E. *Principles of Optics*, Pergamon (1959).

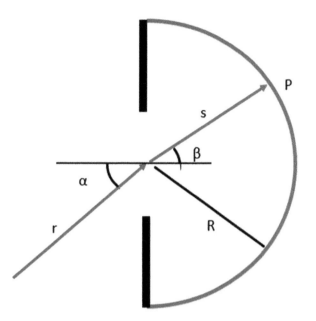

Figure A.1.6.1 Diffraction geometry.

surface normal, respectively, A is the aperture area, and r and s are the distances shown Figure A.1.6.1. Integration occurs over the spread of the wave behind the aperture, through all space to the right of the aperture.

The cosine inclination factor represents the angle of the incoming spherical wave that is incident upon the diffracting aperture or obstruction and the angle at which the light is diffracted. The integration is over the area of the aperture or obstruction and r is the distance in front of it and s the distance behind it.

When the distances are large, the diffraction is called Fraunhofer diffraction, as opposed to Fresnel diffraction. Fortunately, almost all practical problems in diffraction are of the Fraunhofer type. The integral simplifies. The inclination factor may be replaced by $\cos\delta$ and taken out of the integral since it varies very slowly with variation with incident angle that is also small. The range of integration is small so that r and s in the denominator may also be taken out of the integral. Then the expression for diffraction reduces to

$$E(p) = \frac{jA\cos\delta}{\lambda rs} \iint e^{jk(r+s)}dS,$$

where j represents a phase shift (that can usually be ignored), δ is the angle of incidence, and r and s are distances determined by the nature of the problem. It is convenient to write the expression for normal incidence (where $\cos\delta = 1$) as

$$E(p) = \frac{jA}{\lambda rs} \iint e^{jk(r+s)} dS.$$

For a distant source incident perpendicularly upon an optical instrument, Born and Wolf show that the intensity is given by

$$I = A/\lambda^2 = (d/\lambda)^2.$$

The intensity is proportional to the area of the aperture or obstruction divided by the square of the wavelength. That means that on a linear basis the spread of the diffraction pattern is inversely proportional to the number of waves in the linear dimension of the aperture or obstruction.

For a rectangular aperture, the solution of the integral is the sinc function $\sin(x)/x$ and $\text{sinc}(y)$; for a circular aperture, it is the besinc or jinc function $J_1(x)/x$. These are shown in Figure A.1.6.2. The red line is the sinc; the blue is the jinc. They are for a circular aperture with a diameter equal to the side of the square aperture, normalized to 1. Notice that the first zero of the sinc function is at $\pi = 3.14$, whereas the first zero of the besinc function is at 3.83 ($= 1.22 \times 3.14$), which explains the mysterious factor of 1.22 in the diffraction limit equations for circular apertures.

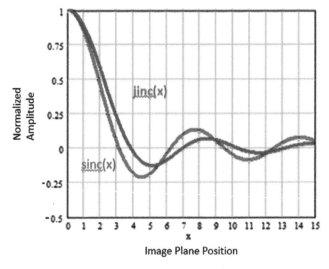

Figure A.1.6.2 Sinc and jinc functions.

It is interesting to speculate a bit about the relative diffraction performance of a square aperture versus a circular one. The first zero of the circular aperture occurs farther from the center than that of a square one, so it would appear that squaring the circle, as shown in Figure A.1.6.3, provides better diffraction performance. Would it be better to cut the sides off all circular telescope mirrors?

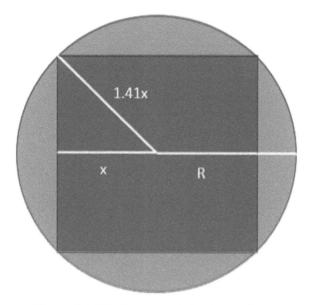

Figure A.1.6.3 Square and circular apertures.

The horizontal and vertical zeroes will occur closer for the square, 3.14 versus 3.83, but the diagonal ones for the square are 1.41 times 3.14 = 4.42. The diffraction patterns for the square are not symmetric. It turns out that about half (56%) of the zeroes of the square occur closer than those of the circle and half are farther. Figure A.1.6.4 shows normalized curves for the circular aperture in blue and the range of diffraction patterns for the square aperture in red. It appears that the diffraction pattern of the circular aperture is about the mean of those for the square aperture. There is a very minimal advantage in squaring the circle, if any.

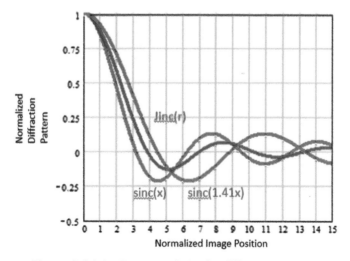

Figure A.1.6.4 Square and circular diffraction patterns.

A.1.7 Scattering

Aerosol scattering, scattering by suspended particles, can be divided into the regions in which the scattering particles are the same size or smaller than the wavelength of the light and those in which they are larger. The first case is called Rayleigh scattering. The second is called Mie scattering.

Rayleigh scattering applies to blue skies and red sunsets. The fraction of light F, that is scattered can be calculated with the following Rayleigh formula, where θ is the angle of scatter, R is the distance from the particle, n is its refractive index, λ is the wavelength of the light, and d is the particle diameter:

$$F = 8\pi^4 \frac{1 + \cos^2 \theta}{R^2} \quad \frac{n^2 - 1}{n^2 + 2} \quad \frac{d^4}{\lambda^4} \quad d^2.$$

A nitrogen molecule has a diameter of 0.155 μm, smaller than the wavelength of blue light (0.4 μm). Oxygen is just about the same, 0.152 μm. The refractive indices of oxygen, nitrogen, and even air are all about 1.005. So, the refractive index term is a small number, about 0.0033. For our atmosphere, the equation reduces to a term in angle $(1 + \cos^2\theta)$, the inverse square of distance, the ratio of particle diameter to wavelength squared, and the size of the particles. The larger the particles are with respect to wavelength, the greater the scattering.

The other situation, in which the particles are larger than the wavelength of the light passing through them, is called Mie scattering.[13] White clouds are **Mie scatterers**. There are expressions that have been derived for a small set of particles including spheres and cylinders. The solutions involve advanced mathematical functions such as spherical Bessel functions and Legendre polynomials. Suffice it to say that Mie calculations are highly complex, requiring the solution of electromagnetic boundary values for spheres and cylinders. Examples can be found in Stratton.[14] They are usually solved by computer codes. Consult Wikipedia for more information. An approximation of the Mie curve for a conducting spherical particle is shown in Figure A.1.7.1. The x axis is the particle diameter divided by the wavelength on a logarithmic scale. The curve shows the approximate fourth power for small particles and oscillates until it becomes normal reflection at a value of about 30 or 40.

---

[13] Mie, G. "Beiträge zur Optik trüber Medien, speziell kolloidaler Metallösungen," *Annalen der Physik.* **330**(3), 377 (1908).
[14] Stratton, J. *Electromagnetic Theory*, McGraw-Hill (1941).

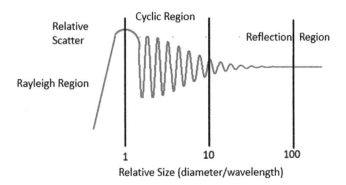

Figure A.1.7.1 Mie scatter by a spherical particle.

Surface scattering depends upon the surface structure, in terms of its microroughness, the distribution of very small heights as a function of position. It has been the subject of intense scrutiny in the design of anti-ballistic missile systems where scattering from sources out of the field of view of the telescope like the sun can really mask the target. The derivations have all come from perturbation theory applied to electromagnetic boundary conditions. One of the best treatments is by Church, and Tacas.[15] It relates the scattering to the spectrum of the surface height distribution. Their expression for the bidirectional reflectivity or BRDF is

$$\text{BRDF} = P(\varphi)R(\theta)S(f),$$

where, as usual, λ is the wavelength of the light, P is a polarization term that is either $\cos^2\varphi$ or $\sin^2\varphi$, R is the ordinary reflectivity of the surface, and S is the power spectral density of the surface height distribution that is a function of the surface spatial frequency f.

A.1.8 Absorption and Emission

I think (hope) the processes described in the main text are clear. Calculation of the various modes in carbon dioxide or the energy levels in sodium are beyond the scope of this book. I think it is worthwhile to review the thought processes that were probably those of Kirchhoff when he derived the fact that emissivity and absorptivity are equal; i.e., a good emitter is a good absorber.

We can imagine two infinitely long objects in radiative equilibrium, as in Figure A.1.8.1. They are at the same temperature and in equilibrium. We assume that neither transmits any power. That can be because they absorb it all and are long enough to do it. To maintain equilibrium, each piece must

---

[15]Church, E. and Tacas, P. "Surface Scattering," *Handbook of Optics*, Bass, M., ed., McGraw-Hill Chapter 7 (1995).

absorb just as much energy as it emits. Its emission equals its absorption. The efficiencies of emission and absorption are equal: $\alpha = \varepsilon$. If one absorbs more than it emits, it will heat up and destroy the equilibrium.

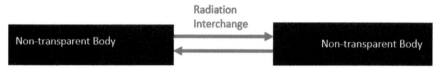

Figure A.1.8.1   Illustrating Kirchhoff's law.

This is also true for spectral quantities. Just imagine a perfect spectral filter between them and repeat the logic. Then, $\alpha(\lambda) = \varepsilon(\lambda)$ at each wavelength but not if the spectral regions are different.

In Section 4.40, the concept of an α over ε ratio, α/ε, is introduced in which that ratio was not equal to 1. Kirchhoff's law applies to equilibrium and the same conditions. That ratio is the absorption of sunlight in the visible divided by the emission of radiation in the infrared. These are not the same conditions, but it is a good measure of radiative cooler performance.

A.1.9 Propagation

The propagation of light (electromagnetic waves) is specified in electromagnetic theory by the Poynting vector **EXH**. It is the cross product of the **E** and **H** fields. Both are perpendicular to the direction of propagation and each other so that the cross-product is perpendicular to them and in the direction of the propagation. It was introduced by John Henry Poynting (it is a shame that his name was not spelled Pointing) in 1844.[16] It is usually denoted by **S** but sometimes by **N** and in boldface capitals to denote that it is a vector. It has the dimensions of watts per square meter. It is the M of radiometry, that is it is power per unit area.

Propagation and the calculation of flux on a detector in an optical system is of two types: (1) if the source is not resolved by the instrument, i.e., if it is smaller than the image of the detector on it, and (2) if it is resolved. The first case is generally called the point source case; the second is the extended source case. No light could come from a true point which has no size. The appellation refers to a source that is smaller than the resolution element of an optical system. It is a useful fiction.

The **point source** calculation starts with the calculation of the radiant intensity of the source. That starts with its temperature and its emissivity. They are usually calculated as $\varepsilon(\lambda)M^{BB}(\lambda)$, the spectral emissivity times the blackbody radiant emittance in watts per square meter. The intensity is found

---

[16]Poynting, J. "On the transfer of energy in the electromagnetic field," *Philo. Trans. Royal Soc. London* **175**, 343–361 (1884).

by multiplying this value by the source area and dividing by the solid angle into which it radiates. If it is an isotropic source, such as a small sphere, it radiates equally into all parts of a sphere. Then, since there are 4π steradians in a sphere, the intensity I is

$$I(\lambda) = \frac{\varepsilon M^{BB}(\lambda)}{4\pi}.$$

We can drop the emissivity, wavelength, and blackbody indications since we are only interested in the geometry:

$$I = \frac{M}{4\pi}.$$

The intensity has units of watts per steradian. The power on the receiver is this radiant intensity times the solid angle Ω the receiver subtends times the transmission losses:

$$P_d = \tau I \Omega_{os} \frac{\tau I A_o \cos\theta}{R^2}.$$

For an **extended** source, the radiance is key. The radiance is power per unit area of the source per solid angle into which it radiates. The radiance is constant as described in the main body and below. It is the same on the receiver as it is from the source. Thus, the power on the detector is given by

$$P_d = \frac{\tau A_s A_o}{R^2} = \tau A_s \Omega_{so} = \frac{\tau A_o A_d}{f^2} = \tau A_o \Omega_{do} = \tau A_d \Omega_{od}.$$

That is quite a mouthful, or rather, a line full. It is five different ways to calculate the power on a detector depending upon what you know. The tau is the total transmission, i.e., that of the atmosphere and the optics. A_s is the projected source area, A_o is the projected optics area, and A_d is the projected detector area. It is a little simpler that way. The solid angles are represented by Ω with subscripts to indicate which. The subscript os means the area of the optics subtended at the source, and so on.

In the main body of the text, it was shown that radiance is constant throughout an optical system by invoking the first and second laws of thermodynamics. Here, it is done in a little more detail with only the first law and some geometry. Consider the geometry of Figure A.1.9.1.

The power in the system must be conserved. The power on each of the surfaces is the same. The power per unit area changes according to the area. The power in terms of the radiance is

$$P = \frac{L_1 A_s A_1}{d_1^2} = \frac{L_2 A_s A_2}{d_2^2} = \frac{L_3 A_s A_3}{d_3^2}.$$

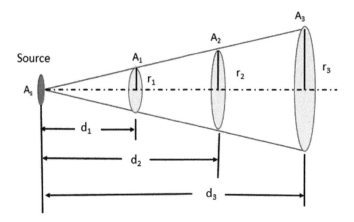

Figure A.1.9.1 Radiance geometry.

It is assumed initially that the radiance may be different on the three different surfaces, i.e., L_1, L_2, and L_3.

The radii and distances form similar triangles so that the surfaces ratios can be written as

$$\frac{A_1}{d_1^2} = \frac{A_2}{d_2^2} = \frac{A_3}{d_3^2},$$

and it is an identity in that $A_s = A_s = A_s$. Therefore, all of the radiances must be equal. I get another QED!

It can also be shown that the reduced radiance L/n^2 is constant through systems such as immersion microscopes that operate in media other than air.

The differential form of a solid angle is

$$d\Omega = \sin\theta\cos\varphi\, d\theta\, d\varphi.$$

Snell's law in air where the first refractive index is assumed to be one is

$$\sin\theta_1 = n\sin\theta_2.$$

The derivative is

$$\cos\theta_1 d\theta_1 = n\cos\theta_2 d\theta_2.$$

By multiplying these equations together and noting the two angles, we find that the relationship for solid angles in two media, $n = 1$ and $n = n$, is

$$d\Omega = n^2 d\Omega.$$

The solid angle in the medium is smaller than that in air by the square of the refractive index as shown in Figure A1.9.2. Therefore, the invariant must be L/n^2 to conserve power. This is sometimes called the reduced radiance.

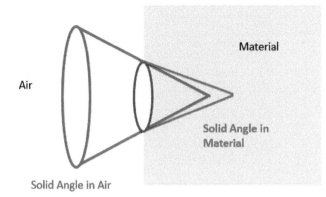

Figure A.1.9.2 Solid angle relations.

A.1.11 Fermat's Principle

This is the principle that light takes the minimum time to go from A to B. That is equivalent to taking the minimum optical path *nd*. It can be stated mathematically as

$$\delta \int nds = \delta \int c/v \ ds = c\delta \int dt = 0.$$

The change in optical path *ns* or the time *t* is zero. That may be any point of inflection, but physical reasoning dictates it is a minimum. It was illustrated in the main text by the distances a photon reflecting from a mirror between two points and via a basketball bounce and that of a pool ball. Here, we give two more mathematical and more complicated examples: one is a swimmer, and the other is an optical ray. The swimmer does not encounter a ray!

Assume a lifeguard wants to get to a drowning swimmer in the shortest time. The geometry is shown in Figure A.1.11.1. She is a distance AB from the shoreline and a distance BF from the shoreline to the swimmer in the water. She runs at a velocity of v_1 and swims at a velocity of v_2. Accepted values for v_1 and v_2 are 10 mph about 8 ft/s and 2 mph about 2.5 ft/s. Based on our knowledge of Fermat's principle and the way lifeguards behave instinctively, she will run farther on the sand than swim in the water. It remains to be seen (calculated) what the distances AD and BD need to be to get there the quickest. It is possible to derive the equations, differentiate them, and set them to zero to find the minimum, but I chose to use a spreadsheet. I have chosen a convenient geometry with no loss of generality. The distance AE is equal to EG. The angle between AE and AG is then 45 degrees, and AEG and CFG are similar equilateral triangles. I have also chosen AB to be 20 feet and BF to be 10 feet. Then FG is 30 feet. The hypotenuses are $20\sqrt{2}$ and $10\sqrt{2}$. The time it takes for her to run and swim are, respectively, $20\sqrt{2}/v_1$ and $10\sqrt{2}/v_2$, or 28.28/8 and 14.14/2.5 = 3.53 + 5.66 = 9.20 seconds.

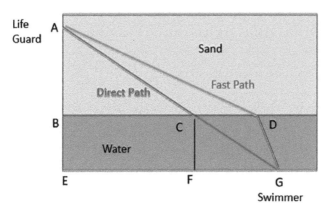

Figure A.1.11.1   Lifeguard geometry.

The following spreadsheet shows the result. The **shortest distance** is for the angle of 45 degrees. The **quickest route** is when the angle is 54.13 degrees and the distance to the shore is 34.13 feet, and in the water only 10.217 feet. This route is in column H. It only takes a total of 8.83 seconds, whereas the direct route would be 9.82 seconds.

	A	B	C	D	E	F	G	H	I	J
1			Lifeguard Rescue							
2										
3	Symbol	Formula				Values				
4	α	assumed	45.00	50.00	52.00	53.00	54.10	54.13	54.14	58.30
5	α	rads	0.79	0.87	0.91	0.93	0.94	0.94	0.94	1.02
6	AB	assumed	20							
7	EG	assumed	30.00							
8	BE	assumed	10.00							
9	v1	assumed	8							
10	v2	assumed	2.25							
11	AD	C6/cos(C5)	28.28	31.11	32.49	33.23	34.11	34.13	34.14	38.06
12	BD	C6xtan(C5)	20.00	23.84	25.60	26.54	27.63	27.66	27.67	32.38
13	FG	C7-C12	10.00	6.16	4.40	3.46	2.37	2.34	2.33	-2.38
14	DG	√(C8²+C13²)	14.14	11.75	10.93	10.58	10.28	10.27	10.27	10.28
15	t1	AD time	3.54	3.89	4.06	4.15	4.26	4.27	4.27	4.76
16	t2	DG time	6.29	5.22	4.86	4.70	4.57	4.56	4.56	4.57
17	t	total tme	9.82	9.11	8.92	8.86	8.83117	8.83115	8.83116	9.33
18		conversion	0.02 rad/deg							

The situation in which light travels through a plane-parallel plate is a bit different but very similar. I have assumed that the direct path angle is 45 degrees to make it simple but with no loss of generality Then, AE = EG, AB = BC, and AC = $\sqrt{2}$ times either one. The geometry is shown in Figure A.1.11.2.

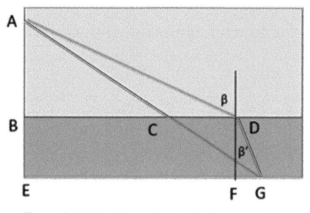

Figure A.1.11.2 Plane-plate optical ray geometry.

The results are shown in the spreadsheet, especially the times printed in red. There is not much difference, but the refracted path does take less time.

	A	B	C	D	E	F	G	H	I	J	K	L
1					**Parallel Plate Ray Paths**							
2						Direct Path ACG				Refracted Path ADG		
3	Symbol	Formula	Value	Units		Formula	Value	Units	Symbol	Formula	Value	Units
4	π	known	3.14159	#								
5	α				AC	√2xD6	28.2	cm	α	assumed	50.23	deg
6	AB	assumed	20	cm	AG	√2x(D6+D7)	42.3	cm	α	C12xK5	0.88	rad
7	BE		10	cm	CG	G6-G5	14.1	cm	AD	A6/cos(K6)	31.27	cm
8	EG		30	cm	t_1		9.4		t_1		10.42	s
9	n	assumed	1.5	#	t_2		7.05		β	J6	0.88	rad
10	v_1	assumed	3	cms^{-1}	t_{direct}	G8+G9	16.45		β'		0.54	rad
11	v_2	assumed	2	cms-1					DG		11.65	cm
12	conversion		0.01745	rad/deg					t_2		5.82	s
13									$t_{refracted}$		16.24	s

A.1.16 Radiometry

One useful and instructive radiometric calculation is the solar constant in both power and photons. The sun may be taken as a blackbody at 5782 K. Its total radiant emittance is 64 megawatts per square meter.

The radius of the sun is 696,330 kilometers, and the distance from the sun to the Earth is 150×10^9 meters, by the Stefan Boltzmann law.

The square of the ratio of the distance from the center of the sun to the Earth compared to radius of the sun is 0.04633.

The solar constant at the top of the Earth's atmosphere is therefore $64 \times 10^6 / 04633 = 1360$ Wm^{-2}.

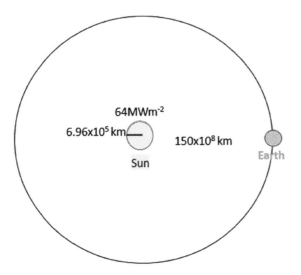

Figure A.1.16.1   The solar constant geometry.

And this is almost exactly what they have measured: 1361 to 1362.[17] It is the inverse square law of propagation at work.

The same geometric ratio can be applied to the photon emittance of the sun, which can be approximated with the use of the average energy of a photon, which is approximately kT, but it is even simpler to apply it to the solar constant:

$$E_q = 1380/kT = 1380/(1.38 \times 10^{-23} \times 5800)$$

$$= 1.72 \times 10^{22} \text{ photons per square meter per second.}$$

[17]Wikipedia. "Solar Constant," https://en.wikipedia.org/wiki/Solar_constant.

Appendix 2
Optics in Nature

It can be shown by the use of Fermat's principle of least time that rainbows only appear at about 42 degrees from the sun.

Fermat's principle in this case states that the path of least time is the path of minimum deviation. Cut that corner as close as you can! Minimum deviation is found by differentiating the total angle of deviation, setting it to zero, and solving.

Figure A.2.9.1 shows that the total angle of deviation D consists of two refraction angles α and four reflection angles β:

$$D = 2\alpha + 4\beta.$$

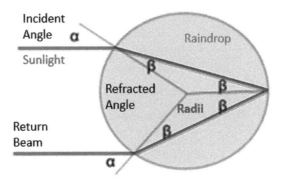

Figure A.2.9.1   Rainbow deviation angle.

Then the derivative of D with respect to the solar angle is taken and set equal to zero so that

$$dD/d\alpha = 2 + 4d\beta/d\alpha = 0.$$

Thus,

$$d\beta/d\alpha = -1/2$$

From Snell's law with the index of air = 1,

$$\sin\alpha = n\sin\beta.$$

The derivative of that equation is

$$\cos\alpha \, d\alpha = n\cos\beta \, d\beta$$
$$\cos\alpha = n\cos\beta \, d\beta/d\alpha = -(n/2)\cos\beta.$$

Squaring both sides gives

$$\cos^2\alpha = (n/2)^2\cos\beta.$$

The trig identity $\sin^2 + \cos^2 = 1$ or $\cos^2 = 1 - \sin^2$ yields

$$(1 - \sin^2\alpha) = (n/2)^2(1 - \sin^2\beta)$$
$$1 - \sin^2\alpha = n^2/4 - (n^2/4)\sin^2\beta/4.$$

Using Snell's law, we get

$$1 - \sin^2\alpha = n^2/4 - (\sin^2\alpha)/4.$$

So

$$4 - 4\sin^2\alpha = n^2 - \sin^2\alpha,$$

and

$$4 - 3\sin^2\alpha = n^2$$
$$3\sin^2\alpha = 4 - n^2.$$

Finally,

$$\alpha = \arcsin\sqrt{[(4 - n^2)/3]}.$$

This can be evaluated using appropriate values of n for the water in the droplets. The angle α is given by this equation. The angle β is obtained from Snell's law. The total deviation angle is $\pi - 2\alpha - 4\beta$, where the π indicates the 180 degree reflection back to the observer.

Figure A.2.9.2 shows the range of viewing angles as a function of the range of wavlengths from 0.4 μm to 0.8 μm in the visible. For these wavelengths, the refractive index of water ranges from about 1.33 to about 1.345, as shown. Although the calculation involves several trig functions, the

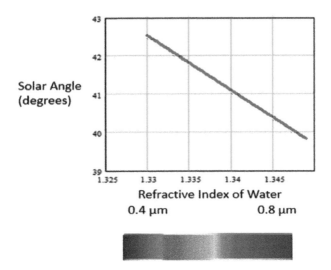

Figure A.2.9.2 Solar angle vs wavelength.

result is very linear because the refractive index does not change much. That is why the colors are bunched and always at 40.2 to 42.5 degrees from the sun angle.

Appendix 3
Components

A.3.3 Blackbodies

This section addresses the different forms of blackbody expressions and the ways they may be calculated, as well as their realization in instrumental form.[1]

Perhaps the most common form of the blackbody equation, the Planck equation, is **flux density** (also called **emittance** or **exitance**) as a function of wavelength with units of watts per square meter or square centimeter:

$$M(\lambda) = \frac{2\pi c^2 h}{\lambda^5(e^{hc/\lambda kT} - 1)} = \frac{c_1}{\lambda^5(e^{c_2/\lambda T} - 1)},$$

where c_1 and c_2 are known as the first and second radiation constants and will be used in the ensuing discussion. The first radiation constant c_1 is usually given in the literature as $3.741771852 \times 10^{-16}$ W·m^2. I prefer 37418 Wcm^{-2}μm^5 to give the flux density in watts per square centimeter and the spectral distribution in terms of wavelength in micrometers. I have never found the additional accuracy useful. The second radiation constant is usually specified as 1.438775×10^{-2} m·K. I usually use it as 14388 μm·K, as I usually work in micrometers and kelvins and do not need the extra accuracy.

A second expression of blackbody radiation is easily obtained. It is the expression for **radiance** as a function of wavelength. For simplicity, I will substitute x for $c_2/\lambda T$ since it always appears that way. Then the radiance is M_λ/π, or

$$L(\lambda) = \frac{M(\lambda)}{\pi} = \frac{2c^2 h}{\lambda^5(e^x - 1)} = \frac{c_1/\pi}{\lambda^5(e^x - 1)}.$$

These expressions can also be written in terms of photons, that is, the radiance and radiant emittance of photons as a function of wavelength or of frequency. Each term is divided by the energy of a photon, hc/λ:

---

[1] Wolfe, W. and Zissis, G. *The Infrared Handbook*, US Government Printing Office (1978).

$$M_q(\lambda) = \frac{2\pi c}{\lambda^4 (e^x - 1)} = \frac{c_1/hc}{\lambda^4(e^x - 1)}.$$

These two functions, emittance and radiance, can also be expressed as functions of frequency. Note that the distributions cannot be equated, only the actual amounts: $M_\lambda \neq M_\nu$; only $M_\lambda d\lambda = M_\nu d\nu$. (One cannot equate watts per area per **wavelength** with watts per area per **cycle per second**.) Thus, $M_\nu = M_\lambda d\lambda/d\nu = M_\lambda c/\nu^2$. Therefore,

$$M_\nu = M_\lambda (d\lambda/d\nu) = M_\lambda(\lambda^2/c).$$

The resultant expression for flux density as a function of radiation frequency is

$$M_\nu = M_\lambda(\lambda^2/c) = \frac{2\pi c^2 h}{\lambda^5 (e^x - 1)} \frac{\lambda^2}{c} = \frac{2\pi ch}{\lambda^3(e^x - 1)} = \frac{2\pi h\nu^3}{c^2(e^x - 1)}.$$

The radiance is determined by dividing by π and equivalent expressions for photon emittance and photon radiance as a function of frequency are obtained by dividing by $h\nu$:

$$M = \frac{2\pi^5 k^4 T^4}{15 h^3 c^2} = \sigma T^4.$$

It may be instructive to see some of these different functions plotted. The functions M_λ and L_λ differ only by the factor of π; they are not both shown. However, $M_{q\lambda}$ and M_λ differ by a wavelength factor. They are both shown as a function of wavelength on the same graph but are normalized. The curves for M and M_q as a function of wavelength in Figure A.3.3.1 show that M_q (in blue) peaks at a shorter wavelength and is flatter at longer wavelengths. This example is just for one value of temperature, 500 K.

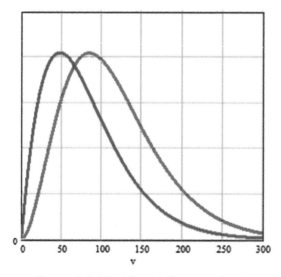

Figure A.3.3.1 M_λ and $M_{q\lambda}$ normalized.

The equivalent functions M_ν and $M_{q\nu}$ as a function of frequency are shown in Figure A.3.3.2, also normalized. The x axis is frequency in teracycles per second (THz). The photon function has its maximum to the left of the power function, as it should be, to lower frequencies and longer wavelengths.

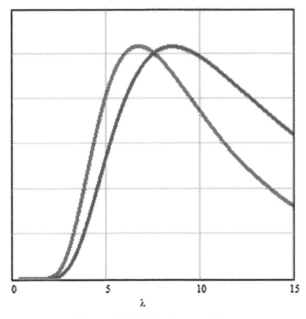

Figure A.3.3.2 $M_{q\nu}$ and $M_{q\nu}$.

We note just a few more properties: the maxima of the different functions and the total integrals, and the total power or photon areal density. They are listed in the table. The maxima are the same for radiance, but the totals are the values given divided by π. The total values for both photons and power are the same whether the distributions are in terms of frequency or wavelength. The total is the total is the total (like roses). It is interesting that for room temperature (about 80 °F or 300 K), everything around us radiates at about a half a kilowatt per square meter. The world is afire, or at least warm, in Tucson in June!

Expression	Maximum	Units	Total	Total	Units
M_λ	$\lambda_{max} = 2898/T$	μm/K	σT^4	$5.6704 \times 10^{-8} T^4$	W/m^2
M_q	$\lambda_{max} =$			$1.5202 \times 10^{11} T^3$	1/sm^2
M_ν	$\nu_{max} = 5.879 \times 10^{10}$	HzK	σT^4	$5.6704 \times 10^{-8} T^4$	W/m^2
$M_{q\nu}$	$\nu_{max} =$	HzK	$\sigma T^4/2.75kT$	$1.5202 \times 10^{11} T^3$	1/sm^2

It may be useful to show just one more representation of these functions on a logarithmic scale. Figure A.3.3.3 shows $M(\lambda)$ for temperatures of 300, 330, and 350 K. This has the nice feature that all the curves are the same shape and the maxima fall on a straight line. The curves are for temperatures of 1000, 1100 and 1200 K.

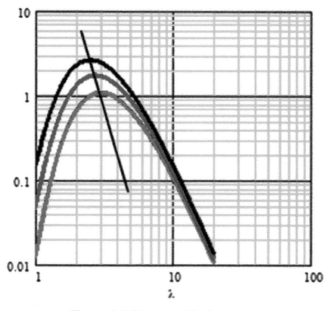

Figure A.3.3.3 Logarithmic curves.

The change in blackbody radiation with respect to temperature is useful in the calculation of NETD for some medical infrared instruments. It can be found by differentiating the Planck function with respect to temperature. It is

$$\frac{dM(\lambda)}{dT} = \frac{xe^x}{e^x - 1} \frac{M(\lambda)}{T}.$$

It is shown in comparison with the blackbody flux density in Figure A.3.3.4. The temperture change curve is shown in blue, and it has been multiplied by 55 to make the comparison clear (again, normalized).

The derivation for the change in radiant emittance with temperature is as follows. The expression for radiant emittance $M(\lambda)$ is

$$M(\lambda) = \frac{c_1}{\lambda^5 (e^x - 1)}.$$

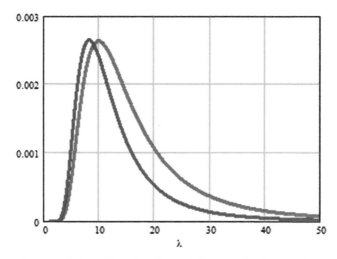

Figure A.3.3.4 Flux density and change with temperature.

The only term in temperature is x, which is $c_2/\lambda T$, so the deifferentiation may be done by the chain rule:

$$\frac{dM}{dT} = \frac{dM}{dx}\frac{dx}{dT}.$$

The differentiation with respect to x is

$$\frac{dM}{dx} = \frac{c_1 e^x}{\lambda^5 (e^x - 1)^2} = \frac{e^x}{e^x - 1} M,$$

and the differentiation of x with respect to T is

$$\frac{dx}{dT} = \frac{d}{dT}\left(\frac{c_2}{\lambda T}\right) = \frac{c_2}{\lambda}\frac{1}{T^2} = \frac{x}{T},$$

The final result, obtained by multiplying them together, is (as claimed)

$$\frac{dM(\lambda)}{dT} = \frac{x e^x}{e^x - 1}\frac{M(\lambda)}{T}.$$

The total radiation from a blackbody is given by the Stefan–Boltzmann expression:

$$M = \sigma T^4,$$

where $\sigma = 2\pi^5 k^4/15 h^3 c^2$. This can be verified by integrating the Planck function. The frequency form cited above is the most convenient:

$$M = \frac{2\pi h}{c^2}\int \frac{v^3}{e^x - 1}\,dv.$$

The integration can be performed by converting the frequency v to the normalized frequency x so that all variables in the integral are the same. The normalized variable x is

$$x = \frac{hv}{kT},$$

so that

$$v = \frac{kT}{h}x,$$

and the derivatives have the same relationship. Thus,

$$M = \frac{2\pi k^4 T^4}{h^3 c^2} \int \frac{x^3}{e^x - 1} dx.$$

The infinite integral is known to be $\pi^4/15$ so that

$$M = \frac{2\pi^5 k^4 T^4}{15\, h3c^2} = \sigma T^4.$$

I will leave it to you to show that, by inserting all the known constant values,

$$\sigma = 5.670374419184429453970996731889230875840122970291130\ldots$$

$$\times 10^{-8}\,\mathrm{W/m^2 K^4}.$$

A similar calculation can be done for the total number of blackbody photons emitted at a specified temperature. It starts with the expression for the number of photons as a function of frequency. That is not given above, but it is the expression in terms of frequency divided by the energy of a photon:

$$M_q = \frac{2\pi}{c^2} \int \frac{v^2}{e^x - 1} dv.$$

The same procedure of replacing the frequency with the normalized frequency to have a uniform integral must be done. That is,

$$v^2 dv = \frac{(kT)^2}{h^2} \frac{kT}{h} x^2 dx = \frac{k^3 T^3}{h^3} x^2 dx.$$

The result is

$$M_q = \frac{2\pi k^3 T^3}{h^3 c^2} \int \frac{x^2}{e^x - 1} dx.$$

All is known except the value of the infinite integral. I have found the value of this infinite integral as 2.404 by numerical integration using Mathcad. Therefore,

$$M_q = \frac{4.808\pi}{h^3 c^2} k^3 T^3.$$

Others have obtained the same result by partial integration and the use of the Riemann zeta function.[2]

The ratio of these two expressions for total blackbody emittance and total blackbody photon emission is $2.70kT$. Just do the division. Then the average energy of a photon is 1.38×10^{-23} times 2.7 times the temperature in kelvin $= 3.73 \times 10^{-23}$ T. And the average energy of a photon from our sun is 2.2×10^{-19} Watt-seconds. It takes a lot of them to generate the solar constant of about 1 kW per square meter—as Carl Sagan used to say, billions and billions.

A.3.5 Detectors

As noted in the main text, there are two types of radiation detectors: thermal and photon.

Thermal detectors operate by changing their resistance R with a change in temperature.[3] They do this better when they have a high coefficient of temperature change, generally denoted by α:

$$\alpha = \frac{1}{R}\frac{dR}{dT}.$$

The change in R, that is, dR, is thus $\alpha R dT$. Sensitive operation requires dT to be large. This is accomplished with a small specific heat c_p and the right geometry. The change in temperature caused by heating is given by

$$dT = \frac{\alpha P t}{m c_p},$$

where αP is the absorbed power, and t is the length of time the radiation is on the detector. The product of P and t is the absorbed heat energy. To maximize this change, the material should have a low specific heat and a small mass. This means it should be thin but have a large area. It should also be fast. So, it is always small. It is also important that the heat stay with the element and not be

---

[2]"Calculating Blackbody Radiance V2," spectralcalc.com.
[3]Kruse, P. *Uncooled Thermal Imaging*, SPIE Press (2001).

conducted away quickly. The rate of conduction is dictated by the lead thermal resistivity and geometry. The thermal energy conducted away is given by

$$Pt = \frac{A}{\rho l}.$$

The electrical leads should have high thermal resistivity ρ, a small area A, and be as long as practical.

All of this dictates the geometry of microbolometer arrays as shown in Figure A.3.5.1. The individual elements have a black coating (it is in the infrared) to maximize α, are reasonably thin, and have areas as large as allowed. They also must be small for high resolution. They are supported by thin leads that are thick enough for support but as long as is practical to maximize the length l. They are shown on the right of the figure.

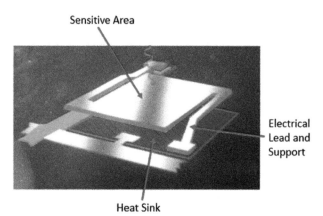

Figure A.3.5.1   Representative bolometer array element.

Detailed considerations by Kruse show that for fast response, these thermal detectors should be small, thin, and largely decoupled from any heat sinks.

Photon detectors detect radiation when a bound electron is freed to the conduction band. These are shown in Figure A.3.5.2. Electrons in the conduction band are free to move and conduct electricity. Those in the valence band are bound to their molecular sites. None can exist in the forbidden gap.

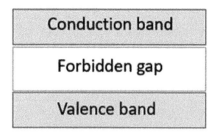

Figure A3.5.2   Bands.

The size of the forbidden gap is dictated by the material and is usually specified in terms of electron volts (eV). The relationship between it and photon energy is 1 eV = 1.24/λ μm. Silicon has a bandgap of 1.14 eV. Therefore, only photons with wavelengths shorter than 1.08 μm can jump the gap. A detector that senses photons with wavelengths as long as 12 μm in the infrared must have a bandgap no larger than 0.1 eV. It is then susceptible to thermal transitions with an energy of kT. Boltzmann's constant is 8.6 × 10^{-5} eVK^{-1}. At room temperature (300 K), these thermal vibrations (phonons) have energies of 0.026 eV, an appreciable fraction of the bandgap. That is why these detectors must be cooled. A typical temperature, such as that of liquid nitrogen, is 77 K. This results in phonon energy of 0.01 eV, or 10% of the gap.

Figure A.3.5.3 shows the specific detectivity of a host of infrared detectors.[4] The photon detectors are those that increase gradually from left to right and have peak values of a little more than 10^{10}. The thermal detectors have values of almost 10^{9} and are spectrally flat. This is because D^{*} is in terms of watts. The thermal detectors respond uniformly to power, but the photon detectors respond uniformly to photons. Photons have decreasing energy with increasing wavelength, hc/λ. As might be expected, visible detectors that would be illustrated to the left of this figure have even higher values of specific detectivity and similar shapes. They increase almost linearly until they reach the wavelength where the photon energy is no longer sufficient to support detection.

---

[4]Wolfe, W. and Zissis, G. eds. *The Infrared Handbook*, US Government Printing Office (1978).

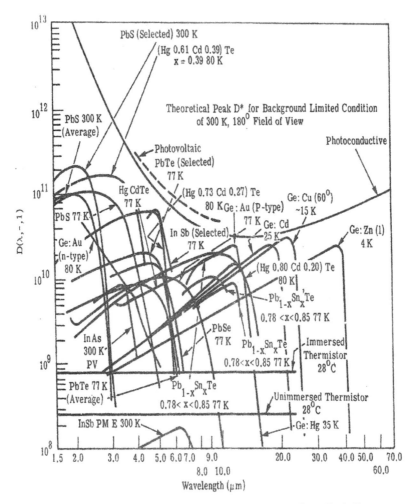

Figure A.3.5.3 Specific detectivities (reprinted from Ref. 4).

A.3.6 Diffraction Gratings

The basic equation for the performance of diffraction gratings is[5]

$$m\lambda = s(\sin\alpha + \sin\beta),$$

where m is the order number, s is the line spacing, α is the angle of incidence, and β is the angle of diffraction. From this, it is easily derived by simple differentiation that the angular dispersion that may be used for instrument design is, assuming quite realistically, a constant angle of incidence:

$$\frac{d\beta}{d\lambda} = \frac{m}{s\cos\beta}.$$

[5]Sawyer, R. *Experimental Spectroscopy*, Prentice Hall (1944).

For normal incidence and the first order, the cosine is 1, and the order is 1. It is just the reciprocal of the line spacing. If the grating has a typical line spacing of 1000 lines per millimeter and the wavelength is 1 μm, the angular dispersion is $0.001/1000 = 10^{-6}$ radians. Two lines that are one micrometer in wavelength will be one microradian apart.

The resolving power $\lambda/d\lambda$ is found in a similar way to be the product of the order number and the total number of lines in the grating.[2] Using the expression for the angular dispersion above, resolving power may be written as

$$\frac{\lambda}{d\lambda} = \frac{\lambda}{d\beta}\frac{d\beta}{d\lambda} = \frac{\lambda}{d\beta}\frac{m}{d\cos\beta}.$$

The term $d\beta$ may be determined from the following geometric considerations shown in Figure A.3.6.1.

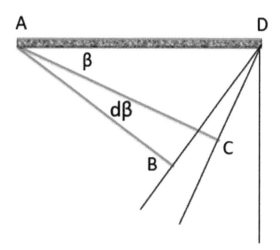

Figure A.3.6.1 Diffraction geometry.

The diffraction grating is AD; and the just resolved beams are AC and AB. They are separated in wavelength by $d\lambda$ and by angle by $d\beta$. The two black lines are respectively perpendicular to the two red ones. By perpendicularity, the angles on the right are β and $d\beta$, as shown. Then, AD = AB cosβ = Nd cosβ, where N is the number of lines in the grating, and d is their spacing. By the same geometry, $d\beta/d\lambda$ is $m/\cos\beta$. Thus,

$$\frac{\lambda}{d\lambda} = Nd\cos\beta\frac{m}{d\cos\beta} = mN.$$

The resolving power of a grating is the total number of lines in it times the order in which it is used. Notice that this is linear as opposed to the nonlinear dispersion of prisms. And it is generally greater than the value for prisms as noted in the main text.

A.3.7 Fibers

The conduction of light in a fiber requires that it be incident at a sufficiently steep angle. Figure A.3.7.1 shows the geometry for the calculation of that limiting angle. It is done three ways in spreadsheets and with Mathcad. The first two are in terms of radians and degrees using the linearized version of the law of refraction; i.e., the angles are proportional to the refractive indices. The third is the full-blown version with sines and radians. It shows that the approximations were not so good and that the input angle is limited a bit more than you might expect from the critical angle. The angles are too big for linearity to hold.

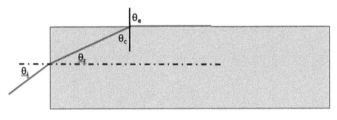

Figure A.3.7.1   Fiber geometry.

The full-blown expression can be derived by considering in order the critical angle, the interior angle, and the incident angle. The critical angle θ_c is given by the inverse sine of 90 degrees divided by the refractive index, which is $1/n$:

$$\theta_c = \arcsin[\sin(\pi/2)/n] = \arcsin(1/n).$$

By virtue of the geometry, the angle interior to the incident angle is the complement of the critical angle:

$$\theta_r = \pi/2 - \theta_c.$$

Then the incident angle is obtained from the law of refraction as the inverse sine of the refractive index times the sine of this angle:

$$\theta_i = \arcsin[n\sin(\theta_r) = \arcsin\{n\sin[\pi/2 - \arcsin(1/n)]\}].$$

This is calculated for a nominal refractive index of 1.5 in the spreadsheet both approximately and correctly in both radians and degrees.

The value can also be calculated for a range of refractive indices in Mathcad. I have done that in two ways, as shown below. One is to calculate each refraction and combine them, i.e., the sequence of *a*(*n*), *b*(*n*), etc. The other is to write a single equation for the whole process, i.e., *y*(*n*), which is sometimes a bit troublesome because it is complicated. The results must be the same—and they are. If you look carefully, you will see a red line covered by a dashed blue line, which agrees with the value of 29.80 from the spreadsheet for an index of 1.5.

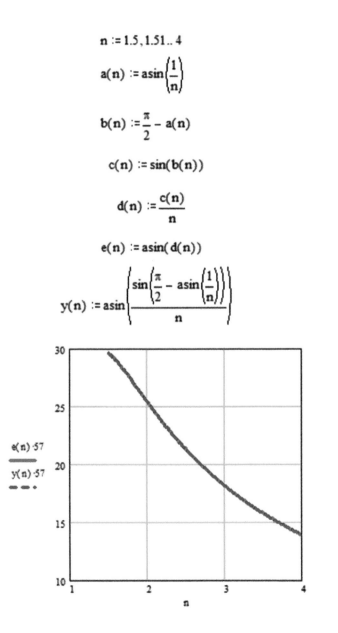

A.3.13 Prisms

As noted in the main text, prisms are used as spectral dispersers and as beam manipulators. They are prism dispersers or prism deviators (colorful or colorless).

Dispersive prisms have long been used in spectrometers. The refraction of a ray through a prism of angle α is shown in Figure A.3.13.1.

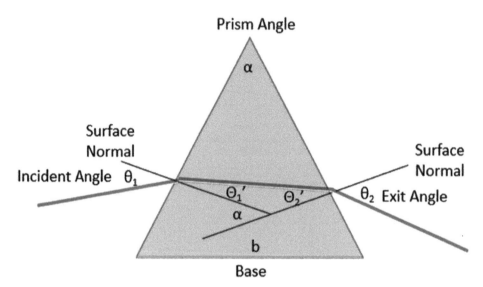

Figure A.3.13.1 Prism geometry.

The total angular deviation by the prism δ is by inspection $\theta_1' + \theta_2'$. The two normals are each perpendicular to the prism sides; so the alternate exterior angle of the primed triangle is α, and therefore

$$\delta = \theta_1 + \theta_2 - \alpha.$$

The total deviation can be calculated using this geometry and Snell's law. The angle θ_1' is obtained from Snell's law:

$$n_1 \sin \theta_1 = n_2 \sin \theta_1',$$

$$\theta_1' = \arcsin((n_1 \sin \theta_1)/n_2).$$

The other interior angle is

$$\theta_2' = \alpha - \theta_1'.$$

The final exiting angle is obtained using Snell's law:

$$n_2 \sin \theta_2' = n_1 \sin \theta_2,$$

$$\theta_2 = \arcsin((n_2 \sin \theta_2')/n_1),$$

$$\theta_2 = \arcsin((n_2 \sin(\alpha - \arcsin(n_1 \sin \theta_1/n_2))/n_1).$$

That may look complicated, but it is the solution for the value of the exit angle in terms of the incident angle and the refractive indices.

It should be no surprise that the minimum deviation occurs when there is complete symmetry, just like the law of reflection in Section 1.4. The geometry looks like that shown in Figure A.3.13.2.

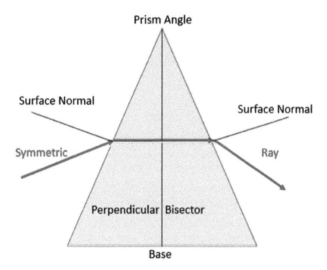

Figure A.3.13.2 Minimum deviation geometry.

I will leave it for you to use the facts that the two θ angles are equal and so are their primes, i.e., $\theta_2 = \theta_1$ and $= \theta_2' = \theta_1'$, to show that the expression for the refractive index is

$$n = \frac{\sin((\alpha + \delta)/2)}{\sin(\alpha/2)}.$$

This is the method used by most to measure the refractive index of a material.[6] Rotate the prism slowly until a particular line changes direction. That is the angle of minimum deviation. Plug in the values for the prism angle and the measured deviation angle. Done. The internal ray is parallel to the base and perpendicular to the perpendicular bisector of the prism. This might have been the inspiration for Ben Platt, who designed the technique for measuring in the infrared[7] where it would be almost impossible to find a

---

[6]Jenkins, F. and White, H. *Fundamentals of Optics*, McGraw-Hill (1957).
[7]Platt, B. "Instruments for Measuring Properties of Infrared Transmitting Optical Materials," Dissertation, The University of Arizona (1976).

minimum deviation angle. Just use the equivalent of half a prism and autocollimate from the front surface to make sure the incident ray is perpendicular to the surface.

The resolving power of a prism $\lambda/d\lambda$ is the thickness of the base of the prism b times the change in the refractive index with wavelength divided by that wavelength range:

$$\frac{\lambda}{d\lambda} = b\frac{dn}{d\lambda}.$$

Note that this is not linear since the refractive index varies with wavelength in a nonlinear fashion. It was one of the factors that helped Herschel find the infrared part of the spectrum (see Section 5.3).

A.3.14 Retroreflectors

I think (hope) the operation of the reflecting cube corner was clear in the main text. The cat's eye reflector is a little more complicated.

For a cat's eye retroreflector to work, it must have the incoming collimated beams focus on the rear of the lens. Although an actual cat's eye is not spherical, many artificial devices are, and the analysis is much simpler. Figure A.3.14.1 shows the geometry. The refractive index and the radius must satisfy the condition that the incoming ray intersects the rear surface as shown. That can be stated in several ways. One is that $(R + s)/h$ must equal the tangent of the angle $c + b$. That angle is determined by the value of the refractive index.

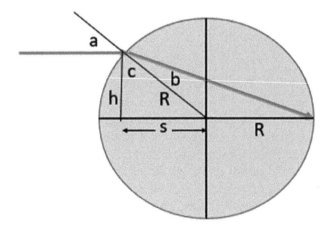

Figure A.3.14.1   Cat's eye reflector geometry.

The calculations are shown in the spreadsheet for a 10 mm radius sphere with a refractive index of two. The first section, rows 3–6, gives the properties of the sphere. The second section, rows 8–10, calculates the required geometry. The third section calculates the position of the ray based on the properties of the sphere. It is almost right: 21.18 versus 18.00. If the refractive index is changed to 1.74, it comes out just about right, and that is closer to a real cat's-eye refractive index.

	A	B	C	D	E	F	G	H	I
1		**Cat's Eye Retroreflector**							
2	Symbol	Meaning	Formula	Value	Units				
3		Sphere Properties							
4	π		known	3.14159	#				
5	R	Radius	assumed	10	mm				
6	h	Ray height	assumed	5	mm				
7		Required Geometry							
8	h	Ray height	assumed	5	mm				
9	s	Distance	√(D5²+D6²)	11.18	mm				
10	R+s	Distance	D5+D9	21.18	mm				
11		Ray Path							
12	n	index	assumed	2	#				
13	a	incident angle	arcsin(D8/D5)	0.52	rad				
14		step	sin(a)/n	0.25	#				
15	b	angle	arcsin(D14)	0.25	rad				
16	c	angle	π/2 -D13	1.05	rad				
17	c+b	angle	D15+D16	1.30	rad				
18	R+s	distance	D6xtan(D17)	18.00	mm				

A variant of this spreadsheet can be used to test tolerances for the design of roadside signs. This one can be used as it is with real-time changes in the refractive index and the radius to see how far the focus moves from its desired point. The tolerances are quite lax. Give it a try.

Appendix 4
Instruments

A.4.3 Automotive Optics

This appendix includes the logic and calculations for a generic infrared night-driving system. The system must be sensitive enough and have enough resolution to detect a person at a safe stopping distance under fair weather conditions with the car moving at the legal speed limit. You need to depend upon my choice of these or do the calculations yourself using the procedure shown in the spreadsheet. Anyone driving over the speed limit shouldn't.

	A	B	C	D	E
1		**Automotive Optics Geometry**			
2	Symbol	Meaning	Formula	Value	Units
3	π	pi	known	3.14159	#
4	d	distance	assumed	800	ft
5	d	distance	D4x12x2.54	24384	cm
6	w	object width	assumed	30	ft
7	w	object widtth	D6x12x2.54	914	cm
8	θ	field angle	2xarctan(D6/2xD6)	0.0375	rad
9	θ	field angle	57xD6	2.14	deg
10	p	object spot size	assumed	1	cm
11	N_h	horizontal pixel #	D7/D10	914	#
12	D_o	optics diameter	assumed	5	cm
13	F#	focal ratio	assumed	3	#
14	f	focal length	D12xD13	15	cm
15	α	resolution angle	D10/D5	4.10E-05	rad
16	s	detector side	D15xD14	6.15E-04	cm

I assume a (reasonable) nighttime driving speed of 60 mph and a normal dry road. The internet provides a calculator that I used.[1] It tells me under those conditions that the stopping distance is 392.2 feet. I rounded that to 400 feet and doubled it for safety. A typical two-lane road is 24 feet wide.[2] I added six feet to include a deer or person about to enter. I assume a resolution spot size on the object of one centimeter because it seems reasonable. I assume 5-centimeter-diameter, F/3 optics as realistic. The spreadsheet shows me that this is all very reasonable: small-diameter optics, a detector 6 μm on a side, an array of not quite 1000×1000, and a small optical field of only about 2 degrees. I could increase the object spot size to 4 centimeters and still be good, but I will stick with this set for illustration purposes and hope for good sensitivity!

The signal-to-noise ratio based on the power on the detector P_d is

$$\text{SNR} = \frac{D^* P_d}{\sqrt{(A_d B)}},$$

where D^* is the specific detectivity, A_d is the detector area, and B is the bandwidth. The SNR has been calculated for this generic night-driving system in two ways: in terms of the object and in terms of the instrument. That is, in object space and in image space. In object space, it is

$$\text{SNR} = \frac{D^* P}{\sqrt{(A_d B)}} = \frac{D^* \tau_o \tau_a L A_o A_t}{R^2 \sqrt{(A_d B)}},$$

where the τ are the atmospheric and optical transmissions, L is the source radiance, and the A are the optics and target areas. In image space, the SNR is

$$\text{SNR} = \frac{D^* P}{\sqrt{(A_d B)}} = \frac{D^* \tau_o \tau_a L A_o A_d}{f^2 \sqrt{(A_d B)}}.$$

They had better be the same! They are if $A_t/R^2 = A_d/f^2$. This is the solid angle of resolution in object space and image space, respectively.

I will use the geometry in the spreadsheet above. I will assume that the bandwidth is the TV rate of 30 frames a second, that the atmospheric transmission is 100% in the assumed good weather and relatively short distance, and that the optics transmission is 50%. I will also assume an array of microbolometers that do not need cooling.

The symbols are shown in columns A and G. The formulas for the calculation are in columns C and H. The values as calculated or assumed are in columns D and I. The units, which must be consistent, are in E and J. The symbol # means that the quantity is unitless, e.g., a number or ratio. The final

---

[1]"Stopping Distance Calculator," omnicalculator.com.
[2]Several sources are available online, for instance, quora.com and fwha.dot.gov.

	A	B	C	D	E	F	G	H	I	J
1					Night Driving					
2			Image Space Calculation					Object Space Calculation		
3			$SNR = \tau_a\tau_o D^* L A_o A_d/f^2 V(A_d B)$					$SNR = \tau_a\tau_o D^* L A_o A_d/R^2 V(A_d B)$		
4		Meaning	Formula	Value	Unit			Formula	Value	Unit
5	π	pi	known	3.14159	#		π	known	3.14159	#
6	τ_a	atmo trans	given	1	#		τ_a	given	1	#
7	τ_o	optics trans	given	0.5	#		τ_o	given	0.5	#
8	D^*	specfic detectivity	given	1.00E+09	$cmW^{-1}s^{-1}$		D^*	given	1.00E+09	$cmW^{-1}s^{-1}$
9	D_o	optics diameter	given	5	cm		D_o	given	5	cm
10	s	detector side	given	6.04E-04	cm		p	given	1	cm
11	F/#	focal ratio	given	3	#		B	given	30	
12	B	bandwidth	given	30	s^{-1}		M	Mathcad	0.017	
13	M	emittance	Mathcad M	0.017	Wcm^{-2}		L	H8/H4	5.41E-03	$wcm^{-2}sr^{-1}$
14	L	radiance	D8/D4	5.41E-03	$wcm^{-2}sr^{-1}$		A_o	C11	19.63	cm^2
15	A_o	optics area	$D4xD9^2/4$	19.63	cm^2		A_d	C16	3.648E-07	
16	A_d	detector area	$D13^2$	3.648E-07	cm		A_s	H11XH11	1	cm^2
17	f	focal length	D9xD11	15	cm		R	given	24834	cm
18	f^2	f squared	$D16^2$	225	cm^2		R^2	$G17^2$	616727556	cm^2
19	SNR	signal to noise		26	#		SNR		26	

Mathcad

$$\lambda := 0.4, 0.41 .. 20$$

$$c1 := 37418 \quad c2 := 14388 \quad T := 300$$

$$x(\lambda) := \frac{c2}{\lambda \cdot T}$$

$$m(\lambda) := \frac{c1}{\lambda^5 \cdot \left(e^{x(\lambda)} - 1\right)}$$

$$M := \int_{8}^{14} m(\lambda)\, d\lambda$$

$$M = 0.017$$

calculation for the SNR is D6 × D7 × D8 × D10 × D1 × D14 / (D17 × D13 × $\sqrt{\text{D18}}$) and is the same in cell I20 for the cells in column I and the Mathcad calculation of radiant emittance is below. The resultant SNR of 27 is modest either way you calculate it. But recall that the stopping distance was doubled, and a small spot size of one centimeter on the object was assumed. These can be relaxed in refined designs that still work and may be cheaper.

I think this is why we now see infrared cameras as a option on many cars and trucks. They work well, have good detectivity, and are now reasonably priced.

A.4.5 Ballistic Missile Interception

The **launch** of a ballistic missile or a satellite by a rocket is accompanied by an enormous, hot plume of water vapor, carbon dioxide, and some carbon particles. That plume is considerably larger than the rocket itself. The Atlas rocket is about 57 meters long by 4 meters in diameter. Call it 60 by 4 m for purposes of this example. As shown in the main text, the plume is a circle with a diameter of about three times the height of the Atlas. That is $\pi/4$ times the square of 6000 centimeters, or 28,274,310 square centimeters. The detection takes place by a satellite in geosynchronous orbit at 35,786 km (call it 35,000 km). Then the maximum angular subtense of the plume is approximately 60 / 30,000,000 = 1.7 microradians. We can assume that it is a point source because it would be just barely resolved by a 1-meter-diameter mirror at a wavelength of 2.5 μm.

The SNR can be calculated in terms of the radiant intensity (for a point source). By definition of the specific detectivity, the SNR is the power P_d on the detector times the specific detectivity D^* divided by the square root of the detector area A_d and noise bandwidth B:

$$\text{SNR} = \frac{D^* P_d}{\sqrt{(A_d B)}}.$$

In this case, the power on the detector is the intensity of the source I, times the solid angle the optics subtend at the source Ω, the transmission of the optics τ_o, and the atmosphere τ_a. The solid angle of the optics at the source is the area of the optics A_o divided by the square of the range R:

$$\text{SNR} = \frac{D^* P_d}{\sqrt{(A_d B)}} = \frac{D^* \tau_a \tau_o I \Omega}{\sqrt{(A_d B)}} = \frac{D^* \tau_a \tau_o I A_o}{R^2 \sqrt{(A_d B)}}.$$

The intensity of radiation emitted by the plume is its total power divided by π, the hemisphere into which its projected area radiates. The evaluation of the radiation from that hot gas is very complicated and beyond the scope of this description.[3] The temperatures vary from almost 1800 K in the core to 300 K at the edges. I choose 1000 K as in intermediate and round value for illustration purposes. I choose the usual and reasonable value of 50% for the transmission of the optics, and the same for the atmosphere. The very hot plume with its broadened spectrum of radiating water vapor will be attenuated by the water vapor in the air. An optics diameter of 10 centimeters is a modest value. An optical speed (F/#) of 3 is also modest. Then the focal length is 30 centimeters, and a detector dimension of 20 μm is standard for either lead sulfide or indium antimonide (as is a specific detectivity of 10^{11}). The intensity is calculated as a blackbody in the spectral band from 2.4 to 2.6 in Mathcad as 0.199 watts per square centimeter. The plume area is a large circle about three times the size of the missile height of 85 feet (~2600 cm). The range is 35,000 km. The detector is 20 μm on a side.

The accompanying spreadsheet shows the results based on the realistic numbers I have chosen to illustrate the calculation. There is a delightful SNR of 2028. That is probably why it could also detect Scud missiles.

If you choose to make your own calculations, refer to column E. That is where the calculations are. They start with the assumed values in row 4 through 8. Then row 9 calculates the source area from the usual equation for the area of a circle. Row 10 inserts the radiant emittance from the Mathcad calculation. Row 11 turns that into an intensity by multiplying by the source area and dividing by π. The assumed optics diameter is in E12, the resultant area in E13 and the linear dimension of the detector in E14. The assumed bandwidth and the known geostationary range are in the next two rows. The final calculation is in cell E17, and its formula, which corresponds to the one at the top is in the last line.

---

[3]Alexeenko, A. et al. "Modeling of Flow and Radiation in the Atlas Plume," *J. Thermophysics and Heat Transfer* **16**, 50 (2002); "IR radiation characteristics of rocket exhaust plumes under varying motor operating conditions," *Chinese J. Aeronautics* **30**(3), 1101–1114 (2017).

	A	B	C	D	E	F
1			**ICBM Launch Detection**			
2			$SNR = \tau_a \tau_o A_s A_o D^* / R^2 \sqrt(A_d B)$			
3	Symbol	Meaning	Formula	Formula	Value	Units
4	π		3.14159	known	3.14159	#
5	τ_a	atmo trans	0.5	assumed	0.5	#
6	τ_o	optics trans	0.5	assumed	0.5	#
7	D*	specific Detectivity	1.00E+11	assumed	1.00E+11	cmvHz/W
8	D_s	source diameter	6000	assumed	6000	cm
9	A_s	source area	$\pi D_s^2/4$	E4xE8^2/4	28274310	cm^2
10	M	emittance		Mathcad	0.243	W
11	I	source intensity	MA_s/π	E9xE10/E4	2187000	W
12	D_o	optics diameter	D_o	assumed	10	cm
13	A_o	optics area	$\pi D_o 2/4$	E4xE12^2/4	78.53975	cm^2
14	d	detector side	0.002	assumed	0.002	cm
15	B	bandwidth	1	assumed	1	Hz
16	R	range	3.70E+09	assumed	3.70E+09	cm
17	SNR	signal to noise			2028	
18			E5XE6XE9XE13XE7/(E16^{2}XE14XE15$^{0.5}$)			

$$\lambda := 0.4, 0.41 .. 5 \qquad T := 1000 \ K$$

$$c_1 := 2 \cdot \pi \cdot c^2 \cdot h \qquad c_2 := \frac{h \cdot c}{k}$$

$$x(\lambda) := \frac{c_2}{\lambda \cdot T} \qquad M(\lambda) := \frac{c_1}{\lambda^5 \cdot \left(e^{x(\lambda)} - 1\right)}$$

$$M := \int_{2.4 \, \mu m}^{2.6 \, \mu m} M(\lambda) \, d\lambda \qquad M = 0.243 \ \frac{W}{cm^2}$$

The interceptions and homing in on an enemy payload during **midcourse** is a bit different. The defensive vehicle is not a satellite but another missile. The infrared target is not the plume but the thermal radiation from the payload vehicle. And the range is much shorter.

The payload, the bad guy, is about 3 meters long and 1 meter in diameter. I will assume that radar has guided our interceptor to within 50 miles (80 kilometers or 8×10^6 centimeters). I will also assume our missile sensor has a 10-centimeter-diameter, F/3 optical aperture and is diffraction limited over its small field of view. The midcourse geometry is detailed in the following spreadsheet.

	A	B	C	D	E
1		**Midcourse Geometry**			
2	Symbol	Meaning	Formula	Value	Units
3	π	pi	known	3.14159	#
4	l	source length	assumed	300	cm
5	d	source diameter	assumed	100	cm
6	B	bandwidth	assumed	1	Hz
7	τ_a	atmo tranmission	assumed	1	#
8	τ_o	optics transmission	assumed	0.5	#
9	R	range	assumed	8.00E+06	cm
10	D_o	optics diameter	assumed	10	cm
11	F/#	F/number	assumed	3	#
12	f	focal length	D10xD11	30	cm
13	λ_{max}	longwavelength	assumed	14	μm
14	λ_{max}	longwavelength	D13/10000	0.0014	cm
15	β	diffraction angle	1.22xD14/D10	1.71E-04	rad
16	α	resolution angle	D15	1.71E-04	rad
17	α_s	source angle	D4/D9	3.75E-05	rad
18	$s = \sqrt{A_d}$	detector side	D15xD12	5.12E-03	cm

This first cut looks pretty good. The optics and detector sizes are good. An array of 1000×10000 detectors will cover an adequate field for tracking the missile. The incoming missile is not resolved at this distance but will be when there is a little closure.

This time, however, it is a gray body made of aluminum or plastic. It will be at about room temperature since it equilibrated with its surroundings on the launch pad and was heated to some extent by atmospheric friction. I will assume that the vehicle is at 300 K with an emissivity of 5%. Mathcad provides the information that a blackbody at 300 K emits 0.015 watts per square centimeter in the 8–14-μm spectral range. This must be multiplied by the emissivity of 0.05 and the area of 3 square meters for the total power of 2.25 mW. We can take at least a second for each look at the target. Therefore, assume a bandwidth of 1 Hertz. The SNR equation is

$$\text{SNR} = \frac{D^* \tau_o A_o l}{R^2 \sqrt{(A_d B)}}.$$

The Midcourse Interception spreadsheet for this scenario is provided below. I had to show the final calculation in cell F20 to fit it in.

That results in a modest SNR just over 2, as shown in the spreadsheet, and it will get larger as the range R gets smaller. We could increase the aperture area to 20 centimeters and thereby increase the SNR to almost 10, but that may not be necessary.

But what about the backgound? The flux density is 10^{-9} watts per square meter—orders of magnitude less. No worries. In effect, there is no background.

As the interceptor approaches, the missile soon becomes an extended source. But the SNR just increases, and soon there is a crash. The missile has hit the missile and smashed it without any explosion. God willing, it will never happen.

	A	B	C	D	E	F	G	H	I
1				**Midcourse Interception**					
2			$SNR = D^* \tau_0 A_0 I / R^2 \sqrt{(A_d B)}$						
3	Symbol	Meaning	Formula	Value	Unit			Mathcad	
4	π	pi	known	3.14159	#			$\lambda := 0.4, 0.41 .. 20$	
5	D^*	specific detectivity	assumed	1.00E+10	cms-1				
6	τ_a	atmo transmission	assumed	1	#		$c1 := 37418$ $c2 := 14388$ $T := 300$		
7	τ_0	optics transmission	assumed	0.5	#				
8	D_0	optics diameter	assumed	10	cm			$x(\lambda) := \dfrac{c2}{\lambda \cdot T}$	
9	R	range	assumed	8.00E+06	cm				
10	F/#	focal ratio	assumed	3	#			$m(\lambda) := \dfrac{c1}{\lambda^5 \cdot \left(e^{x(\lambda)} - 1\right)}$	
11	α	resolution angle	assumed	1.71E-04	rad				
12	B	bandwidth	assumed	1	s^{-1}				
13	A_s	sorce area	assumed	3.00E+04	cm^2			$M := \displaystyle\int_8^{14} m(\lambda)\, d\lambda$	
14	e	source emissivity	assumed	0.05	#				
15	f	focal length	D7xD9	30	cm				
16	s=√A$_d$	detector side	D14xD10	5.13E-03	cm			$M = 0.017$	
17	A_0	optics area	D4xD7^2/4	78.54	cm^2				
18	M	source emittance	Mathcad	0.017	Wcm^{-2}				
19	P	power	D13xD11	25.5	W				
20	I	intensity	D15/4/D4	2.03	Wsr^{-1}				
21	SNR	signal to noise	see cell F21	2.43		D5xD6xD19xD18/(D8^2D15xD11)			

A.4.14 Emissometers

The calculation of emissivity with the Gier–Dunkle emissometer was promised in Section 4.14, so here it is.

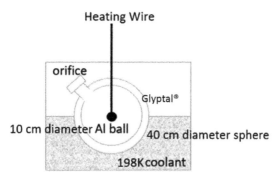

Heating Wire

orifice

Glyptal®

10 cm diameter Al ball

40 cm diameter sphere

198K coolant

Figure A.14.1 Emissometer schematic.

The unknown material is coated on a 10-centimeter-diameter aluminum sphere. It is suspended inside a 40-centimeter sphere maintained at $-75°C$ (198 K). The inside of the chamber is coated with Glyptal® black, a very black commercial electricity-conducting enamel. The wire that suspends the unknown material conducts electricity to it to heat it. The temperatures of both spheres are monitored with thermocouples. The power transfer between the two spheres is only radiative and conductive along the wire since the instrument is evacuated. There is no convection. The only conduction is in the wire. The power transfer at equilibrium is

$$dP = VI = \varepsilon_1 A_1 \sigma T_1^4 - \varepsilon_2 A_2 \sigma T_2^4 - (kA_w/l)(T_2 - T_1),$$

where dP is the power transfer between the two spheres, V is the applied voltage, I is the applied current, the ϵ are the emissivities, the A are the areas, and the T are the temperatures. The conductance term includes the thermal conductance of the wire k. the area of the wire A_w, and its length l. These are all known or measurable quantities. The temperatures are measured with the thermocouples. The areas are calculated by a knowledge of their diameters. The conductance can be calculated from the conductivity and the area and length of the wire, and it can also be measured independently.

The emissivity of the unknown ϵ_1 can be found from all the known quantities, the measured temperatures and input power, the emissivity and size of the sphere, and the Stefan–Boltzmann constant. The uncertainty can be calculated as the rms error in the areas, temperatures, Glyptal emissivity, wire conductance, and the electrical inputs. They will be small. My guess is a few percent total.

A.4.19 Human Eye

The sensitivity of our human eyes was alleged in the main text to be good enough to detect a single photon. The calculation follows. It relies on published data about the sensitivity of the eye in other units. The calculation is for scotopic vision using just the rods which are the most sensitive. But color and structure will not be perceived.

The published data[4] indicate that the human eye, when completely dark adapted, can detect as little as 0.75×10^{-6} cdm^{-2}. That is less than a millionth of a candela per square meter. Is that a single photon?

The spreadsheet gives the answer. It can only be approximate since eyes vary so much. I have used all the reported values for the sensitivity in terms of candelas, the geometry of the eye, and the rods. I also used red light because we see it better in the dark. The result is that we can sense (maybe not see) between one and two photons a second with a completely dark-adapted eye.

This is supported by the information in the reference.[1] The calculations are as follows.

The reported sensitivity in candelas per square centimeter of 0.75 micro-candelas per square meter is a luminance specification equivalent to lumens per solid angle and area. My first step is to convert it to radiance by way of the scotopic sensitivity of the eye. Note that 1700 lumens per watt is the same as 1700 lumens per steradian per square meter per watts per steradian per square meter. Then the task is to use the properties of the eye to convert the value to watts. Finally, a photon rate and the number detected during one integration time can be determined.

These are approximate calculations, but they are about right and do show what a fantastic sensor the human eye is.

An article in *Nature Communications* reports the experimental detection of a single photon.[5] It reports that it is not so much a matter of seeing the photon but sensing it.

---

[4]Poelman, D., Avci, N., and Smet, P. "Measured luminance and visual appearance of multi-color persistent phosphors," *Opt. Express* **17**(1), 358 (2009).

[5]Tinsley, J., Molodtsov, M., Prevedel, R. et al. "Direct detection of a single photon by humans," *Nat. Commun.* **7**, 12172 (2016).

◢	A	B	C	D	E
1		**Detecting a Single Photon**			
2	**Symbol**	**Meaning**	**Formula**	**Value**	**Units**
3	π	pi	known	3.14159	#
4	S	Sensitivity	reference	7.50E-11	$cdcm^{-2}$
5	Vs	Scotopic Efficacy	known	1700	lmW^{-1}
6	L	Radiance	D4/D5	4.41E-14	$Wcm^{-2}sr^{-1}$
7	d	Eye diameter	approximate	2.5	cm
8	l	Eye depth	approximate	2.5	cm
9	A	Eye aperture area	$D3xD6^2/4$	4.91	cm^2
10	d_{rod}	Rod diameter	approximate	0.0002	cm
11	A_{rod}	Rod area	$D3xD10^2/4$	3.14159E-08	cm^2
12	G	Etendue	$D9xD11/D8^2$	0.00008	cm^2sr
13	Pd	Detected power	D6xD12	3.53E-18	W
14	λ	Photon wavelength	assumed	0.8	μm
15	λ	Photon wavelength	converted	8.00E-07	m
16	c	Light speed	known	3.00E+08	ms^{-1}
17	v	Photon frequency	D16/D15	3.75E+14	Hz
18	h	Planck's constant	known	6.63E-34	JHz^{-1}
19	E_q	Photon energy	D17xD18	2.48E-19	J = Ws
20		Photon rate detected	D13/D19	14.20	s^{-1}
21	t	Integration time	approimate	0.10	s
22	N	Photons detected	D20xD21	1.42	#

A.4.21 Interferometers

The transmission of a Fabry–Perot (FP) interferometer is the same as the plane-parallel plate and can be inferred from Figure A.4.21.1. You can view this as a single pane of glass or the air space between two FP etalons.

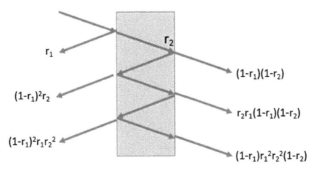

Figure A.4.21.1 Multiple reflections.

The total transmission due to all multiple reflections is

$$\tau = (1 - r_1)(1 - r_2)(1 + r_1 r_2 + r_1^2 r_2^2 + \ldots) = (1 - r_1)(1 - r_2)/(1 - r_1 r_2).$$

If the two reflections are equal, as is often the case with a plate of glass in air and a Fabry–Perot that has air between two glass plates, then the expression reduces to

$$\tau = (1 - r)^2/(1 - r^2) = (1 - r)/(1 + r).$$

This transmission is shown in Figure A.4.21.2.

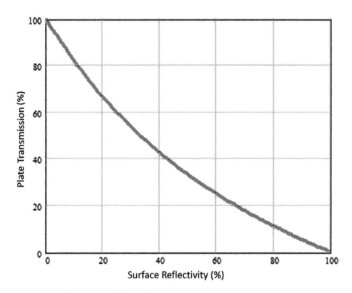

Figure 4.21.2   Transmission vs reflectance.

The resolving power (RP) $\lambda/d\lambda$ of a Fabry–Perot interferometer, which is used for high-spectral-resolution measurements, is

$$RP = \lambda/d\lambda = m\pi\sqrt{r/(1 - r)},$$

where π is its usual value, r is the reflectance, and m is the interference order number. The order number may be taken as $2d/\lambda$, i.e., the number of half waves in the etalon. Then,

$$RP = \frac{\lambda}{d\lambda} = \frac{2\pi d r^{1/2}}{\lambda(1 - r)}.$$

The resolving power is increased by increasing the reflectance. Figure A.4.21.3 shows how the resolving power depends on reflectance (normalized to 100%)

and why it is good to have high-reflectance etalons in a Fabry–Perot even if they decrease the transmission. A compromise between 60% and 80% seems appropriate. That is where the transmission shown in Figure 4.21.2 is down to 10 to 20 percent but the resolving power, shown in Figure 4.21.3 is rising rapidly.

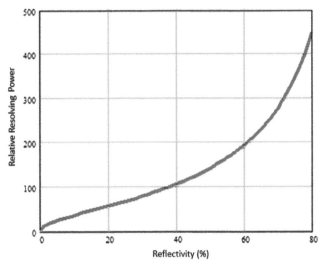

Figure A.4.21.3   Resolving power vs reflectance.

A.4.22 Laser Damage

This section describes in more detail how both intentional and accidental laser damage may occur. The first part calculates the approximate time it takes for a laser weapon to penetrate the aluminum fuselage of an aircraft. Assume any hostile aircraft you like. The second part is about eye damage from laser pointers.

Assume, optimistically, a one-megawatt **laser weapon** operating at 10 μm and a one-meter-diameter mirror to form the beam. The assumed range is 10 kilometers, or 6 miles. The figures are shown in the spreadsheet. These choices will illustrate the situation. Most of the entries in the spreadsheet come from these assumptions. The values are all converted to centimeters to be consistent. The beam spread in cell D9 is based on the diffraction limit.

	A	B	C	D	E
1		**Laser Weapon Geometry**			
2	**Symbol**	**Meaning**	**Formula**	**Value**	**Units**
3	π	pi	known	3.14159	#
4	P	laser power	assumed	1.00E+06	W
5	D_o	optics diameter	assumed	1	m
6	D_o	optics diameter	100xD5	100	cm
7	λ	wavelength	assumed	10	μm
8	λ	wavelength	D7/10000	0.001	cm
9	α	beam spread	1.22xD8/D6	1.22E-05	rad
10	R	range	assumed	10	km
11	R	range	10^5xD10	1.00E+06	cm
12	b	beam diameter	D11xD9	12.2	cm
13	A_b	beam area	D3xD12^2/4	117	cm^2

The significant result is that for these assumptions, all the laser power is on an area of approximately 117 square centimeters. It will proceed to melt the skin based on the weight, specific heat, and melting temperature.

Melting aluminum requires that the spot rises to a temperature of 933 K. Attainment of this value is determined by the power incident upon it, how much is absorbed, the mass of the absorbing volume, and the specific heat of aluminum.

The melting temperature of aluminum is 933 K, its specific heat capacity is 0.9 Joules per gram degree Kelvin, and its density is 2.7 g/cc. The definition of specific heat is the amount of energy E it takes to raise one gram of mass m by one degree Kelvin dT. The energy E is the power P times the time of exposure t. Thus,

$$c_p = \frac{E}{mdT} = \frac{Pt}{mdT}.$$

The time of exposure is then, by rearrangement,

$$t = \frac{c_p \rho A x dT}{P},$$

where the mass m is the density ρ times the area times the thickness x. The power P is the incident power; most of it is reflected. Only about 5 percent is absorbed, i.e., 50000 watts. The temperature needs to go from an ambient of about 300 K to the melting temperature of 933 K, for a change of 633 K. We can now plug and play. The values are $c_p = 0.9$ Ws/gK; $\rho = 2.7$ g/cc;

$A = 78.5 \text{ cm}^2$; $x = 1$ cm; and $P = 10^6$ watts. So the time the laser takes to do damage, based on these assumptions, is

$$t = \frac{0.9 \times 2.7 \times 78.5 \times 1 \times 633}{50000} = 2.414.$$

The time it takes for a 1-megawatt laser to melt an aluminum fuselage is almost 2.5 seconds. A tracker must keep the beam on the moving target for that long. A jet traveling at Mach 2 goes about 5000 feet in that time. A 1-megawatt laser is optimistic. A 0.1-megawatt laser will take ten times as long, and so on. It will take even longer with an iron target that has a larger density. I think that it is correct to assume these values for a fighter jet or bomber. I do not know how it works for a re-entry vehicle, but the concepts are the same.

Eye damage is another form of laser damage. We can get some estimates of how dangerous laser pointers might be as an example of this kind of damage. The maximum permissible exposure for lasers emitting in the visible range varies from about 1 to 10 milliwatts per square centimeter for almost any exposure time.[6] The eye surface is about 1 centimeter in diameter and about 1 square centimeter in area. These are approximate. Thus, a 1-milliwatt well-collimated pointer puts about 1 milliwatt per square centimeter of power density on the eye. That is in the danger range. Fortunately, there have been few such events reported. The same reference points out that red pointers of less than 5 mW with an exposure of less than 0.25 seconds are no danger. Green lasers, with a color to which the eye is more sensitive, are slightly more trouble. The normal eye blink is 0.4 seconds. Hopefully, a laser-caused blink is faster.

If you are the speaker, use a red pointer and keep it aimed at the screen. If you are part of the audience, keep your eyes away. The reflection off the screen is okay but not the direct beam.

A.4.26 Medical Thermographs

As noted in the main text, infrared cameras (thermographs) have a variety of medical uses. These include breast cancer screening, malingering, circulation problems, and determination of the extent of necrotic tissue around a burn. The designs require adequate thermal and spatial resolution as well as relatively fast reaction times. Modern devices use arrays of either thermal or photon detectors and operate at TV rates of about 30 frames per second. In the examples cited below and in general, a temperature difference of about 0.01 K is required to be sensed over an object field of about 25 by 25 centimeters and a spatial resolution of about 1 square millimeter. The criterion is the minimum resolvable temperature difference (MRTD).[7] The

---

[6]Wikipedia. "Laser Safety," https://en.wikipedia.org/wiki/Laser_safety.
[7]Lloyd, J. *Thermal Imaging* Systems, Plenum (1978).

full-blown MRTD involves the noise equivalent temperature difference (NETD), the modulation transfer function (MTF), and a few visibility factors such as eyeball integration time, narrowband spatial filtering, and viewing conditions, but these examples will only include the NETD and MTF. I consider two designs: one using photon detectors, and one using micro-bolometers. I describe a breast cancer screening camera. Other modalities are less demanding or equivalent. Shoulder pain, lower-back pain, circulation, or necrotic tissue determination have about the same requirements, but leg and arm examinations have smaller fields of view and will be covered easily by such an instrument.

As with all such design problems, the first order of business is the required geometry. Most doctors use the camera at an object distance of about 30 centimeters. Most women have a chest area of about 25×25 centimeters. An appropriate resolution is 1 millimeter, but smaller is better. The geometry spreadsheet shows these figures and the fact that something is wrong. Some things are not acceptable. The field of view, cell D14, is much too large, and so is the detector size, cell D12. These can both be helped by increasing the object distance. A doctor often uses a camera like this at a larger distance of at least 100 centimeters.

The next spreadsheet shows the geometry for the next iteration, an improved and reasonable arrangement.

	A	B	C	D	E
1		**Medical Thermograph Geometry**			
2	Symbol	Meaning	Formula	Value	Units
3	π	pi	known	3.14159	#
4	o	object distance	assumed	30	cm
5	p	object pixel	assumed	0.1	cm
6	h	object height	assumed	25	cm
7	w	object width	assumed	25	cm
8	D_o	optics diameter	assumed	10	cm
9	F/#	focal ratio	assumed	3	#
10	f	focal length	D8xD9	30	cm
11	α	resolution angle	D5/D4	0.003333	rad
12	s=√Ad	detector size	D10xD11	0.1	cm
13	θ	field angle	2arctan(D6/D4)	1.39	rad
14	θ	field angle	2arctan(D6/D4)	79.20	deg

▲	A	B	C	D	E
1		**Medical Thermograph Geometry**			
2	Symbol	Meaning	Formula	Value	Units
3	π	pi	known	3.14159	#
4	o	object distance	assumed	100	cm
5	p	object pixel	assumed	0.1	cm
6	h	object height	assumed	25	cm
7	w	object width	assumed	25	cm
8	D_o	optics diameter	assumed	10	cm
9	F/#	focal ratio	assumed	3	#
10	f	focal length	D8xD9	30	cm
11	α	resolution angle	D5/D4	0.001	rad
12	s=√Ad	detector size	D10xD11	0.03	cm
13	θ	field angle	2arctan(D6/D4)	0.49	rad
14	θ	field angle	2arctan(D6/D4)	27.93	deg

This is acceptable. The object distance in D4 is okay. The angular field of view in D14 is reasonable, and the detector size in D12, is consonant with available detector arrays. This also illustrates the procedure by which such instruments are designed: by guess and by golly; by educated guess and some good experience; and by educated trial and error.

The next step is sensitivity analysis. Can the system detect temperature differences that are small enough? This is determined by calculating the NETD, the temperature difference that provides a signal equal to the noise level. It is obtained from the SNR. In fact, it is the temperature difference that yields an SNR equal to 1. The SNR in terms of the change of radiance, with respect to temperature dL/dT, is

$$\text{SNR} = \frac{D^* \tau_a \tau_o A_d (dL/dT) dT}{f^2 \sqrt{(A_d B)}}.$$

The NETD is obtained by setting the SNR equal to 1 and solving for the temperature difference:

$$\text{NETD} = dT \Big|_{\text{SNR}=1} = \frac{d^2 \sqrt{(A_d B)}}{D^* \tau_a \tau L A_s A_o dL/dT} = \frac{f^2 \sqrt{(A_d B)}}{D^* \tau_a \tau_o A_p A_d dL/dT}.$$

This equation is calculated in the spreadsheet.

	A	B	C	D	E	F	G	H	I	J
1			**Medical Thermograph**							
2		$B^{0.5}f^2/(\tau_a\tau_o A_o A_d^{0.5}D^*dL/dT)$								
3		Formula	Value	Unit	Value					
4	π	known	3.14159	#	3.14159					
5	d	assumed	30	cm	30			Mathcad		
6	L	assumed	30	cm	30			$\lambda := 0.4, 0.41 .. 20$		
7	W	assumed	30	cm	30					
8	D^*	assumed	1.00E+10	cms^{-1}W-1	1.00E+09		$c1 := 37417$		$c2 := 14388$	
9	τ_a	assumed	1	#	1					
10	τ_o	assumed	0.5	#	0.5		$T := 300$		$x(\lambda) := \dfrac{c2}{\lambda\cdot T}$	
11	l	assumed	0.1	cm	0.1					
12	B	assumed	30	s^{-1}	30		$N(\lambda) := \dfrac{c1}{\lambda^5\cdot\left(e^{x(\lambda)}-1\right)}$		$f(\lambda) := \dfrac{x(\lambda)\,e^{x(\lambda)}}{e^{x(\lambda)}-1}\dfrac{1}{T}$	
13	D_o	assumed	10	cm	10					
14	F/#	assumed	3	#	3					
15	f	C11xC12	30	cm	30		$M := \displaystyle\int_8^{14} N(\lambda)\,d\lambda$		$dM := \displaystyle\int_8^{14} f(\lambda)\cdot N(\lambda)\,d\lambda$	
16	α_s	C9/C5	0.003333	rad	0.003333					
17	s=√Ad	C15xC16	0.1	cm	0.1					
18	√B	C10*0.5	5.48	s$^{-1/2}$	5.48		$M = 0.015$		$dM = 2.312\cdot10^{-4}$	
19	f^2	C13^2	900	cm2	900					
20	A_o	C4xC13^2/4	15.70795	cm2	15.70795					
21	dM/dT	Mathcad	2.30E-04	Wcm^{-2}K^{-1}	2.30E-04					
22	dL/dT	C21/C4	7.32E-05	Wcm^{-2}K^{-1}sr^{-1}	7.32E-05					
23	NETD		0.008573	K	0.08573					

A word of explanation is probably in order, maybe a few. Column A contains all the symbols that are used in the calculation. Column B contains the formula or origin of the value. The first 14 are all assumed and discussed in the text above. Cell B15 is the first one calculated. It is the focal length that is the assumed aperture diameter times the assumed focal ratio. It is shown as C11 × C12 because the actual calculation is the value in C11 times the value in C12. Column C is those calculations. Column D is the units, and column E is for the microbolometer system.

The calculation for the microbolometer array is in column E, but the result is obvious for the assumptions I have made. The only difference is the detectivity, a factor of ten. The result is that the NETD for microbolometers is 0.086 K instead of 0.0086. They are both just fine.

The MRTD is the NETD divided by the MTF. It was shown in Section A.1.13 that the MTF of circular optics is given by

$$\text{MTF} = \frac{2}{\pi}[\arccos(\xi) - \xi(1 - \xi^2)^{1/2}],$$

where ξ is the cutoff frequency = $1/\lambda F\#$. In this case, $\xi = 1/3 \times 12$ μm = $1/36$ μm^{-1} = 280 pixels per centimeter, 28 per millimeter. It is not the limiting resolution.

Figure A.4.26.1 shows the MTF of the optics, and Figure A4.26.2 shows the MRTD without the additional human factors that will improve it by a factor of about two. The MTF is the typical one for a circularly symmetric system. The MRTD starts with the NETD of 0.086 K at zero frequency and increases to 2 K at 250 cycles per centimeter. It reached up to infinity at the cutoff frequency of 280. That means, for instance, that it requires a temperature difference of 2 degrees to barely see the difference between two spots 1/28th of a millimeter apart (a tenth of a degree if they are $1/10^{th}$ millimeter apart). That is still just fine.

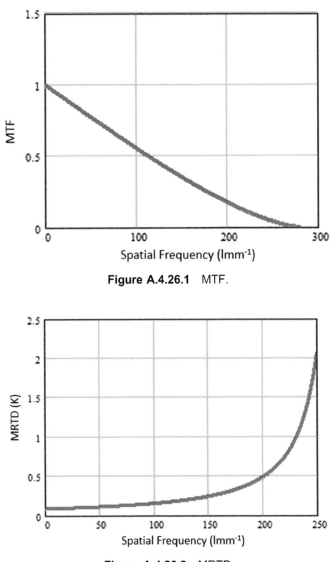

Figure A.4.26.1   MTF.

Figure A.4.26.2   MRTD.

A.4.29 Multispectral Imagers

As noted in the text, there are two major applications of multispectral imagers. One is the use in satellites or aircraft to assess crops, cities, and littoral areas. The other is for medical purposes. This section includes descriptions of the two kinds of devices and general performance estimates.

Satellite devices detect the sunlight as it is reflected from crops and seas and cities. They are extended source sensors since they make images. They detect the radiance reflected from the object in a narrow spectral band. If they operate in the infrared, they also detect radiation emitted by the Earth. The detection of emitted, thermal radiation is dealt with in other sections; this one concentrates on reflected sunlight.

As is the case with all these examples, we start with the geometry. This is in the first spreadsheet. It starts with some very general data like the orbital altitude and period. Some assumptions are made about the scan pattern, the desired size of the spot on the ground, the ground spatial distance (GSD), and the scan width. This immediately devolves to total coverage and timeliness. A wider scan width means fewer scans are necessary to cover the Earth, but a wider scan width requires more difficult optics. It also means that the GSD becomes larger as it is projected on an assumed flat Earth. I have chosen 10 kilometers for illustration purposes. It will take several days for such a scanner to cover the entire Earth, but things such as crops do not change that rapidly.

If the system measures reflected sunlight only sunlit scans are useful. These polar orbiters make about 16 orbits in a day, during which time the Earth rotates 360 degrees. These calculations are above my paygrade. I will deal with the geometry of the system in a single orbit. I will assume that we want a GSD of 1 meter and a swath width of 1 kilometer for illustration purposes. I also assumed fairly large optics to get this very good ground resolution.

▲	A	B	C	D	E
1			**Orbital Geometry**		
2	Symbol	Meaning	Formula	Value	Units
3	h	orbital altitude	assumed	800	km
4	C_e	Earth circumference	known	40000	km
5	C_e	Earth circumference	known	40000000	m
6	t	orbital period	known	90	min
7	t	orbital period	known	5400	s
8	v	orbital speed	C4/C6	7407	ms^{-1}
9	GSD	line width	assumed	1	m
10	t_l	line time	D9/D10	0.000135	s
11	B	bandwidth	1/2D8	3704	s^{-1}
12	α_0	GSD nadir angle	D9/D3/1000	1.25E-06	rad
13	S	swath width	assumed	10	km
14	D_0	optics diameter	assumed	100	cm
15	β_1	diffraction at 1 um	1.22x0.0001/D15	1.22E-06	rad
16	F/#	focal ratio	assumed	3	#
17	f	focal length	C13xC14	300	cm
18	s = VA$_d$	detector side	D12xD18	3.75E-04	cm

These are reasonable values for the optics, detector, GSD, and bandwidth. Experience is a good teacher. The next order of business is to see if the system is sensitive enough. Many investigators calculate the noise equivalent reflectivity, and we do that as well. It is a good indication of how well you can obtain a useful reflected image of various objects. And it is often good (not always) to be one with the crowd.

The SNR is

$$\text{SNR} = \frac{D^* P_d}{\sqrt{(A_d B)}},$$

where D^* is the specific detectivity, P_d is the power on the detector, A_d is the detector area, and B is the electrical bandwidth. The power on the detector in terms of ground radiance L is

$$P_d = \frac{D^* \tau_a \tau_o L A_s A_o}{R^2 \sqrt{A_d B}} = \frac{D^* \tau_o \tau_a L A_d A_o}{f^2 \sqrt{A_d B}},$$

where the τ's are the atmospheric and optical transmissions. The A's are the source, optics, and detector areas, and B is the noise bandwidth. The radiance

L is the radiance reflected by the foliage from the solar irradiance. In a detailed, precise calculation, that would have to be the part of the spectrum incident from the sun, the solar spectral irradiance, times the BRDF. That accounts for the solar incidence angle and the angle of viewing. For purposes of this example, we can assume that it is perfectly diffuse with a directional-hemispherical reflectivity given by ρ and that the sun is directly overhead. It is a reasonable assumption that makes the reflectivity ρ/π.

We could emphasize that it is in a narrow spectral band by writing $L_\lambda d\lambda$, but it is enough to keep in mind that L is the radiance in a narrow spectral band.

The SNR is then

$$\mathrm{SNR} = \frac{\tau_a \tau_o \rho E A_s A_o D^*}{\pi R^2 \sqrt{(A_d B)}} = \frac{\tau_a \tau_o \rho E A_d A_o D^*}{\pi f^2 \sqrt{(A_d B)}}.$$

We can obtain the noise equivalent reflectivity, NEρ, or NERD, by setting the SNR equal to one and solving for the reflectivity, ρ.

$$\mathrm{NERD} = \rho \Big|_{\mathrm{SNR}=1} = \frac{\pi f^2 \sqrt{(A_d B)}}{\tau_a \tau_o E A_d A_o D^*} = \frac{\pi R^2 \sqrt{(A_d B)}}{\tau_a \tau_o E A_s A_o D^*}.$$

This value is calculated in the next spreadsheet using the information from the geometric considerations above. The assumed irradiance is the known solar constant of $1 \ \mathrm{kWm}^{-2} = 0.1 \ \mathrm{Wcm}^{-2}$.

I have assumed that the atmospheric transmission is 50%, the optics transmission is 50%, the solar irradiance at the Earth's surface is distributed like a 5000 K blackbody, and the D^* is 10^{11} in the visible. For simplicity, I also assumed that the spectral bandwidth is 10 percent of the total solar irradiance. In a real case, one needs to be more precise. This is a generic example. It does show that this kind of spectral examination of the Earth from satellites is very practical. It has been used for years.

The spreadsheet shows how to make this calculation. Most of it should be clear, but note that it is important to keep track of dimensions. In particular, the specific detectivity D^* is given in terms of watts, the bandwidth is usually stated in hertz but has the dimension of reciprocal seconds, and the square root of the detector area is in **centimeters**. A good procedure is to keep all linear dimensions in centimeters.

	A	B	C	D	E
1	**Multispectral Satellite Sensor**				
2	NERD = $\pi f^2 \sqrt{(A_d B)}/\tau_a \tau_o E A_d A_o D*$				
3	Symbol	Meaning	Formula	Value	Units
4	π	pi	known	3.14159	#
5	D*	speciic deectivity	assumed	1.00E+11	cm√HzW^{-1}
6	τ_a	atmo transmission	assumed	0.5	#
7	τ_o	optics transmission	assumed	0.5	#
8	D_o	optics diameter	assumed	100	cm
9	E	solar irradiance	assumed	0.1	Wcm^{-2}
10	fraction	spectral part	assumed	0.1	#
11	F/#	focal ratio	assumed	5	cm
12	B	bandwidth	above	3704	HZ
13	GSD	ground spot	assumed	1.00E+02	cm
14	R	altitude	assumed	5.00E+07	cm
15	α	resolution angle	C14/C15	2.00E-06	rad
16	f	focal length	C8xC11	500	cm
17	√Ad	detector side	C15xC17	1.00E-03	cm
18	A_o	optics area	C4xC8^2/4	7854	cm^2
19	A_d	detector area	C13^2	1.00E-06	cm^2
20	NERD	noise equivalent ρ	see B21	0.0001	%
21		D4xD11^{2}xD13xD12$^{0.5}$/(D6xD7xD9xD10xD15xD14xD5)			

We can also consider as an example the **multispectral cancer imager** that is under development at the Wyant College by my colleagues. The prototype consists of eight different spectral bands. Each is illuminated by a quartet of LEDs in the selected wavelengths. Each LED emits at least 1 milliwatt. This illumination covers an area of about 5 centimeters in diameter at an object distance of about 30 centimeters from the patient. The resolution spot on the patient is 0.1 mm = 0.01 cm. That arrangement results in an irradiance of $(0.004 \text{ W}/\pi D^2/4) = 0.004/19.6 = 0.0002 \text{ Wcm}^{-2}$. It was inspired by a squamous cell carcinoma on my hand.

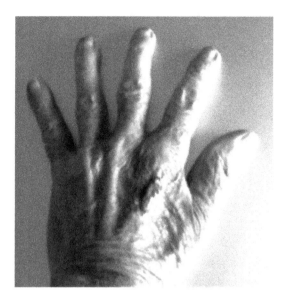

Figure A.4.29.1 A squeamish squamous.

The geometry is summarized in the spreadsheet. It looks pretty good. It should—it closely approximates what we are doing. The working distance is what all doctors or their assistants do with most cameras. The pixel of one millimeter on the object is good resolution, and the total field of view is small enough that a good camera lens will suffice. The detector size is realistic for silicon detectors, and so is the array size.

	A	B	C	D	E
1		**Multispectral Cancer Geometry**			
2	Symbol	Meaning	Formula	Value	Units
3	π	pi	known	3.14159	#
4	d	object distance	assumed	30	cm
5	l	object length	assumed	7	cm
6	w	object width	assumed	7	cm
7	p	object pixel	assumed	0.01	cm
8	α	pixel angle	D6/D3	3.33E-04	rad
9	θ	total field	2xarctan(D5/2/D4)	0.23	rad
10	θ	total field	D9x57	13.24	deg
11	D_o	optics diameter	assumed	5	cm
12	β	diffraction angle	1.22x0.0001/D11	2.44E-05	rad
13	N_h	length pixel #	D5/D7	700	#
14	N_v	width pixel #	D6/D7	700	#

The next step in the design or evaluation is sensitivity (in this case, whether the system can detect small enough differences in the reflectivity of the skin in narrow spectral bands). This is described in the next spreadsheet in terms of the NERD that was described above.

In the current design, there are eight sets of LED illuminating quartets that radiate at different wavelengths throughout the visible and near infrared. We can assume that each diode has a power of one milliwatt. Then each color will provide 4 milliwatts over an area of almost 50 square centimeters for an irradiance of 0.00008 Wcm^{-2}. That is part of the input.

	A	B	C	D	E
2	NERD = $\pi f^2 v(A_d B)/\tau_a \tau_o EA_d A_o D^*$				
3	Symbol	Meaning	Formula	Value	Units
4	π	pi	known	3.14159	#
5	D^*	specific detectivity	assumed	1.00E+11	$cm\sqrt{Hz}W^{-1}$
6	τ_a	atmo transmission	assumed	1	#
7	τ_o	optics transmission	assumed	0.5	#
8	D_o	optics diameter	assumed	5	cm
9	E	irradiance	assumed	0.00008	Wcm^{-2}
10	d	object distance	assumed	30	cm
11	p	object pixel size	assumed	0.01	cm
12	F/#	focal ratio	assumed	3	#
13	B	bandwidth	assumed	30	Hz
14	α	resoltion angle	K11/K10	0.000333	rad
15	f	focal length	K8XK12	15	cm
16	s =$\sqrt{Ad}$	detector size	K14xK15	0.01	cm
17	A_o	optics area	$K4xK8^2/4$	19.63494	cm^2
18	NERD	noise equivalent p	see B19	0.13	%
19		$D4xD15^2xD16xD13^2/(D6xD7xD9xD16^2xD17xD5)$				

It is clear that there is sufficient sensitivity to detect differences on the order of fractions of a percent in the reflection from the skin or other tissue. The assumption of a diffuse reflection is also quite valid—just look at the sunlight reflected from your skin at different angles. It is all about the same. The remaining uncertainty is if there is enough variation between normal and malignant tissue and if there is enough constancy in this variation. That is, as is often written, TBD.

As we go to press, a preliminary multispectral image has the same normal/malignant pattern as that obtained with the standard Mohs technique.

A.4.39 Radiative Coolers

Purdue University researchers have claimed they developed a paint with properties that make it a better cooler than air conditioners. They describe an acrylic paint with calcium carbonate filler that obtains a solar reflectance of 98.1% and an emissivity of 94% in the sky window. They further claim that a 1000-square-foot roof with this paint on it provides 10 kW of cooling.[8]

It is both interesting and instructive to evaluate these claims. The inputs, of course, are the spectrum of the paint and the solar input. The output is the thermal radiation from the Earth.

The reflectance spectrum of calcite,[9] calcium carbonate, can be approximated by a constant value according to their claim and the data shown below as 98.1% from 0.4 to 2.2 μm. The transmission is 1.9%.

The ratio of the emittance from a blackbody at 5800 K from 0.4 to 2.2 um to the total is 0.829 (Mathcad). The solar constant, the total irradiance on the Earth's surface from the sun, is 1000 Wm^{-2}. Therefore, the input is 0.019 times 0.829 × 1000 watts per square meter, or 15.75 watts per square meter.

The radiation input in the region 2.2 to 10 μm is 0.046 of the solar constant, or 46 Wm^{-2}. Therefore, the total radiation absorbed by the roof is approximately 62 Wm^{-2}.

The output is the radiation from the paint in the long wave part of the spectrum; it appears that the paint has an emissivity of 0.94 in this region. We can assume a temperature of 300 K. The ratio of that blackbody emission to the total is 0.376, so the emission is 0.94 × 0.376 × 1000 = 353 Wm^{-2}.

The net exchange is 353 − 62 = 291 Wm^{-2} out. It does cool.

A 1000-square-foot roof is 92.9 square meters. It therefore radiates 291 × 92.9 = 27033 watts, or 27 kilowatts.

These are approximate calculations, but even if the numbers change a little, it seems that the Purdue claims are valid. My remaining question is the maintenance. Any dirt or smudges will nullify these good results. How clean will I have to keep my roof?

---

[8]Several sites can be found by searching on Purdue white paint.
[9]Gaffey, S. "Spectral reflectance of-carbonate minerals in the visible and near infrared (0.35–2.55 microns): calcite, aragonite, and dolomite," *Am. Mineralogist* **71**, 151 (1986).

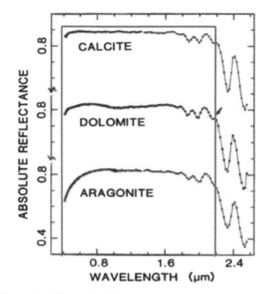

Figure A.4.39.1 Calcite spectrum and approximation.

A.4.40 Radiometers

The operation of the Nichols radiometer[10] can be understood based on the calculations of the momentum provided by the photons emitted by the sun.

The momentum of each photon is h/λ. For a yellow photon in the middle of the visible range, it is 1.325×10^{-24} gms^{-1}. The photon solar constant is approximately the solar constant divided by $2.7\ kT$. That is 1000 Wm^{-2} divided by 1.38×10^{-23} times 5900, which is 5×10^{17} photons per second per square centimeter. The momentum on a typical vane of such a radiometer is 0.6×10^{-6} grams per second each second.

That is why the vane assembly needs to be mounted on an almost frictionless pedestal and why the experiment is difficult.

A.4.47 Spectrometers

As noted in Section 4.47, spectrometers are devices that record or display spectra. The standard ones, with slits or arrays, use either a prism or a grating to disperse the spectrum. The Fourier transform infrared (FTIR) spectrometer records an interferogram and transforms it to a spectrum with computer programs.

Spectrometers have different spectral resolutions and sensitivities.

The FTIR has both a throughput and multiplex advantage, also called Jacquinot and Felgett after their originators. An approximate calculation of

---

[10]Nichols, E. and Hull, G. "A preliminary communication on the pressure of heat and light radiation," *Physical Review. Series I. American Physical Society* **13**(5), 307–320 (1901).

the throughput advantage of the FTIR compares a slit with a circle, assuming the rest of the systems are comparable and have the same optical speed and therefore solid angle. If the circle (the optical element, lens, or mirror) has a radius of R, then its area is $\pi D^2/4$, and the slit is D times its width w, which must be small for good spectral resolution. The advantage is $\pi D/4w$. A typical value of w is about 0.01D. The advantage is then $100\pi D$. If D is only 5, the advantage is 1571.

The multiplex advantage can be estimated with a couple assumptions. Assume the spectral resolution is 0.1 μm and the total spectrum goes from 1 μm to 20 μm and the scan takes 1 second. That is 200 samples. The time for each sample is 1/200 = 0.005 seconds. The bandwidth is the reciprocal of twice the dwell time, $1/2t_d$, and is therefore 100. Since SNR depends on the square root of the bandwidth, the advantage is 10.

The resolving power of a prism spectrometer is the distance the light goes through the prism b times the total change in refractive index dn over the spectral region divided by the spectral extent $d\lambda$:

$$\frac{\lambda}{d\lambda} = \frac{bdn}{d\lambda}.$$

Typical values are 5 centimeters for b, 0.0003 for dn, and 0.00004 cm for the visible $d\lambda$. The resulting value is 5 × 0.0003/0.00004 = 38.

The resolution of a grating is the number of lines times the order number. I assume first order for simplicity. A typical grating has at least 1000 lines. The RP is 1000.

I think two things are obvious: (1) an FTIR is more efficient than a prism or grating spectrometer, and (2) a grating spectrometer is more efficient than one that uses a prism.

A.4.55 Television Sets

It was stated in Section 4.55 that some 4k television sets are now at the limits of visual acuity. This is the support for that.

Assume a large 75-inch TV set across the room. It is 12 feet away with a 75-inch diagonal. These 4k sets are specified as having 3840 pixels across[11] with a 1.91 to 1 aspect ratio.[12] The relationship between the angular resolution of the eye and the angular pixel size calculated in the spreadsheet. I calculated its horizontal width of the 75-inch set based on an aspect ratio of 1.91 to 1 based on the reference. The horizontal pixel number came from the same reference, but the resolution of the eye came from several other references. It is generally accepted as 28 arcseconds.

---

[11]"What is 4K resolution?" techradar.com
[12]"What's the difference between 4K and UHD?" extremetech.com.

The result is that the pixel subtends an angle smaller than the resolution of the eye. But remember that the eye is a variable thing. This can only be an approximate calculation that the pixel of a large, 4k TV set is about at the limit of human acuity. For these assumptions, a pixel subtends 0.00012 radians (cell D11), whereas the commonly accepted resolution of the eye is 0.000136 radians (cell D14). The pixel is less than 9/10ths the resolution of the eye.

	A	B	C	D	E
1		**Television Set Acuity**			
2	Symbol	Meaning	Formula	Value	Units
3	d	TV distance	assumed	12	ft
4	d	TV distance	12xD3	144	in
5		TV diagonal	assumed	75	in
6		Aspect ratio	reference	1.91	#
7		Aspect angle	arctan(D6)	1.09	rad
8	w	Horizontal width	sin(D7)	66.44	in
9	N_h	Horizontal pixels	assumed	3840	#
10	p	Pixel size	D8/D9	0.0173	in
11	α	Pixel angle	D10/D4	0.00012	rad
12	β	Eye resolution	reference	28.00000	arcsec
13	β	Eye resolution	D12/3600	0.007778	deg
14	β	Eye resolution	D13xD15	0.000136	rad
15		conversion		0.017453	rad/deg

A.4.57 Underground Object Detection

Several different underground objects were discussed in Section 4.58. Here is the quantification of the detection of two types of land mines: anti-personnel and anti-tank military mines. The other applications, cities, pipes, etc. can be extrapolated from these.

The typical installation I assume by way of illustration is a personnel or anti-tank mine with two different types of explosives (TNT and RDX) in a plastic container buried in sand. I compare the rise in temperature between such a mine and sand of the same volume. Both receive the same solar input for the same amount of time and are assumed to be the same depth for comparison, that is, a disc of normal sand versus a mine of the same dimensions.

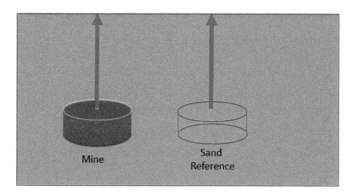

Figure A.4.57.1 Underground mine assumptions.

The temperature increase is governed by the specific heat capacity c_p of each disc. The specific heat capacity is defined as the amount of heat energy E it takes to raise a given amount of mass m by one degree temperature dT:

$$c_p = \frac{E}{m\,dT}.$$

This equation can be inverted to obtain the increase in temperature:

$$dT\frac{E}{mc_p} = \frac{E}{\rho V c_p}.$$

The energy on both the mine and the sand is the same: the solar input over time. Equal volumes are to be compared. Only the differences in the products of the specific heat and the density need to be considered. These are listed in the spreadsheet. I think it is obvious that, no matter what the composition of the mine, the sand will be different. Even the lowest product for a pure plastic disc is about 30% different from sand. The data are from several places on the net, mostly from NIST.

	A	B	C	D
1	**Mine Detection**			
2	**Material**	**Density**	**Specific Heat**	**Product**
3		g/cm3	JgK	
4	TNT	1.65	1.07	1.77
5	RDX	1.82	1.01	1.84
6	Tetryl	1.73	1.01	1.75
7	HMX	1.91	1.00	1.91
8	Plastic	1.00	1.67	1.67
9	Loose sand	1.44	0.83	1.20

Appendix 6
Foundations Introduction

This appendix describes some of the foundational scientific concepts that may not be familiar to every reader. It starts with a section on dimensions. In optics, the dimensions range from the very minute, atomic sizes to the very large, astronomical ones. And they are usually specified by the SI system. The full name of the SI system is the Système international (d'unités), or in English, the International System of Units. It uses meters for lengths, areas, and volumes; kilograms for mass; and seconds for time. It is therefore often referred to as the mks or metric system. Most scientific dimensions are metric. The SI system also uses the candela for light and the Kelvin temperature scale.

The third section describes the four different temperature scales. Again, most scientific endeavors use the Celsius and Kelvin scales that are unfamiliar to most Americans. And hardly anyone uses the absolute version of the Fahrenheit scale. What is it called again?

Conic sections are used for the shape of telescope mirrors and lenses, and they deserve a little description. Why are they called conic sections? In short, because they are planar cuts through a cone. But there is more.

Spectra and their manipulation by Fourier series and transforms are critical to the understanding of many optical applications.

In a few of the mathematical appendices, some relatively advanced math is used, such as vector differential calculus and matrices. These are briefly described.

Finally, the basic ideas of fields, functions, and variables in math are discussed for the non-scientific reader.

A.6.1 Numbers and Dimensions

Most Americans are used to the English or common system of units. The only other countries that are not metric are the United Kingdom, Myanmar, and Liberia. The British Imperial system involves measures such as the imperial gallon, which is 20% larger than an American gallon, but the Imperial system will not be considered here. Most of this book uses the metric system. It was adopted by most countries in 1960 and by the United States in 1975 via the Metric Conversion Act. It cited the SI system as the **preferred** way of doing business, **not the only** way. That is why I have two sets of sockets in my wrench set. This appendix makes the connections between SI and US units. I have included only the dimensions and conversions needed here. I have often listed both in the text for convenience.

Some useful **approximations** are

- One centimeter is about 0.4 inches.
- One meter is a little more than a yard.
- One kilometer is about 0.6 miles.
- One liter is about one quart.
- One micrometer is about 0.4 millionths of an inch.
- One nanometer is 1000 times smaller than a micrometer.

Linear Dimensions

The basic linear unit in the SI system is the meter. In the US system, it is the yard (not the foot). There are some weird units, such as the link, rod, and furlong, that will not be included here. The Ängstrom is no longer recommended as a small unit of measurement. The conversions from metric units to US units are

- 1 micrometer = 0.001 millimeters = 0.00000393701 inch
- 1 millimeter = 0.001 meter = 0.0393701inch
- 1 centimeter = 0.01 meter = 0.393701 inch
- 1 meter = 1.09361 yard
- 1 kilometer = 1093.613 yards
- 1 kilometer = 1000 meters = 0.621371 mile

The inverse conversion, from US to metric, are

- 1 inch = 25,400 micrometers
- 1 inch = 25.4 millimeters
- 1 inch = 2.54 centimeters
- 1 yard = 0.9144 meter
- 1 mile = 1.60934 kilometers

Area Dimensions

All the linear dimensions squared are applicable. In addition,

- 1 hectare = 10,000 square meters = 2.471 acres

Volume Dimensions

- 1 liter = 1.06 quarts
- 1 stere = 1 cubic meter = 1.30794 cubic yards = 61,023 cubic inches
- 1 cc = 1 cubic centimeter = 1 milliliter
- 1 gallon = 3.79 liters

Weight Dimensions

- 1 gram = 0.0353 ounces
- 1 kilogram = 2.2 pounds
- 1 pound = 0.045 kilograms
- 1 ounce = 28.35 grams

Time is of the essence and is the same in both systems.

Some Computer Measures

A bit is the basic unit of information. It is a binary digit, a one or a zero. Yes or no. On or off.

- Eight bits is a byte.
- Four bits is just a nibble.
- Sixteen bits is not a gulp!
- Information rate is specified in baud, one byte per second.

The use of scientific representation of numbers is useful especially for very large and very small numbers. The procedure is to represent the total number by a multiplier and an exponent of ten. The number 12 million (12,000,000) is written as either 12×10^6 or 1.2×10^7. The exponent indicates how many places there are behind the decimal point. A similar procedure is used for decimal fractions. The number 0.00012 is written as 1.2×10^{-4}. The −4 indicates how many decimal places should be in front of the multiplier. These values have mathematical bases. Each multiplier is multiplied by ten to the right power to represent the number. They are usually written the way I just have, but they can also be written as 1.2E+10 or 1.20E−6.

The following is a table of numerical prefixes and their meanings.

Numerical Prefixes			
Prefix	Common Name	Value	Value
atto	quadrillionth	0.0000000000001	1.00E-12
nano	trillionth	0.000000001	1.00E-09
micro	millionth	0.000001	1.00E-06
milli	thousandth	0.001	1.00E-03
centi	hundredth	0.01	1.00E-02
deci	tenth	0.1	1.00E-01
uni	one	1	1.00E+00
deka	ten	10	1.00E+01
hect	hundred	100	1.00E+02
kilo	thousand	1000	1.00E+03
mega	million	1,000,000	1.00E+06
giga	billion	1,000,000,000	1.00E+09
tera	trillion	1,000,000,000,000	1.00E+12

A.6.2 Temperatures

I confess that I am amtriguous about temperature scales. I am an American who lives everyday with the Fahrenheit scale. But I am a scientist, and I work with the Kelvin and Celsius scales. I work with all three. I know that room temperature is about 72 °F and body temperature is about 98.6 °F. I remember that room temperature is about 300 K, but I do not remember body temperature in kelvins. I have described different phenomena in this book in Fahrenheit, centigrade, and Kelvin scales. Their conversions are in the tables below. I consider the absolute Fahrenheit scale, the Rankine, as obscene.

A few common temperatures in these three scales are listed in Table 6.2.1.

Table 6.2.1 Common temperature values.

Condition	Fahrenheit	Centigrade	Kelvin
Freezing water	32	0	273
Boiling water	212	100	373
Ambient	80	27	300
Body	98.6	37	310
Freezing nitrogen	−320	−196	77

As mentioned, the four temperature scales are Fahrenheit, Rankine, Celsius, and Kelvin. The first two and the second two are pairs; i.e., Rankine is the absolute version of Fahrenheit, and Kelvin is the absolute version of Celsius. The Fahrenheit scale, the one Americans are used to, defines the freezing point of water as 32 degrees and the boiling point as 212 degrees. There are therefore 180 degrees between them. The absolute-zero temperature on the Fahrenheit scale, when all molecular motion stops, is -459.67 degrees. That obscene scale is not used in this book.

The Celsius scale, also known as the centigrade scale, defines the freezing point of water as 0 degrees and its boiling point as 100 degrees. That 100-degree spread is why it is sometimes called **centi**grade. Each degree is a 100th of the total from freezing to boiling. It is also why a centigrade degree is larger than a Fahrenheit one by $180/100 = 1.8$. The corresponding absolute scale is the Kelvin scale, in which the degree is the same size as in the centigrade scale, but absolute zero is -273 degrees centigrade. Figure A.6.2.1 shows the conversion from Fahrenheit to centigrade.

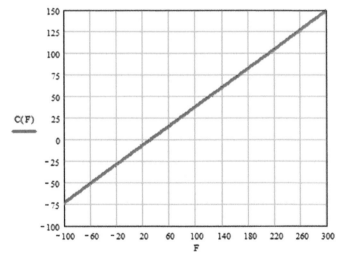

Figure A.6.2.1 Fahrenheit-to-centigrade conversion.

Figure A.6.2.2 shows the conversion the other way around, i.e., from centigrade to Fahrenheit.

Figure A.6.2.3 shows the conversion from Fahrenheit to Kelvin.

Finally, Figure A.6.2.4 shows the conversion from Kelvin to Fahrenheit.

F(C)

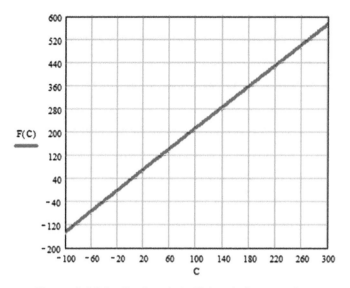

Figure A.6.2.2   Centigrade-to-Fahrenheit conversion.

K(F)

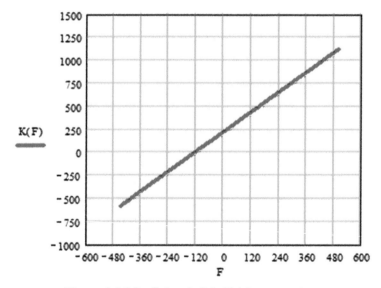

Figure A.6.2.3   Fahrenheit-to-Kelvin conversion.

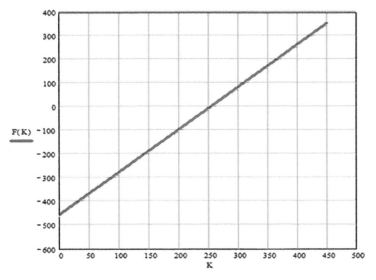

Figure A.6.2.4 Kelvin-to-Fahrenheit conversion.

A.6.3 Conic Sections

Conic sections were referred to in the section on mirrors. Mirrors are described as paraboloidal, ellipsoidal, hyperboloidal, and spherical. These are the rotationally symmetric versions. They are the solid, rotational versions of their plane cousins, parabolas, ellipses, hyperbolas, and circles.

They are called conic sections, and Figure A.6.3.1 shows why. In it, the planar slices through the light blue cone define the conic sections of circle, ellipse, parabola and hyperbola. When the cone has a plane cut through it parallel to the base, it forms a circle, shown in blue. If the plane slices it diagonally, at an angle to the perpendicular, it forms an ellipse, shown in green. If it slices diagonally and goes through the base, it forms a parabola, in yellow. Finally, if it slices vertically and goes through the base, it forms an hyperbola, shown in purple. Their solid equivalents are the sphere, paraboloid, ellipsoid, and hyerboloid. Mirror technicians use both terms for the figures of their mirrors.

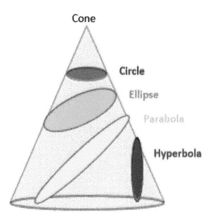

Figure A.6.3.1 Conic sections.

A.6.4 Spectra and Fourier Transforms

Spectra are central to much of the understanding of optical concepts. We all have a general idea of what a spectrum is. It is the distribution of colors behind a prism or in a rainbow. But it has a more technical meaning and use than that. A second important concept is the relationship a spectrum has to its Fourier transform. This is especially important in understanding spectrometers of that type. I will illustrate it by my imaginary rambling in the morning.

I get up and go to the bathroom, 10 feet away. Then to the sink, another 10 feet away, then back to the closet, 20 feet away. I won't bore you with the rest of my movements, but I make many of them. We all do: to the kitchen, the garage, the back yard, etc. They are all certain distances at different times. So I can plot them, as in Figure A6.4.1. This graph is a timeline of my movements of how far I moved when I did.

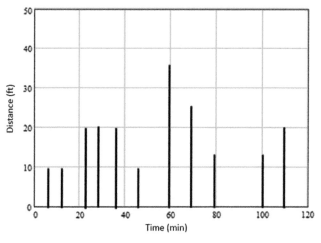

Figure A.6.4.1 Movement timeline.

This same information can be presented in a different way: a plot of how often I made a move of a certain distance. That is a spectrum. It would look like Figure A.6.4.2.

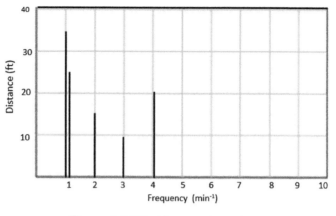

Figure A.6.4.2 Movement spectrum.

Joseph Fourier, in 1822, showed the relationship between these two representations. He showed that any function can be represented by a series of sine waves, its spectrum. These are typically time and frequency relations.

There is a distinct mathematical relationship, but here is an example. It is the representation of a square wave by a series of sine waves. The square wave is shown in Figure A.6.4.3.

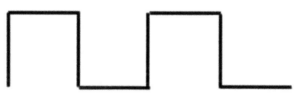

Figure A.6.4.3 Square wave.

The first approximation would be a sine wave of the same frequency and about the same amplitude, as shown in Figure A.6.4.4.

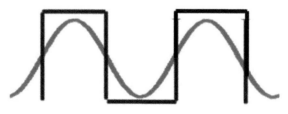

Figure A.6.4.4 First approximation.

This is a reasonable approximation, but it is not so good at the corners. Therefore, add a sine wave that is three times the frequency and smaller amplitude to fill in the corners. It then looks like Figure A.6.4.5: the corners have filled in a bit, although there is now a dip in the middle of the peak.

Figure A.6.4.5 Next approximation.

It should not be hard to imagine that adding more such terms will result in a sufficiently accurate representation of the square wave. The math tells us what to do: it is the sum of the sine of the same frequency and $4/\pi$ times the amplitude of the square wave plus one third times $4/\pi$ times the sine of three times the frequency plus one fifth of $4/\pi$ times the sine of five times the frequency, and so on for all odd numbers. The factor $4/\pi$ may look strange, but it adjusts the amplitude. Figure A.6.4.6 shows the approximation with five terms. Higher frequencies with smaller amplitudes will fill in the rest.

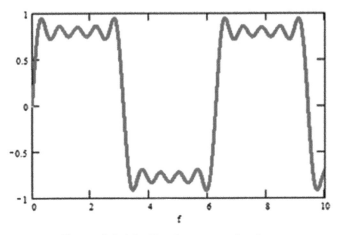

Figure A.6.4.6 Five-term approximation.

The spectrum of that square wave is shown in Figure A.6.4.7 for ten terms. For completeness, the number of terms is actually infinite. I did not have the time to plot those, but you can see that the higher frequencies have lower amplitudes that approach the same value. That smooths out the center of the square wave.

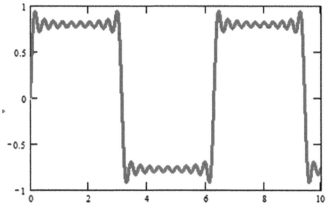

Figure A.6.4.7 Ten-term approximation.

The small peaks at the corners are known as Gibb's phenomena and will not be discussed further. The approximations are good.

The Fourier transform that is the mathematical method to go from one form to the other is

$$S(f) \int f(t) e^{i\omega t} dt,$$

where $S(f)$ is the spectrum, and $f(t)$ is the time function.

A.6.5 Mathematical Operations

In several sections of these appendices some relatively advanced mathematical operations are employed. The following explanations are for those who have some competence in the field but may need a reminder.

Several vector calculus operations are used in the appendix on light, A1.1. These are the del, divergence, curl and del squared operations, which are often called del, del dot, del cross, and del squared for obvious reasons and are written, respectively, as

$$\nabla, \nabla \cdot, \nabla \times, \text{ and } \nabla^2.$$

The del operator is a vector differential operator defined as

$$\nabla = \hat{x} \frac{\partial}{\partial x} + \hat{y} \frac{\partial}{\partial y} + \hat{z} \frac{\partial}{\partial z}.$$

It is the three unit vectors, x hat, y hat, and z hat times their respective partial derivatives.

The divergence is the dot product of the del operator with a vector, **V**. It is a dot product, so it is a scalar:

$$\nabla \cdot \mathbf{V} = \frac{\partial V_x}{\partial x} + \frac{\partial V_y}{\partial y} + \frac{\partial V_z}{\partial z}.$$

The curl is the cross product of del with its corresponding vector. It is a cross cross-product so it is a vector.

$$\nabla \times \mathbf{V} = \left(\frac{\partial V_z}{\partial y} - \frac{\partial V_y}{\partial z}\right)\hat{x} + \left(\frac{\partial V_x}{\partial z} - \frac{\partial V_z}{\partial V_x}\right)\hat{y} + \left(\frac{\partial V_x}{\partial y} - \frac{\partial V_y}{\partial x}\right)\hat{z}.$$

Del squared is del dot del, the dot product of del with itself:

$$\nabla^2 \mathbf{V} = \nabla \cdot \nabla = \frac{\partial^2 V_x}{\partial x^2} + \frac{\partial^2 V_y}{\partial y^2} + \frac{\partial^2 V_z}{\partial z^2}.$$

The vector identity used in the derivation of the wave equation is

$$\nabla \times \nabla \times V = \nabla^2 \mathbf{V} - \nabla \cdot v.$$

In the appendix on polarization, Section A1.2, several matrices are used. They are arrays of numbers that can be manipulated in specific ways. In that section matrices with one column and two rows, one column and four rows, as well as 2 by 2 and 4 by 4 matrices are multiplied together. This section shows how it is done. Consider first a 2×2 matrix multiplying a 2×2 matrix with elements a_{ij} and b_{ij}:

$$\begin{bmatrix} a_{11} & a_{12} \\ a_{21} & a_{22} \end{bmatrix} \times \begin{bmatrix} b_{11} & b_{12} \\ b_{21} & b_{22} \end{bmatrix} = \begin{bmatrix} c_{11} & c_{12} \\ c_{21} & c_{22} \end{bmatrix}.$$

The element c_{11} is the sum of $a_{11}b_{11}$ and $a_{11}b_{21}$:

$$c_{11} = a_{11}b_{11} + a_{12}b_{21}.$$

The first term in the resultant **c** matrix is the sum of the products of the terms in the first row of the first matrix and the first column of the second. Mathematicians call this the inner product. I think you can view it as laying the row on the column and multiplying the pairs of numbers and adding them.

The entry in the second row of the second column in the **c** matrix, c_{22}, is the sum of the product of the elements in the second row of the **a** matrix and the elements in the second column of the **b** matrix:

$$c_{22} = a_{21}b_{12} + a_{22}b_{22}.$$

And for good measure, here is a visual way to remember this. Consider c_{12}, the element that is in the first row and the second column. The first row in **a** is a_{11} and a_{12}. The second column of **b** is b_{12} and b_{22}. View them like this:

$$\begin{bmatrix} a_{11} & a_{12} & b_{12} \\ & & b_{22} \end{bmatrix}.$$

Then multiply and add them by rotating the first row to make it a column and then multiplying the pairs.

$$\begin{bmatrix} a_{11} & b_{12} \\ a_{21} & b_{22} \end{bmatrix} \quad \begin{bmatrix} a_{11}b_{12} \\ + \\ a_{12}b_{22} \end{bmatrix}.$$

And the answer is

$$c_{12} = a_{11}b_{12} + a_{12}b_{22}.$$

For the more mathematically inclined, the general product matrix element c_{ij} can be written as

$$c_{ij} = \Sigma a_{ik}b_{kj},$$

where k is the running index.

The vector operator curl, $\nabla\mathbf{X}$, can be formed as a matrix, just to combine the two concepts:

$$\mathbf{W} = \begin{bmatrix} \hat{x} & \hat{y} & \hat{z} \\ \partial/\partial x & \partial/\partial y & \partial/\partial z \\ V_x & V_y & V_z \end{bmatrix}.$$

The solution to the wave equation was shown to be in terms of sines or cosines. Some of the manipulations have been in terms of exponentials. There is a relationship and a convention. The relationship is

$$e^{j\varphi} = \cos\varphi + j\sin\varphi.$$

We can use the exponential form to represent a sine or cosine if we are careful. Only linear processes are valid with this maneuver and we need to keep in mind that the exponential form represents only the sine. We can write $\sin(\omega t - kr)$ as $e^{j(\omega t - kr)}$ with it understood that it means $\text{Re}(e^{j(\omega t - kr)})$. It is often easier to manipulate the exponential form than the trigonometric forms. Another way to write $e^{j(\omega t - kr)}$ is $\exp j(\omega t - kr)$.

A.6.6 Fields, Functions, and Variables

Fields, functions, and both independent and dependent variables are second nature to mathematically experienced people. They have special meanings to them and me. This appendix explains those special meanings that may not be familiar to the non-scientific reader.

Fields are regions of influence. In politics and business, people have their fields of influence. In science, certain forces have theirs. Iron filings held a few inches from a good magnet will be drawn to it. The magnetic force has an influence of attraction at some distance from it. A ball thrown up in the air comes back down because the gravitational force of the Earth pulls it down. An electron can move another electron at some distance up and down if it moves up and down by virtue of the repulsion of like charges. Light is an electromagnetic field, an influence at a distance. A field is the area over which some kind of force has an influence.

A **function** can be a particular use. The function of a dishwasher is to wash dishes. But in a mathematical sense and in this book, a function describes how a **dependent variable** relates to an **independent one**. For instance, the height of the terrain in your neighborhood varies. The height is a function of position. Position is the independent variable; height is the dependent one. The function describes how the height varies from place to place. That height function might be represented as $h(x)$.

Two-dimensional functions are often plotted as a graph with the independent variable on the x axis (the horizontal one) and the dependent variable on the y axis (the vertical one), as shown in Figures A.6.6.1 and A.6.6.2.

Figure A.6.6.1   Example of a function.

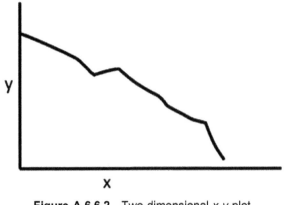

Figure A.6.6.2   Two-dimensional *x-y* plot.

Three-dimensional functions are plotted on the *z* axis, with the *x* and *y* axes plotted in perspective. Figure A.6.6.3 shows a cone that rests on the *x-y* plane and rises along the *z* axis.

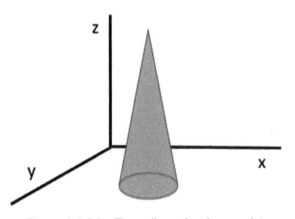

Figure A.6.6.3   Three-dimensional. *x,y,z* plot.

Three functions that are used frequently in this book are the sine, cosine, and exponential. The sine is a function that is repetitious. Its amplitude varies from zero to plus one back through zero to minus one and back to zero. Figure A.6.6.4 shows one cycle. The first maximum is at $\pi/2$ radians, or 90 degrees. The first zero is at π radians, or 180 degrees. Minus one occurs at $3\pi/2$, or 270 degrees, and one full phase ends at 2π, or 360 degrees, with an amplitude of zero. This process is then repeated for as many cycles as are appropriate to the problem at hand. The cosine is the sine shifted by 90 degrees, or $\pi/2$ radians. The sine is in red in Figure A.6.6.5; the cosine is in blue. The first zero of the cosine occurs at $\pi/2$, where the sine is maximum. They are complementary functions in the sense that one is the other shifted by 90 degrees, or $\pi/2$.

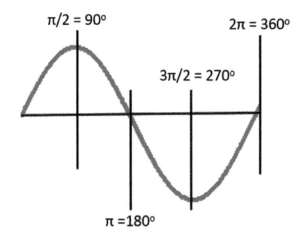

Figure A.6.6.4   Single-cycle sine function.

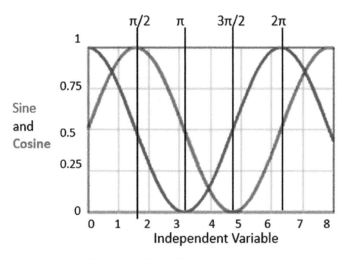

Figure A.6.6.5   Sine and cosine.

The cosine is the sine shifted by 90 degrees, or $\pi/2$ radians. The sine is in red in Figure A.6.6.5; the cosine is in blue. The first zero of the cosine occurs at $\pi/2$, where the sine is maximum. They are complementary functions in the sense that one is the other shifted by 90 degrees, or $\pi/2$.

A complex number is a number that includes both a real number and an imaginary number. The imaginary number itself is $\sqrt{-1}$, which I have specified as j. Other authors use i. It seems that most physicists use i and most engineers use j. I am both, so I made an arbitrary choice. A general imaginary number is the real number times the unitary imaginary number. That is, an imaginary number might be $5j$. Then a complex number is the combination of a real one and an imaginary one, such as $5 + 3j$. This is easier than writing $5 + 3\sqrt{-1}$.

The exponential function is just e^x, the number 2.718 raised to the power x. But it is used often in this book as e raised to an imaginary number. We can write

$$e^{j\theta} = \cos\theta + j\sin\theta.$$

Note that imaginary numbers are just as real as real numbers. When they were first introduced, there was skepticism about them, and people did a lot of imagining. The name stuck. Do not be discouraged if this is somewhat obscure. I took an entire three-hour class called "functions of complex variables." I remember a little bit of it.

A bonus bit of trivia for those who are curious: a discussion of the mysterious number $e = 2.718$ and the relation of the exponential to sines and cosines.

It was Leonard Euler who came up with the funny number e. It is probably named after him. We can derive it in terms of compound interest. We all know the power of compound interest, but do we all realize that the more it is compounded, the faster it accumulates?

The standard formula for the annual amount realized when compounded yearly is

$$R = A(1 + r)^n,$$

where R is the return after n years at an interest rate of r, and A is the original amount.

Another version is the return if the interest is compounded more often than annually, say, quarterly or monthly. The interest rate must be reduced by the number of times it is applied. The expression then is

$$R + A(1 + r/n)^n,$$

where n is the number of times it is compounded. It is the exponent of the parens and the divisor of the interest rate. If you invest 100 cents at 8%, at the end of the year with annual compounding you will have earned eight cents:

$$R = 100(1 + 0.08)^1 = 108.$$

If you can have the amount compounded quarterly, it will be

$$R = 100(1 + 0.02)^4 = 108.24,$$

which is almost a quarter cent more. Monthly, it would be 30 cents.

This concept may have been what set Euler to thinking about what happens if you compound daily or even every minute or second. That is, you let n in the expression go to infinity. The answer is 2.718281828459045 2353602874713527... That is the value of e as far as I could go, but it can almost always be truncated to 2.718. Doing that incurs a 1% error. Using the next digit, that is, 2.7183, gives an error of 0.06%. It is hard to imagine you will need better accuracy than that.

Figure A.6.6.6 shows much of this in practical terms. An annual interest rate of 8% compounded annually is 8%. If it is compounded monthly, it becomes 8.3%, and if compounded daily, it is 8.3278%. Compounding monthly is worthwhile, but there is hardly any additional advantage after that.

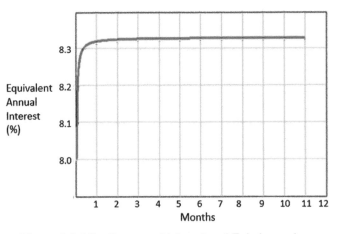

Figure A.6.6.6   Compound interest and Euler's number.

An explanation of how the exponential e^{jx} can stand for the complex sum of sines and cosines can be done in terms of their series representations. The expression to be proven is

$$e^{jx} = \cos x - j \sin x.$$

The series expression for e^{jx} is

$$e^{ix} = 1 + jx + \frac{j^2 x^2}{2!} + \frac{j^3 x^3}{3!} + \frac{j^4 x^4}{4!} + \frac{j^5 x^5}{5!} + \frac{j^6 x^6}{6!} + \frac{j^7 x^7}{7!} + \ldots$$

That is equivalent to

$$e^{ix} = 1 + jx - \frac{x^2}{2!} - \frac{jx^3}{3!} + \frac{x^4}{4!} + \frac{jx^5}{5!} - \frac{x^6}{6!} - \frac{jx^7}{7!} + \ldots$$

The second version is obtained by observing that each even power of j is –1 and each odd power is –j.

The series expression for $\sin(jx)$ is

$$\sin(jx) = j\left(x - \frac{x^3}{3!} + \frac{x^5}{5!} - \frac{x^7}{7!} + \ldots\right) = j\sin(x).$$

The series expression for $\cos(jx)$ is

$$\cos(x) = 1 - \frac{x^2}{2!} + \frac{x^4}{4!} - \frac{x^6}{6!} + \ldots$$

If all the even powers are separated from the odd powers, the expression is

$$e^{jx} = \cos(x) + j\sin(x).$$

Quod erat demonstandum!

Finis, mirabile dictum!